INSTRUCTOR'S MANUAL
to accompany

CHEMISTRY
Science of Change
third edition

Oxtoby ~ Freeman ~ Block

Wade A. Freeman
The University of Illinois at Chicago

SAUNDERS GOLDEN SUNBURST SERIES

Saunders College Publishing
Harcourt Brace College Publishers

Fort Worth   Philadelphia   San Diego   New York   Orlando   Austin
San Antonio   Toronto   Montreal   London   Sydney   Tokyo

Oxtoby, Freeman, Black: Instructor's Manual to accompany *Chemistry: Science of Change, 3e*. Freeman.

ISBN 0-03-020467-4

789 021 7654321

Contents

Answers to Even-Numbered Problems, Chapter 1

1-2 At room conditions, "absolute" alcohol (pure alcohol) is a homogeneous liquid. It is a substance, a compound of three elements. Milk is heterogeneous. It is a mixture of an aqueous phase and a fat phase. This is true even of "homogenized" milk, in which the droplets of fat suspended in the aqueous phase are smaller than in raw milk, but still large enough to be seen with an optical microscope. Copper wire is homogeneous. It is an elemental substance; the physical form of a sample (wire, rod, lump, etc.) is immaterial to the classification. Ordinary scales of rust often contain unreacted iron and oxides of iron other than Fe_2O_3 and are therefore heterogeneous and mixtures. The chemical name barium bromide implies a pure compound. A sample of barium bromide under ordinary conditions is homogeneous. At its melting point (847°C), barium bromide could be heterogeneous (solid and liquid phases coexisting). Concrete is heterogeneous and a mixture. Baking soda and baking powder are both heterogeneous mixtures. Baking soda is very nearly pure $NaHCO_3$, but some impurities are present; baking powder is $NaHCO_3$ mixed with an acidic salt (either $Ca(HPO_4)_2 \cdot H_2O$ or a mixture of $NaAl(SO_4)_2$ with $Ca(HPO_4)_2 \cdot H_2O$).

1-4 Spring water direct from an artesian well is by definition pure spring water, because it is water right out of a spring. Triply distilled water is more pure in the chemical sense that it contains fewer materials that are different from H_2O.

1-6 Numerous examples are possible. A solution of sugar in water is a homogeneous mixture. If left open to the room it would give sugar crystals plus water as the water evaporated. Ultimately, only the sugar crystals would remain. Thus, this example starts homogeneous, become heterogenous and goes back to being homogeneous.

1-8 The answer is $\boxed{\text{elements}}$. Although all of the formulas (H_2, N_2, etc.) are for diatomic molecules, compounds require the presence of two different elements.

1-10 Let the mass of the total sample be x. Then $0.944x = 125.0$ g. Solving gives $x = \boxed{132.4 \text{ g}}$.

1-12 A likely explanation is that some copper reacts with a compound or element in the air to form a solid compound or solid compounds of greater mass. That is, the coin $\boxed{\text{chemically incorporates mass}}$ from the surrounding air.

1-14 If decomposition into potassium chloride and oxygen is the only reaction undergone by the $KClO_3$, then the mass of $KClO_3$ remaining must equal the original mass minus the masses of the two products: $100.0 \text{ g} - 36.9 \text{ g} - 57.3 \text{ g} = \boxed{5.8 \text{ g}}$.

1-16 The ratio of the mass of tellurium to the mass of hafnium in this compound is

$$\frac{\text{mass Te}}{\text{mass Hf}} = \frac{31.5 \text{ g Te}}{25.0 \text{ g Hf}} = 1.26\frac{\text{g Te}}{\text{g Hf}}$$

Because the compound from the rock is identical, it contains Te and Hf in the same ratio.

$$\text{mass Te} = 0.125 \text{ g Hf} \times \left(1.26\frac{\text{g Te}}{\text{g Hf}}\right) = \boxed{0.158 \text{ g Te}}$$

The compound may of course contain other elements as well.

1-18 (a) The chemical formula of the ternary compound is $\boxed{Na_4P_2O_7}$.

(b) The formula is $\boxed{N_2O_5}$. Formulas with fractional subscripts (such as $NO_{2.5}$ or $NO_{2\,1/2}$) are usually avoided.

(c) The formula is $\boxed{FeC_{10}H_{10}}$ (the compound is perhaps ferrocene, $Fe(C_5H_5)_2$).

1-20 (a) The law of multiple proportions states that the masses of an element that combine with a fixed mass of another element in a series of compounds stand in the ratio of small integers. The ratio of the masses of S per 1.0000 g of H in the two compounds is

$$\frac{15.907 \text{ g S}}{31.813 \text{ g S}} = 0.50001$$

which is consistent with the law (it is the ratio of 1 to 2). The ratio of the masses of O per 1.0000 g of H in the two compounds is $31.747/23.810 = 1.3333$, which is the ratio of 4 to 3. The recognition of ratios from their decimal equivalents is hard for many students (see Appendix C-3 for exercises).

(b) The first compound contains 1.0000 g of H per 23.810 g of O. It also contains 31.813 g of S per 1.000 g of H. Hence it must contain 31.813 g of S for every 23.810 g of O. The desired ratios are

$$\frac{31.813 \text{ g S}}{23.810 \text{ g O}} = \boxed{1.3361\frac{\text{g S}}{1.000 \text{ g O}}} \quad \text{and} \quad \frac{1.0000 \text{ g H}}{23.810 \text{ g O}} = \boxed{0.041999\frac{\text{g H}}{1.000 \text{ g O}}}$$

Similar ratios can be formed for the second compound:

$$\frac{15.907 \text{ g S}}{31.747 \text{ g O}} = \boxed{0.50106\frac{\text{g S}}{1.000 \text{ g O}}} \quad \text{and} \quad \frac{1.0000 \text{ g H}}{23.810 \text{ g O}} = \boxed{0.031499\frac{\text{g H}}{1.000 \text{ g O}}}$$

(c) The second compound is 1/2 as rich in sulfur per gram of hydrogen as the first, but 4/3 richer in oxygen. Its formula therefore equals the formula of the first, but with the subscript of the S multiplied by 1/2 and the subscript of the O multiplied by 4/3: $H_2S_{1/2 \times 2}O_{4/3 \times 3}$. This works out to $\boxed{H_2SO_4}$.

1-22 (a) In each case the mass of fluorine that combines with 1.0000 g of iodine is simply the mass percentage of fluorine divided by the mass percentage of iodine. This is best shown by considering samples of the compounds that have masses of 100.000 g. The masses contributed by each element in the compounds are then easily computed.

Compound 1	13.021 g F/86.979 g I	0.14970 g F/ g I
Compound 2	30.993 g F/69.007 g I	0.44913 g F/ g I
Compound 3	42.809 g F/57.191 g I	0.74853 g F/ g I
Compound 4	51.171 g F/48.829 g I	1.04796 g F/ g I

(b) The law of multiple proportions involves the ratio of the numbers in the last column. Simply divide all four by the smallest. The results are: 1.0000 for compound 1; 3.0002 for compound 2; 5.0002 for compound 3; 7.0004 for compound 4. These equal the small whole numbers $\boxed{1, 3, 5, \text{ and } 7}$ within the precision of the data.

1-24 The only products are gaseous N_2 and gaseous H_2. From the formula of the starting compound there must be twice as many molecules of H_2 as of N_2 in the products. The law of combining volumes (or, in this case, the law of "uncombining" volumes) holds that the volume of hydrogen is twice the volume of nitrogen as long as the temperature and pressure remain unchanged. The answer is $\boxed{27.4 \text{ mL}}$.

1-26 Answering requires application of the law of combining volumes. It is *not* necessary to know how to balance a chemical equation to conclude that for every one volume of methanol vapor that burns, one volume of carbon dioxide and two volumes of water vapor form. The formulas of carbon dioxide (CO_2) and water (H_2O) *are* required. Both appear in the chapter. The correct answers are $\boxed{2.0 \text{ L}}$ of carbon dioxide and $\boxed{4.0 \text{ L}}$ of water vapor.

1-28 The volumes of the reacting N_2 and O_2 are in the ratio

$$\frac{11.71 \text{ mL}}{17.56 \text{ mL}} = 0.6668 = \frac{2}{3}$$

The volume of the single, gaseous product is the same as the volume of the reacting N_2; hence: $2 \text{ vol } N_2 + 3 \text{ vol } O_2 \longrightarrow 2 \text{ vol product}$. Because all of the oxygen is incorporated into molecules of product, the ratio of N to O in the product must equal $\boxed{2 \text{ to } 3}$. The answer is also obtained by balancing the equation $2 N_2 + 3 O_2 \longrightarrow 2 N_2O_3$, but only the law of combining volumes is actually needed.

1-30 For any atom, $A = Z + N$

Symbol	Z	N	A	No. of electrons
$^{228}\text{Ra}^{2+}$	$\boxed{88}$	$\boxed{140}$	$\boxed{228}$	$\boxed{86}$
$^{127}\text{I}^-$	53	$\boxed{74}$	127	54
^{40}Ca	$\boxed{20}$	20	40	20
N^{3-}	$\boxed{7}$	8	$\boxed{15}$	$\boxed{10}$

1-32 Divide the mass of the proton by the mass of the electron

$$\frac{m_{\text{proton}}}{m_{\text{electron}}} = \frac{1.672623 \times 10^{-27} \text{ kg}}{9.109390 \times 10^{-31} \text{ kg}} = 1.836153 \times 10^3$$

The answer therefore is just over $\boxed{1836}$ electrons.

1-34 The atoms of the isotope of element 112 that was produced contain $\boxed{112 \text{ protons}}$, $\boxed{165 \text{ neutrons}}$ and $\boxed{112 \text{ electrons}}$. See text page 628 for more on the synthesis of element 112.

1-36 (a) Promethium has $Z = 61$. The ratio of the number of neutrons to the number of protons in Pm-145 is $\boxed{2.377 \text{ to } 1}$.

(b) A single Pm^{3+} ion contains $\boxed{5}$ electrons.

1-38 Take relative atomic masses from the table on the inside back cover of the text. The atomic mass of uranium exceeds that of phosphorus by a factor of $238.0289/30.973761 = 7.684856$; if an atom of P has a mass of 10 paulings, then an atom of U has a mass of $\boxed{76.84856 \text{ paulings}}$.

1-40 The relative atomic mass of naturally-occurring chromium is the average of the relative masses of the isotopes weighted by their fractional abundances:

$$A_{Cr} = \sum_{i=1}^{n} A_i p_i$$
$$= (49.9461)0.0431 + (51.9405)0.8376 + (52.9407)0.0955 + (53.9389)0.0238$$
$$= \boxed{51.998}$$

1-42 The fractional abundances of the two isotopes of iridium are *not* known, but their atomic masses are. Also, the atomic mass of naturally occurring iridium is known. Let the fractional abundance of the lighter isotope equal p. Then that of the heavier isotopes is $1 - p$. Substitute these values into the equation for the weighted average and solve for p:

$$A_{Ir} = \sum_{i=1}^{n} A_i p_i$$
$$192.22 = 190.96058p + 192.96292(1 - p)$$
$$p = \boxed{0.37}$$

Note that the difference between 192.22 and 192.96292 is precise only to the hundredths place, so this answer is correctly given to only two significant figures.

1-44 (a) The mass spectrum has $\boxed{\text{three peaks}}$: one for H_2^+, one for HD^+, and one for D_2^+.

(b) The relative atomic masses of the three peaks are the relative masses of the three varieties of hydrogen molecules: $\boxed{2.01566, 3.02194, \text{ and } 4.02822}$.

(c) The $\boxed{H_2^+}$ peak will be the largest by far because the H-atoms are so abundant in the sample. The smallest peak will be the D_2^+ peak. The sizes of the peaks are in the ratio $0.9997 : 0.00030 : 2.3 \times 10^{-8}$.

1-46 Note the similarity between this problem, which treats a chemical system, and problem 1-45, which deals with chocolate-covered cherries. The relative atomic mass of the indium is

$$A_{In} = \sum_{i=1}^{n} p_i A_i = 0.0428(112.904) + (1.0000 - 0.0428)(114.904) = \boxed{114.82}$$

1-48 The point of the problem is the proper handling of nested parentheses in chemical formulas when computing molecular masses, formula masses, and molar masses. The answers have no units because they are relative masses:

(a)	(b)	(c)	(d)	(e)
$[Ag(NH_3)_2]Cl$	$Ca_3[Co(CO_3)_3]_2$	OsO_4	H_2SO_4	$Ca_3Al_2(SiO_4)_3$
177.38	598.16	254.2	98.08	450.45

1-50 There are N_0 gold atoms in a mole of gold, each with a diameter of 2.88×10^{-10} m. The length of the line is therefore:

$$(6.022 \times 10^{23} \text{ atom}) \left(\frac{2.88 \times 10^{-10} \text{ m}}{\text{atom}} \right) = \boxed{1.73 \times 10^{14} \text{ m}}$$

This is over 1100 times the distance from the earth to the sun!

1-52 Comparing the masses of the samples requires expressing their amounts in the same unit of mass. In the case of the SF_4, conversion necessary from the chemical amount to mass (using the molar mass). In the cases of the Cl_2O_7 and Ar, a preliminary conversion from the given number of particles to chemical amount is also necessary. The amount of CH_4 is already in terms of a unit of mass (grams). The results are:

$$\boxed{SF_4 \; (114.5 \text{ g}) \quad < \quad CH_4 \; (117 \text{ g}) \quad < \quad Cl_2O_7 \; (264 \text{ g}) \quad < \quad Ar \; (2766 \text{ g})}$$

1-54 First, establish how many atoms there are in 1.000 mol of vitamin A:

$$1.000 \text{ mol vitamin A} \times \left(\frac{51 N_0 \text{ atoms}}{1 \text{ mole}} \right) = 51.00 N_0 \text{ atoms}$$

Next, use this figure to compute the chemical amount of vitamin A_2:

$$51.00 N_0 \text{ atoms} \times \left(\frac{1 \text{ mol vitamin } A_2}{49 N_0 \text{ atoms}} \right) = \boxed{1.041 \text{ mol vitamin } A_2}$$

1-56 The 100 million atoms of fluorine will weigh $18.998403/12.00000$ times the mass of 100 million atoms of ^{12}C. Avogadro's number of ^{12}C atoms weighs exactly 12 g, so 100 million atoms of ^{12}C weighs

$$m_C = (100 \times 10^6 \text{ atoms}) \times \left(\frac{12.00000 \text{ g}}{6.022137 \times 10^{23} \text{ atoms}} \right) = 1.992648 \times 10^{-15} \text{ g}$$

The mass of the fluorine is

$$m_F = 1.992648 \times 10^{-15} \text{ g C} \times \left(\frac{18.998403 \text{ g F}}{12.00000 \text{ g C}} \right) = \boxed{3.154761 \times 10^{-15} \text{ g F}}$$

The answer can be obtained in atomic mass units by an exactly similar computation starting with the fact that one atom of ^{12}C weighs exactly 12 u:

$$m_C = (100 \times 10^6 \text{ atoms}) \times \left(\frac{12.00000 \text{ u C}}{1} \text{ atom C} \right) = 1.200000 \times 10^9 \text{ u}$$

The mass of the fluorine is

$$m_F = 1.200000 \times 10^9 \text{ u C} \times \left(\frac{18.998403 \text{ u F}}{12.00000 \text{ u C}} \right) = \boxed{1.8998403 \times 10^9 \text{ u F}}$$

1-58 (a) The molar mass of vanillin $C_8H_8O_3$ is 152.14 g mol^{-1}. The chemical amount of vanillin equals its mass divided by its molar mass

$$n_{C_8H_8O_3} = 10.00 \text{ g} \times \left(\frac{1 \text{ mol}}{152.14 \text{ g}} \right) = \boxed{0.06573 \text{ mol}}$$

(b)

$$N_{C_8H_8O_3} = 0.06573 \text{ mol} \times \left(\frac{6.022137 \times 10^{23} \text{ molecule}}{1 \text{ mol}} \right) = \boxed{3.958 \times 10^{22} \text{ molecule}}$$

(c) The mass of one molecule of vanillin in atomic mass units is simply the relative molecular mass on the accepted scale expressed in atomic mass units: $\boxed{152.14 \text{ u}}$.

(d) One gram equals Avogadro's number of atomic mass units:

$$N_{C_8H_8O_3} = 152.14 \text{ u} \times \left(\frac{1 \text{ g}}{6.022137 \times 10^{23} \text{ u}} \right) = \boxed{2.5263 \times 10^{-22} \text{ g}}$$

1-60 Using the chemical formula and the table of atomic masses, TiO_2 is 59.94% Ti by mass. This is the relative mass of the single titanium in the formula divided by the relative mass of the entire formula unit and multiplied by 100%.

(a) In 91.2 g of TiO_2, the mass of titanium is $0.5994 \times 91.2 \text{ g} = \boxed{54.7 \text{ g Ti}}$.

(b) The mass of titanium in 91.2 tons of TiO_2 is $\boxed{54.7 \text{ tons}}$. It is computed similarly.

1-62 Weigh out some reasonable, small mass of screws. Count the screws in this sample. Multiply the count by the mass of the entire lot of screws expressed in the unit of mass you chose. Thus, if you weigh out one pound of screws and find that it contains 170 screws, the lot must have $250 \times 170 = 42500$ screws. The analogy to using the mole in chemical calculations is clear.

1-64 The concentration of lead in the patient's blood is given in units of moles per milliliter. The following series of conversions takes this to the unit that the EPA uses:

$$\frac{1.50 \times 10^{-8} \text{ mol Pb}}{1.00 \text{ mL blood}} \times \left(\frac{100 \text{ mL blood}}{1 \text{ dL blood}}\right) \times \left(\frac{207.2 \text{ g Pb}}{1 \text{ mol Pb}}\right) \times \left(\frac{10^6 \ \mu\text{g Pb}}{1 \text{ g Pb}}\right)$$

$$= \boxed{311 \ \mu\text{g Pb dL}^{-1}}$$

This is more than ten times the maximum safe level.

1-66

$$9.33 \text{ cm}^3 \text{ CH}_3\text{OH} \times \left(\frac{0.7914 \text{ g CH}_3\text{OH}}{1.000 \text{ cm}^3 \text{ CH}_3\text{OH}}\right) = \boxed{7.38 \text{ g CH}_3\text{OH}}$$

$$7.38 \text{ g CH}_3\text{OH} \times \left(\frac{1.00 \text{ mol CH}_3\text{OH}}{32.042 \text{ g CH}_3\text{OH}}\right) = \boxed{0.230 \text{ mol CH}_3\text{OH}}$$

1-68 The density is "turned upside-down" and used as a unit factor:

$$V_{\text{Re}} = 1.000 \text{ g Re} \times \left(\frac{1.000 \text{ cm}^3 \text{ Re}}{21.02 \text{ g Re}}\right) = \boxed{0.04757 \text{ cm}^3 \text{ Re}}$$

1-70

$$10.0 \text{ cm}^3 \text{ Au} \times \left(\frac{19.32 \text{ g Au}}{1 \text{ cm}^3 \text{ Au}}\right) \times \left(\frac{1 \text{ oz}}{31.1035 \text{ g Au}}\right) \times \left(\frac{\$400}{1 \text{ oz}}\right) = \boxed{\$2480}$$

1-72 Start with the reciprocal of the density of lithium:

$$\left(\frac{1 \text{ cm}^3 \text{ Li}}{0.534 \text{ g Li}}\right) \times \left(\frac{6.941 \text{ g Li}}{1 \text{ mol Li}}\right) \times \left(\frac{1 \text{ mol Li}}{6.022137 \times 10^{23} \text{ atom Li}}\right) = \boxed{\frac{2.16 \times 10^{-23} \text{ cm}^3}{\text{atom Li}}}$$

1-74 (a) The conclusion van Helmont reached was $\boxed{\text{not invalid}}$. Much of the matter making up the 164 lb of wood, bark and roots in fact came from carbon dioxide absorbed by the growing willow tree from the air.

(b) $\boxed{\text{Yes}}$, Van Helmot was using the law of conservation of matter, as shown by his attempt to strike a balance of masses in the experiment and the care he used to exclude dust and other solids from the growing tree.

1-76 In Crispix, the ratio of rice to corn is fixed at $1 : 1$ by the structure of the flake. Thus $\boxed{\text{Crispix more like a compound}}$ than Raisin Bran, which contains bran flakes and raisins in proportions that can be varied arbitrarily. $\boxed{\text{Raisin Bran is more like a mixture.}}$

1-78 The word "molecule" in this statement is misused to mean "substance." It is of course no big feat to synthesize more than seven million different molecules—they could be molecules of the same substance, for example water.

1-80 Consider a 100.00-g sample of this binary compound. It contains 78.06 g of Ni and 21.94 g of O. This is 78.06 g/58.69 g mol^{-1} = 1.330 mol of Ni and 21.94 g/15.9994 g mol^{-1} = 1.371 mol of O. The ratio of these two chemical amounts is 1.031 to 1. If the data are truly precise to four significant figures, this is almost certainly a $\boxed{\text{nonstoichiometric}}$ compound. The answer is qualified because "Ni$_{1000}$O$_{1031}$" is a conceivable stoichiometric formulation. These subscript are whole numbers, but they are hardly small whole numbers.

1-82 If the relative mass of the oxygen atom were half of its actual value, then the number of oxygen atoms represented in a chemical formula involving O would have to be twice as great, in order to keep the mass of oxygen represented by the formula unchanged.

(a) H$_2$O would become H$_2$O$_2$, which gives the empirical formula $\boxed{\text{HO}}$.

(b) H$_2$O$_2$ would become H$_2$O$_4$, which gives the empirical formula $\boxed{\text{HO}_2}$.

1-84 (a) Davy understood that it is always possible that some further reaction or procedure might succeed in further decomposing what had been thought to be an element.

(b) Isotopes are different forms of a single element that are extremely difficult to separate chemically. Their discovery showed that most of the elements of the 19th century in fact *were* capable of further decomposition (into two or more isotopes).

1-86 First, compute the fractional abundances of the two isotopes in the desired mixture. Let x equal the fraction of ^{121}Sb. Then $1 - x$ is the fraction of ^{123}Sb. The weighted-average mass is 121.5000 so

$$121.5000 = (120.9038)x + (122.9041)(1 - x)$$

Solution for x gives 0.701945 as the fractional abundance of ^{121}Sb, and 0.298055 as the fractional abundance of ^{123}Sb. The problem calls for the largest possible amount of mixture, so plan to use the whole supply of ^{121}Sb (the computation shows that it is the more abundant isotope in the mixture). Compute the chemical amount of ^{121}Sb in the 100.00 g of pure isotope, and reason that the required number of moles of ^{123}Sb is 0.298055/0.701945 times this. From the chemical amount of ^{123}Sb comes the mass of ^{123}Sb to mix in. The detailed set-up is:

$$m_{123\text{Sb}} = 100.00 \text{ g } ^{121}\text{Sb} \times \left(\frac{1 \text{ mol } ^{121}\text{Sb}}{120.9038 \text{ g } ^{121}\text{Sb}}\right) \times \left(\frac{0.298055 \text{ mol } ^{123}\text{Sb}}{0.701945 \text{ mol } ^{121}\text{Sb}}\right)$$

$$\times \left(\frac{122.9041 \text{ g } ^{123}\text{Sb}}{1 \text{ mol } ^{123}\text{Sb}}\right) = \boxed{43.164 \text{ g } ^{123}\text{Sb}}$$

1-88

$$1.16 \times 10^{22} \text{ molec.} \times \left(\frac{1 \text{ mol}}{6.022 \times 10^{23} \text{ molec.}}\right) \times \left(\frac{29.0 \text{ g}}{1 \text{ mol}}\right) = \boxed{0.559 \text{ g of air}}$$

1-90 (a) The molar mass has units, no units appear in the statement.

(b) There is ambiguity. Is 1.0 mol of Br_2 or 1.0 mol of Br intended?

(c) The term "multiple proportions" has no place here. The proportion of the elements in water to 2 to 1 on a mole basis.

(d) This is the mass of one *mole* of ^{12}C atoms, not the mass of one atom.

(e) A mole of water contains 6.022×10^{23} *molecules* of water, not that number of atoms. Each H_2O molecule contains three atoms, so there are actually 18.066×10^{23} atoms in a mole of water.

1-92 The 1.00 mol of silicon has a mass of 28.086 g. The number of atoms of "dopant" germanium that has been added is

$$28.086 \text{ g Si} \times \left(\frac{0.62 \times 10^{-7} \text{ g Ge}}{100 \text{ g Si}}\right) \times \left(\frac{1 \text{ mol Ge}}{72.61 \text{ g Ge}}\right)$$

$$\times \left(\frac{6.022 \times 10^{23} \text{ atoms Ge}}{1 \text{ mol Ge}}\right) = \boxed{1.4 \times 10^{14} \text{ atoms Ge}}$$

1-94 (a) The volume occupied by a mole of any substance can be calculated from its density and molar mass. For argon under room conditions:

$$V_{\text{Ar}} = 1.00 \text{ mol Ar} \times \left(\frac{39.948 \text{ g Ar}}{1 \text{ mol Ar}}\right) \times \left(\frac{1 \text{ L Ar}}{1.64 \text{ g Ar}}\right) = \boxed{\frac{24.4 \text{ L Ar}}{\text{mol Ar}}}$$

(b) The volume per mole of liquid argon is only $\boxed{0.0285 \text{ L}}$. The calculation is similar to the preceding.

(c) Because $0.0285 \text{ L} \approx 1/1000 \times 24.4 \text{ L}$, the percentage of empty space in gaseous argon under room conditions is about $\boxed{99.9\%}$.

1-96 Convert from volume to mass (using density), then to chemical amount, then to the number of NaCl formula units, and finally to the number of atoms:

$$N_{\text{atoms}} = 1.000 \text{ cm}^3 \text{ NaCl} \times \left(\frac{2.165 \text{ g NaCl}}{1 \text{ cm}^3 \text{ NaCl}}\right) \times \left(\frac{1 \text{ mol NaCl}}{58.4428 \text{ g NaCl}}\right)$$

$$\times \left(\frac{6.022 \times 10^{23} \text{ NaCl units}}{1 \text{ mol NaCl}}\right) \times \left(\frac{2 \text{ atoms}}{1 \text{ NaCl unit}}\right) = \boxed{4.462 \times 10^{22} \text{ atoms}}$$

Some might quibble that there are no atoms in NaCl, but instead a collection of ions. Actually, the difference between a lattice of Na^+ and Cl^- ions in NaCl and a lattice of Na and Cl atoms is experimentally imperceptible.

1-98 The volume and mass of the neutron star are:

$$V_{\text{star}} = \frac{4\pi}{3}r^3 = \frac{4\pi}{3}(2.0 \times 10^6 \text{ cm})^3 = 3.35 \times 10^{19} \text{ cm}^3$$

$$m_{\text{star}} = 6.0 \times 10^{56} \text{ neutron} \times \left(\frac{1.009 \text{ u}}{\text{neutron}}\right) \times \left(\frac{1 \text{ u}}{6.02 \times 10^{23} \text{ g}}\right) = 1.01 \times 10^{33} \text{ g}$$

The density is:

$$d_{\text{star}} = \frac{m_{\text{star}}}{V_{\text{star}}} = \frac{1.01 \times 10^{33} \text{ g}}{3.35 \times 10^{19} \text{ cm}^3} = \boxed{3.0 \times 10^{13} \text{ g cm}^{-3}}$$

The density of the ^{232}Th nucleus equals its mass divided by its volume. The mass of the ^{232}Th atom could be looked up in text Table 15-1 (text page 614). For this comparison, approximating the mass as 232 u, which is the mass of 232 protons and neutrons at about 1 u each, is sufficient:

$$d_{\text{nucleus}} = \frac{m_{\text{nucleus}}}{V_{\text{nucleus}}} = \frac{232 \text{ u}}{3.1565 \times 10^{-36} \text{ cm}^3} \times \left(\frac{1 \text{ g}}{6.022 \times 10^{23} \text{ u}}\right)$$

$$= \boxed{12.2 \times 10^{13} \text{ g cm}^{-3}}$$

This density is close to the density of the neutron star.

Answers to Even-Numbered Problems, Chapter 2

2-2 The empirical formula is $\boxed{C_6H_7}$. This formula uses the smallest possible whole-number subscripts.

2-4 The molar mass of $C_{44}H_{54}N_4O_{10}$ is $\boxed{798.934 \text{ g mol}^{-1}}$. The empirical formula is $\boxed{C_{22}H_{27}N_2O_5}$, obtained by factoring out and then canceling the common factor of 2 in the subscripts in the molecular formula.

2-6 The formula indicates 4 mol of NH_3 per formula unit of the compound. Hence:

$$n_{NH_3} = 10.9 \text{ mg ZnCl}_2 \cdot 4NH_3 \times \left(\frac{1 \text{ mmol ZnCl}_2 \cdot 4NH_3}{204.41 \text{ mg ZnCl}_2 \cdot 4NH_3} \right)$$

$$\times \left(\frac{4 \text{ mmol NH}_3}{1 \text{ mmol ZnCl}_2 \cdot 4NH_3} \right) = 0.213 \text{ mmol NH}_3 = \boxed{2.13 \times 10^{-4} \text{ mol NH}_3}$$

2-8 (a) The molar mass of AZT is $283.244 \text{ g mol}^{-1}$. Then

$$n_{AZT} = 60.5 \text{ kg body} \times \left(\frac{3.00 \text{ mg AZT}}{1 \text{ kg body}} \right) \times \left(\frac{10^{-3} \text{ g AZT}}{1 \text{ mg AZT}} \right)$$

$$\times \left(\frac{1 \text{ mol AZT}}{283.244 \text{ g}} \right) = \boxed{6.41 \times 10^{-4} \text{ mol AZT}}$$

(b) The problem asks for the chemical amount, but the chemical amount in the dose cannot be determined without knowing how much a mole weighs. The empirical formula does not supply this information; the molecular formula does.

2-10 The instructional value in the problem comes when students notice that the percentages that they compute for the two compounds are identical—the two compounds have the same empirical formula. The percentages of boron in the two are also equal to the mass percentage of boron in "BCl." Arithmetic is minimized by doing the computation for the 1-to-1 case:

$$\% \text{ B} = \frac{\text{relative mass B}}{\text{relative mass BCl}} \times 100\% = \frac{10.81}{10.81 + 35.453} \times 100\% = \boxed{23.37\%}$$

2-12 The mass percentage of fluorine in each of the compounds can certainly be calculated, and the results used to get the required order. It is quicker however just to estimate the relative fluorine content of each compound. Thus, HF is certainly the richest in fluorine by mass because the only other atom is the very light hydrogen atom—there is only 1 unit of non-fluorine mass per fluorine atom. The non-fluorine mass per fluorine atom in C_6HF_5 is about $(6 \times 12)/5 \approx 14$; in BrF it is 79.9; in UF_6 it is $238/6 \approx 40$. The desired order is therefore

$$\boxed{\text{BrF} \quad < \quad UF_6 \quad < \quad C_6HF_5 \quad < \quad \text{HF}}$$

2-14 (a) Use the molecular formula of estradiol $C_{18}H_{24}O_2$ as follows:

$$\% \; C = \frac{18(12.011)}{18(12.011) + 24(1.00794) + 2(15.9994)} \times 100\% = \boxed{79.37\%}$$

(b) Testosterone contains more carbon atoms per molecule than the female sex hormone, but it also contains additional *non*-carbon atoms (four hydrogen atoms). The difference in formula between the two is "CH_4." This fragment is almost exactly $12/16$ or 75% carbon by mass; adding it to a molecule brings the overall percentage of carbon in that molecule closer to 75%.

2-16 The molar mass of acetaminophen (*p*-acetylaminophenol) is $151.165 \; \text{g mol}^{-1}$, the sum of the contributions to the mass by the elements given in the formula $C_8H_9NO_2$. The mass percentages are:

$$\boxed{\text{C, } 63.56\%; \; \text{H, } 6.001\%; \; \text{N, } 9.266\%; \; \text{O, } 21.17\%}$$

2-18 The magnesium in the compound remains magnesium even if its symbol is written twice in this formula. Rewrite the formula as $Mg_3Si_2O_6F_2$, and the computation is straightforward. The molecular mass is 263.079 and the atomic mass of the three magnesium atoms is 72.915. The mass percentage of magnesium is then $\boxed{27.716\%}$.

2-20 The pharmacist mixes 286 g of one compound with 150 g of another, using water as a blending agent. After all the water is driven away, the mixture obviously weighs 436 g. The first compound (Na_2CO_3) contains a fixed proportion of carbon as does the second ($C_2H_5NO_2$). The mass of carbon in the Na_2CO_3 is $(12.011/105.989) \times 286$ g and the mass of carbon in the $C_2H_5NO_2$ is $(2 \times 12.011/75.067) \times 150$ g where the 105.989 and 75.067 are the respective molecular masses of the compounds. The mass of carbon in the mixture is the sum of these two masses. It equals 80.41 g. The mass percentage of carbon in the mixture is this mass divided the mass of the mixture (436 g) and multiplied by 100%. It is $\boxed{18.4\%}$.

2-22 The masses of bromine, carbon and hydrogen add up to the mass of the sample. Therefore, the compound contains only these three elements. Convert each of the elemental masses to chemical amount by dividing by the molar mass of the element: there are 0.4748 mol of Br, 0.1582 mole of C, and 0.1587 mol of H.[1] The three chemical amounts stand in the ratio of 3.00 to 1.00 to 1. Within the precision of the data this ratio is 3 to 1 to 1. The empirical formula is $\boxed{Br_3CH}$.

2-24 Exactly 100 g of this compound contains 75.26 g of Ag and 24.74 g of Cl. Divide each of these masses by the molar mass of the element to obtain the number of moles of that element per 100 g of compound:

$$\frac{75.26 \text{ g Ag}}{107.8682 \text{ g mol}^{-1}} = 0.6977 \text{ mol Ag} \quad \text{and} \quad \frac{24.74 \text{ g Cl}}{35.453 \text{ g mol}^{-1}} = 0.6978 \text{ mol Cl}$$

The chemical amounts of the two elements are essentially equal. Therefore the empirical formula is $\boxed{AgCl}$.

2-26 The compound contains only two elements, so exactly 100 g sample would furnish 1.6907 g of oxygen and 98.3093 g of cesium. Convert both of these masses to chemical amounts by dividing them by the molar masses of the element. There are 0.739694 mol of Cs and 0.105673 mol of O. The ratio of these molar amounts is 6.9999 to 1. Within the precision of the data the formula is $\boxed{Cs_7O}$.

Students often ask whether they should use the molar mass of O_2 to compute the chemical amount of oxygen. After all, gaseous O_2 is formed. A good answer is that they may do so if they wish, but they must then also use O_2 in the formulas that result. It is easy to confirm in this problem that the chemical amount of gaseous O_2 that evolves is 0.0528364 mol. The ratio of 0.739694 mol to 0.0528364 mol is 14.000 to 1. The chemical formula by this method is then $Cs_{14}(O_2)_1$—which is the same as Cs_7O.

2-28 Proceeding exactly as in problem 2-24 gives these empirical formulas of the five compounds (in alphabetical order): $\boxed{CO_2;\ CO;\ C_4O_3;\ C_3O_2;\ C_5O_2}$. The data are given to four significant figures, so the ratios of the chemical amounts should be computed to this degree of precision in each of the five cases. Students tend to round-off such ratios far too much; rounding the ratio 4/3 to 3/2, for example. All five of these compositions exist. The first two are common. The third is mellitic anhydride and has the molecular formula $C_{12}O_9$. The last two, which have molecular formulas identical to their empirical formulas, are rare suboxides of carbon.

---

[1]Nonsignificant figures appear in these intermediate values for the sake of greater precision in the final result.

2-30 Take the pewter bowl to weigh 100.0 g. It then contains 92.1 g of tin, 6.1 g of antimony and 1.8 g of copper. Dividing each of these masses by the molar mass of the element gives the chemical amounts of the three: 0.7758 mol of Sn, 0.0501 mole of Sb, and 0.0283 mol of Cu. The *total* number of moles of metals is then 0.8542 mol. The fraction of the moles of metals that are from the Cu is $(0.0283/0.8542) = 0.0331$. The number of atoms of each metal is directly proportional to its number of moles. Also, the total number of atoms of metals is directly proportional to the total number of moles of metals. The copper therefore comprises $\boxed{33}$ out of every 1000 atoms in the bowl.

2-32 87.41 g of the compound contains 51.83 g of I, 15.97 g of K, and, by difference, 19.61 g of O. Convert each of these masses to chemical amount by dividing by the proper elemental molar mass: 0.4084 mol of I, 0.4085 mol of K and 1.2257 mol of O. The ratio of these chemical amounts is 1.000 to 1.000 to 3.001; the chemical formula is $\boxed{KIO_3}$.

2-34 Determine the chemical amounts of the three elements in the original sample:

$$2.3890 \text{ g CaO} \times \left(\frac{1 \text{ mol CaO}}{56.077 \text{ g CaO}}\right) \times \left(\frac{1 \text{ mol Ca}}{1 \text{ CaO}}\right) = 0.042602 \text{ mol Ca}$$

$$1.8760 \text{ g CO}_2 \times \left(\frac{1 \text{ mol CO}_2}{44.0098 \text{ g CO}_2}\right) \times \left(\frac{1 \text{ mol C}}{1 \text{ CO}_2}\right) = 0.042627 \text{ mol C}$$

$$3.9214 \text{ g NO}_2 \times \left(\frac{1 \text{ mol NO}_2}{46.0055 \text{ g NO}_2}\right) \times \left(\frac{1 \text{ mol N}}{1 \text{ NO}_2}\right) = 0.085238 \text{ mol N}$$

The empirical formula is then:

$$\text{"Ca}_{0.042602}\text{C}_{0.042627}\text{N}_{0.085238}\text{"} \implies \boxed{CaCN_2}$$

2-36 The 369.78 mg of carbon dioxide contains a definite mass of carbon because CO_2 has a definite composition:

$$369.78 \text{ mg CO}_2 \times \left(\frac{12.011 \text{ mg C}}{44.0098 \text{ mg CO}_2}\right) = 100.919 \text{ mg C}$$

A similar computation based on the formula of water gives the mass of hydrogen in the original sample. It is 3.6311 mg H. The total mass of the sample was 317.4 mg. Everything in the sample that was neither C nor H was Cl. Hence the mass of Cl in the sample was $317.4 - 3.631 - 100.92 = 212.85$ mg Cl. Converting the masses of the three elements to chemical amounts reveals that the sample contained 6.004 mmol of Cl, 8.402 mmol of C and 3.602 mmol of H. Dividing each chemical amount by the smallest among them gives these ratios: 1.6667 mol Cl to 2.332 mol C to 1 mol H. These numbers equal 5/3, 7/3 and 3/3 within the precision of the data. The empirical formula is $\boxed{Cl_5C_7H_3}$.

2-38 The molar mass computed for the formula $C_{10}H_{12}N_2Cu_3I_5$ is 985.38 g mol^{-1}. This number goes into the quoted molar mass 12.00 times. Hence, the molecular formula is 12 times the empirical formula: $\boxed{C_{120}H_{144}N_{24}Cu_{36}I_{60}}$. This problem becomes difficult when students refuse to believe that a molecular formula may have such big subscripts. Yet the compound is real.

2-40 (a) $2\,Al + 3\,FeO \longrightarrow Al_2O_3 + 3\,Fe$

(b) $3\,Br_2 + I_2 \longrightarrow 2\,IBr_3$

(c) $3\,PbO + 2\,NH_3 \longrightarrow 3\,Pb + N_2 + 3\,H_2O$

(d) $Cl_2O_7 + H_2O \longrightarrow 2\,HClO_4$

(e) $2\,CaC_2 + 3\,CO_2 \longrightarrow 2\,CaCO_3$

2-42 (a) $2\,Al + 6\,HCl \longrightarrow 2\,AlCl_3 + 3\,H_2$

(b) $4\,NH_3 + 5\,O_2 \longrightarrow 4\,NO + 6\,H_2O$

(c) $2\,Fe + O_2 + 2\,H_2O \longrightarrow 2\,Fe(OH)_2$

(d) $2\,HSbCl_4 + 3\,H_2S \longrightarrow Sb_2S_3 + 8\,HCl$

(e) $2\,Al + Cr_2O_3 \longrightarrow Al_2O_3 + 2\,Cr$

(f) $XeF_4 + 2\,H_2O \longrightarrow Xe + O_2 + 4\,HF$

(g) $(NH_4)_2Cr_2O_7 \longrightarrow N_2 + Cr_2O_3 + 4\,H_2O$

(h) $NaBH_4 + 2\,H_2O \longrightarrow NaBO_2 + 4\,H_2$

2-44 (a) $C_8H_8 + 10\,O_2 \longrightarrow 8\,CO_2 + 4\,H_2O$

(b) $Mn_2O_7 + CaO \longrightarrow Ca(MnO_4)_2$

(c) $4\,Mg + 3\,N_2 \longrightarrow 2\,Mg_2N_3$

(d) $4\,Fe_3O_4 + O_2 \longrightarrow 6\,Fe_2O_3$

(e) $2\,MgO + C \longrightarrow 2\,Mg + CO_2$

2-46 (a) This is the first step in the two-step process for the production of zinc considered in Example 2-13 (text page 60):

$$2\,ZnS + 3\,O_2 \longrightarrow 2\,ZnO + 2\,SO_2$$

(b) The point here is that a reaction that does not occur can still be represented by a respectable balanced equation:

$$ZnS + 2\,O_2 \longrightarrow ZnO + SO_3$$

2-48 The law of conservation of mass is fulfilled in every case.

(a) One molecule of CO_2 (mass 44.01 u) reacts with one atom of C (mass 12.01 u) to give 2 molecules of CO (mass 2 times 28.01 u): $\boxed{56.02 \text{ u in; } 56.02 \text{ u out}}$.

(b) Two molecules of C_6H_{14} (mass 172.4 u) react with 19 molecules of O_2 (mass 608.0 u) to give CO_2 (528.1 u) and water (252.2 u): $\boxed{780.4 \text{ u in; } 780.3 \text{ u out}}$. Sometimes, as here, rounding off causes an apparent deviation from the law of conservation of mass. In fact, the law is fulfilled within the precision i implied by given the masses to ± 0.1 u.

(c) One molecule of Mn_2O_7 (mass 221.9 u) and one molecule of water (mass 18.0 u) give two molecules of $HMnO_4$ (mass 239.9 u): $\boxed{239.9 \text{ u in; } 239.9 \text{ u out}}$.

2-50 According to the balanced equation, 2 mol of NH_3 gives 2 mol of HCN and consumes 3 mol of O_2 in doing so. Therefore, 6.48 mol of NH_3 gives $\boxed{6.48 \text{ mol}}$ of HCN and consumes $3/2 \times 6.48 = \boxed{9.72}$ mol of O_2 in doing so.

2-52 In all parts, the computation follows the pattern outlined in text Figure 2-6 (text page 52).

$$\text{(a)} \quad 1.000\,\text{g}\,O_2 \times \left(\frac{1\,\text{mol}\,O_2}{32.00\,\text{g}\,O_2}\right) \times \left(\frac{2\,\text{mol}\,C_5H_{12}}{16\,\text{mol}\,O_2}\right) \times \left(\frac{72.151\,\text{g}\,C_5H_{12}}{1\,\text{mol}\,C_5H_{12}}\right)$$

$$= \boxed{0.2818\,\text{g}\,C_5H_{12}}$$

$$\text{(b)} \quad 1.000\,\text{g}\,H_2O \times \left(\frac{1\,\text{mol}\,H_2O}{18.0153\,\text{g}\,H_2O}\right) \times \left(\frac{1\,\text{mol}\,XeF_4}{2\,\text{mol}\,H_2O}\right) \times \left(\frac{207.28\,\text{g}\,XeF_4}{1\,\text{mol}\,XeF_4}\right)$$

$$= \boxed{5.753\,\text{g}\,XeF_4}$$

(c) $1.000 \text{ g KOH} \times \left(\dfrac{1 \text{ mol KOH}}{56.106 \text{ g KOH}}\right) \times \left(\dfrac{1 \text{ mol K}_2\text{Cr}_2\text{O}_7}{2 \text{ mol KOH}}\right)$

$$\times \left(\dfrac{294.185 \text{ g K}_2\text{Cr}_2\text{O}_7}{1 \text{ mol K}_2\text{Cr}_2\text{O}_7}\right) = \boxed{2.622 \text{ g K}_2\text{Cr}_2\text{O}_7}$$

(d) $1.000 \text{ g H}_2\text{O}_2 \times \left(\dfrac{1 \text{ mol H}_2\text{O}_2}{34.0147 \text{ g H}_2\text{O}_2}\right) \times \left(\dfrac{1 \text{ mol N}_2\text{H}_4}{2 \text{ mol H}_2\text{O}_2}\right) \times \left(\dfrac{32.0451 \text{ g N}_2\text{H}_4}{1 \text{ mol N}_2\text{H}_4}\right)$

$$= \boxed{0.4711 \text{ g N}_2\text{H}_4}$$

2-54

$$m_{\text{CS}_2} = 67.2 \text{ g S} \times \left(\dfrac{1 \text{ mol S}}{32.06 \text{ g S}}\right) \times \left(\dfrac{1 \text{ mol CS}_2}{4 \text{ mol S}}\right) \times \left(\dfrac{76.139 \text{ g CS}_2}{1 \text{ mol CS}_2}\right) = \boxed{39.9 \text{ g CS}_2}$$

2-56

$$m_{\text{Ca}_3(\text{PO}_4)_3} = 69.8 \text{ g P}_4 \times \left(\dfrac{1 \text{ mol P}_4}{123.896 \text{ g P}_4}\right) \times \left(\dfrac{2 \text{ mol Ca}_3(\text{PO}_4)_2}{1 \text{ mol P}_4}\right)$$

$$\times \left(\dfrac{310.177 \text{ g Ca}_3(\text{PO}_4)_2}{1 \text{ mol Ca}_3(\text{PO}_4)_2}\right) = \boxed{349 \text{ g Ca}_3(\text{PO}_4)_2}$$

$$m_{\text{CaSiO}_3} = 69.8 \text{ g P}_4 \times \left(\dfrac{1 \text{ mol P}_4}{123.896 \text{ g P}_4}\right) \times \left(\dfrac{6 \text{ mol CaSiO}_3}{1 \text{ mol P}_4}\right) \times \left(\dfrac{116.16 \text{ g CaSiO}_3}{1 \text{ mol CaSiO}_3}\right)$$

$$= \boxed{393 \text{ g CaSiO}_3}$$

2-58 The equation furnishes the ratio 5 mol O_2 to 4 mol NO. The same ratio applies to the volumes of the gases as long as the temperature and pressure are constant. Hence $5/4 \times 16.4 = \boxed{20.5 \text{ L}}$ of O_2 is needed.

2-60 Whatever the details of the reactions that give $Pt_2C_{10}H_{18}N_2S_2O_6$, the chemical amount that forms cannot exceed one half of the chemical amount of platinum because the compound contains two atoms of Pt per molecule. Thus:

$$1.406 \text{ g Pt} \times \left(\dfrac{1 \text{ mol Pt}}{195.08 \text{ g Pt}}\right) \times \left(\dfrac{1 \text{ mol Pt}_2\text{C}_{10}\text{H}_{18}\text{N}_2\text{S}_2\text{O}_6}{2 \text{ mol Pt}}\right)$$

$$\times \left(\dfrac{716.555 \text{ g Pt}_2\text{C}_{10}\text{H}_{18}\text{N}_2\text{S}_2\text{O}_6}{1 \text{ mol Pt}_2\text{C}_{10}\text{H}_{18}\text{N}_2\text{S}_2\text{O}_6}\right) = \boxed{2.582 \text{ g Pt}_2\text{C}_{10}\text{H}_{18}\text{N}_2\text{S}_2\text{O}_6}$$

2-62 By subtraction, 3.802 g of H_2O was driven off by heating the $Na_2SiO_3 \cdot 9H_2O$. This is equivalent to 0.21104 mol of H_2O (taking the molar mass ($\mathcal{M}$) of water to equal 18.0153 g mol^{-1}). Meanwhile the original sample consisted of 0.035186 mol of $Na_2SiO_3 \cdot 9H_2O$ ($\mathcal{M} = 284.2008$ g mol^{-1}). The ratio of these two chemical amounts is 5.998 to 1, which means that 6.00 mol of water was lost during the heating for every 1.00 mol of compound that was present before the heating. The formula of the residue is $\boxed{Na_2SiO_3 \cdot 3H_2O}$.

2-64 Balance an equation to represent the reaction:

$$Si_4H_{10} + 13/2\,O_2 \longrightarrow 4\,SiO_2 + 5\,H_2O$$

Use the balanced equation to help generate a series of chemical conversion factors:

$$m_{SiO_2} = 25.0 \text{ cm}^3 \text{ Si}_4\text{H}_{10} \times \left(\frac{0.825 \text{ g Si}_4\text{H}_{10}}{1 \text{ cm}^3 \text{ Si}_4\text{H}_{10}}\right) \times \left(\frac{1 \text{ mol Si}_4\text{H}_{10}}{122.4194 \text{ g Si}_4\text{H}_{10}}\right)$$

$$\times \left(\frac{4 \text{ mol SiO}_2}{1 \text{ mol Si}_4\text{H}_{10}}\right) \times \left(\frac{60.0843 \text{ g SiO}_2}{1 \text{ mol SiO}_2}\right) = \boxed{40.5 \text{ g SiO}_2}$$

2-66 Let Wi stand for a wing, Fu for a fuselage, En for an engine, and Wh for a wheel. In terms of these symbols, the production of a kit is represented

$$2\,Wi + 1\,Fu + 4\,En + 6\,Wh \longrightarrow 1\,kit$$

A quick way to identify the limiting reactant in a chemical reaction is to divide the number of moles of each reactant by its coefficient in the balanced equation and then to compare the numbers. The smallest number identifies the limiting reactant. Doing this with the "reaction" shows that the engines run out first in the making of the kits. Because there are $31,008/4 = 7,752$ sets of engines, there can be only $\boxed{7,752 \text{ kits}}$. There are 1,750 wings, 961 fuselages, and 1,631 wheels left over.

2-68 Compute how many reams of copy stock and how many reams of cover stock are available. This is done by dividing the mass of each material by the mass per ream.

$$5.00 \text{ kg cover} \times \left(\frac{1 \text{ ream cover}}{3.071 \text{ kg cover}}\right) = 1.628 \text{ ream cover}$$

$$50.00 \text{ kg copy} \times \left(\frac{1 \text{ ream copy}}{1.628 \text{ kg copy}}\right) = 30.71 \text{ ream copy}$$

The materials are assembled according to the equation:

$$1 \text{ cover sheet} + 16 \text{ copy sheets} \longrightarrow 1 \text{ booklet}$$

With unlimited copy stock, 1.628 reams of booklets could be made. With unlimited cover stock, $30.71/16 = 1.919$ reams of booklets could be made. Output is limited by the supply of copy stock; it cannot exceed 1.628 reams of booklets. At 500 booklets per ream of booklets, this equals $\boxed{814 \text{ booklets}}$.

The assembly of the booklets is analogous to a chemical reaction. The ream is analogous to the chemical mole, and the masses per ream are analogous to molar masses.

2-70 The silicon and carbon are present in equal masses, but not in equal chemical amounts: n_{Si} is *less* than n_C because the molar mass of silicon is *greater* than that of carbon. The two elements react in a 1 to 1 molar ratio, so the $\boxed{\text{silicon}}$ runs out first. The chemical amount of silicon is $2.00 \text{ g}/28.086 \text{ g mol}^{-1} = 0.0712 \text{ mol}$. From the balanced equation, this is also the chemical amount of SiC formed; the mass of 0.0712 mol of SiC is $\boxed{2.86 \text{ g}}$.

2-72 (a) The 2.135 g of HCl ($\mathcal{M} = 36.461 \text{ g mol}^{-1}$) amounts to 0.05856 mol of HCl, and the 11.563 g of $AgNO_3$ ($\mathcal{M} = 169.873 \text{ g mol}^{-1}$) amounts to 0.06807 mol of $AgNO_3$. The two react in a 1 to 1 ratio, so the $\boxed{AgNO_3}$ is in excess.

(b) The unreacted $AgNO_3$ amounts to $0.06807 - 0.05856 = 0.00951 \text{ mol}$. This equals $\boxed{1.62 \text{ g}}$ of $AgNO_3$.

2-74 The $AgNO_3$ and NaCl react in a 1 to 1 molar ratio to give AgCl, which precipitates, and $NaNO_3$, which stays in solution. The chemical amount of $AgNO_3$ is 0.63500 mol, and the chemical amount of NaCl is slightly less, 0.60663 mol. Hence the NaCl is the limiting reactant. The theoretical yield of AgCl is computed based on the amount of the limiting reactant:

$$35.453 \text{ g NaCl} \times \left(\frac{1 \text{ mol NaCl}}{58.4428 \text{ g NaCl}} \right) \times \left(\frac{1 \text{ mol AgCl}}{1 \text{ mol NaCl}} \right)$$
$$\times \left(\frac{143.323 \text{ g AgCl}}{1 \text{ mol AgCl}} \right) = \boxed{86.944 \text{ g AgCl}}$$

2-76 Assume that all of the nitrogen in the 2.33 L of N_2 ends up in the N_2O. If so, the maximum possible yield of gaseous N_2O would equal 2.33 L. Now, assume that all of the oxygen in the 2.70 L of O_2 ends up in the N_2O. The maximum yield of gaseous N_2O would then equal 5.40 L, because two moles of N_2O, as shown by its formula, can be formed from every one mole of O_2. Obviously, N_2 is the limiting reactant. When it is all consumed, a maximum of $\boxed{2.33 \text{ L}}$ of N_2O has formed.

2-78

$$7.39 \text{ kg TiO}_2 \times \left(\frac{1 \text{ kmol TiO}_2}{79.8788 \text{ kg TiO}_2}\right) \times \left(\frac{1 \text{ kmol TiCl}_4}{1 \text{ kmol TiO}_2}\right)$$

$$\times \left(\frac{189.692 \text{ kg TiCl}_4}{1 \text{ kmol TiCl}_4}\right) = \boxed{17.55 \text{ kg TiCl}_4}$$

$$\% \text{ yield} = \left(\frac{\text{actual}}{\text{theoretical}}\right) \times 100\% = \left(\frac{14.24 \text{ kg TiCl}_4}{17.55 \text{ kg TiCl}_4}\right) \times 100\% = \boxed{81.1\%}$$

2-80 The yield of the product is less than the theoretical, and this fact must be reckoned with. A very good approach is to use the percentage yield to construct an additional unit-factor (the second factor in the following):

$$125 \text{ g Si}_3\text{N}_4 \text{ isolated} \times \left(\frac{100 \text{ g Si}_3\text{N}_4 \text{ formed}}{95.0 \text{ g Si}_3\text{N}_4 \text{ isolated}}\right) \times \left(\frac{1 \text{ mol Si}_3\text{N}_4}{140.2835 \text{ g Si}_3\text{N}_4 \text{ formed}}\right)$$

$$\times \left(\frac{3 \text{ mol Si}}{1 \text{ mol Si}_3\text{N}_4}\right) \times \left(\frac{28.0855 \text{ g Si}}{1 \text{ mol Si}}\right) = \boxed{79.0 \text{ g Si}}$$

2-82 First calculate the number of moles of TCDD absorbed by a kilogram of the body using the molar mass of TCDD (321.97 g mol^{-1}) computed according to the molecular formula given in problem 2-83:

$$\frac{107,000 \text{ ng TCDD}}{1 \text{ kg body}} \times \left(\frac{10^{-9} \text{ g}}{1 \text{ ng}}\right) \times \left(\frac{1 \text{ mol TCDD}}{321.97 \text{ g TCDD}}\right)$$

$$\times \left(\frac{0.2 \text{ g TCDD absorbed}}{1 \text{ g TCDD applied}}\right) = \frac{6.65 \times 10^{-8} \text{ mol TCDD absorbed}}{1 \text{ kg body}}$$

The kilogram of body mass consists approximately of 55.5 mol of H_2O. The fraction of TCDD molecules is the number of moles of TCDD in the kilogram divided by the total number of moles. It is $\boxed{1.2 \times 10^{-9}}$.

2-84 The mass of calcium in the annual production of portland cement is 46 percent of 800,000,000 metric tons, or 368,000,000 metric tons. By the formula of lime (CaO), calcium contributes $(40.08/56.08) = 0.715$ of the mass of lime. The mass of lime is therefore $368,000,000/0.715 = \boxed{500,000,000 \text{ metric ton}}$.

2-86 Hydrogen (H_2) and oxygen (O_2) combine in the mass ratio $2.01/15.9994$ to form water. Since 16.0 g of oxygen requires 2.0 g of hydrogen, 8.0 g of oxgyen requires 1.0 g of hydrogen. It follows that 8.0 lb of oxygen requires $\boxed{1.0 \text{ lb}}$ of hydrogen.

2-88 The problem requires more significant figures than similar previous computations, but follows the same principles. The molar mass of the compound is 13 932.2. The mass percentages are:

$$\boxed{C,\ 59.571\%;\ H,\ 6.497\%;\ N,\ 12.567\%;\ O,\ 18.833\%;\ S,\ 2.532\%}$$

2-90 If the compound is ErB_{69} it is 18.32% erbium by mass. If the compound is ErB_{61} it is 20.23% erbium by mass. The range is $\boxed{18.32\text{-}20.23\%}$.

2-92 All of the molecular formulas are integral multiples of the formula C_5H_8; they have a $\boxed{\text{moleular subunit in common}}$. (All of these compounds are terpenes: a diterpene, triterpene, and tetraterpene).

2-94 Calculate the empirical formula of djenkolic acid from the mass percentages. First, the chemical amounts of the component elements in 100 g of djenkolic acid:

$$33.06 \text{ g C} \times \frac{1 \text{ mol C}}{12.01 \text{ g C}} = 2.753 \text{ mol C} \qquad 5.55 \text{ g H} \times \frac{1 \text{ mol H}}{1.008 \text{ g H}} = 5.506 \text{ mol C}$$

$$11.01 \text{ g N} \times \frac{1 \text{ mol N}}{14.01 \text{ g N}} = 0.7859 \text{ mol N} \qquad 25.16 \text{ g O} \times \frac{1 \text{ mol O}}{16.00 \text{ g O}} = 1.572 \text{ mol O}$$

$$25.22 \text{ g S} \times \frac{1 \text{ mol S}}{32.07 \text{ g S}} = 0.7864 \text{ mol S}$$

This gives the formula "$C_{2.753}H_{5.506}N_{0.7859}O_{1.573}S_{0.7864}$." Dividing each subscript by 0.7859 , which is the smallest subscript, re-expresses the relative number of atoms as $C_{3.503}H_{7.006}N_1O_{2.002}S_{1.001}$. which is very close to $\boxed{C_7H_{14}N_2O_4S_2}$. This last formula is the molecular formula because it represents two atoms of S as required by the problem. The molar mass of the compound is $\boxed{254.45 \text{ g mol}^{-1}}$.

2-96 The mass percentages of the elements in the two oxides are given to six significant figures. The precision in the calculation is limited by the fact that the molar mass of tungsten is known to only five significant digits. There is 0.43132 mol of W in a 100.000 g sample of the white oxide, and 1.29395 mol of O in the same sample. The ratio of these chemical amounts is 3.0000—the empirical formula is $\boxed{WO_3}$. In the blue oxide the chemical amounts in a 100.000 g sample are 0.43975 mol W and 1.19709 mol of O. The ratio of these numbers is 2.7222. This turns out to equal the ratio of 49 to 18, within the precision of the data. Hence the formula $\boxed{W_{18}O_{49}}$ is a correct answer. The blue oxide is really a nonstoichiometric compound, however.

2-98 (a) "Binary" means that the only elements in the three compounds are oxygen and M. The first compound contains 13.38 g of O for every 86.62 g of M, which is $\boxed{0.15447 \text{ g}}$ of O per gram of M. Similarly, the second and third compounds have $\boxed{0.1029 \text{ g}}$ of O and $\boxed{0.07721 \text{ g}}$ of O per gram of M, respectively.

(b) If the first compound of MO_2, then the second is "$MO_{4/3}$" because the second compound has almost exactly 2/3 the oxygen per quantity of M as the first. This formula is improved by multiplying both subscripts by 3 to clear the fraction. The answer is $\boxed{M_3O_4}$. The third compound is $\boxed{MO}$ if the first is MO_2 because it has almost exactly 1/2 as much oxygen per quantity of M as the first.

(c) Compute the amount of metal that combines with 2 mol (2×15.9994 g) of O in the first compound. It is 207.2 g M. This means that the relative atomic mass of M is 207.2. because 1 mol of M is combined in this compound with 2 mol of O. Scanning the periodic table for metallic elements of the proper relative atomic mass establishes $\boxed{\text{lead}}$ as the metal.

2-100 The given information can be summarized in the equation

$$(C_6N_2O_3)_n + 3\,H_2O \longrightarrow (C_3HNO_2)_{2n}$$

where the subscript n is the number of times the empirical formula is repeated in the molecular formula. A $2n$ appears on the right because this is the only way that C and N will be balanced. There are $(3n + 3)$ O's on the left and $4n$ O's on the right. Obviously $n = 3$. The molecular formula is then $\boxed{C_{18}N_6O_9}$.

The molar mass is $\boxed{444.23 \text{ g mol}^{-1}}$.

2-102 The equations given in this problem do not represent single chemical changes but rather the sum of the two chemical changes:

$$C + O_2 \longrightarrow CO_2 \quad \text{and} \quad 2\,C + O_2 \longrightarrow 2\,CO$$

The first equation is formed by multiplying the first equation by 4 and the second equation by 3 and adding them; the second equation comes from multiplying the first by 5 and the second by 2 and adding them. An infinite number of other "balanced equations" can be formed by picking different multiplying factors.

2-104 The general formula C_nH_{2n} implies that all alkenes have the same fractions of carbon and of hydrogen by mass. This fraction appears as the first unit-factor in the following two equations. All the carbon from an alkene ends up incorporated in CO_2 and all of the hydrogen ends up incorporated in H_2O if an alkene is completely burned. The second unit-factors derive from the compositions by mass of these two products:

$$1.000 \text{ g alkene} \times \left(\frac{12.011 \text{ g C}}{14.027 \text{ g alkene}}\right) \times \left(\frac{44.01 \text{ g CO}_2}{12.011 \text{ g C}}\right) = \boxed{3.138 \text{ g CO}_2}$$

$$1.000 \text{ g alkene} \times \left(\frac{2.0159 \text{ g H}}{14.027 \text{ g alkene}}\right) \times \left(\frac{18.015 \text{ g H}_2\text{O}}{2.0159 \text{ g H}}\right) = \boxed{1.284 \text{ g H}_2\text{O}}$$

It is also possible to write the general equation for the burning of an alkene in oxygen as

$$C_nH_{2n} + 3/2\,n\,O_2 \longrightarrow n\,CO_2 + n\,H_2O$$

and solve a stoichiometry problem: unknown n's appear in the molar mass of the alkene and in the number of moles of CO_2, but they cancel out.

2-106 (a) Heating the hydrated lanthanum phosphate breaks it down to H_2O, La_2O_3 and P_4O_{10}. Note that the sum of the masses of the three product compounds *does* equal the 1.000 g of compound originally present. Compute the masses of P and O in the 0.2922 g of P_4O_{10} using the known proportions of the elements in that compound. There is 0.1275 g of P and 0.1647 g of O. By similar means, there are 0.5719 g of La and 0.0988 g of O in the La_2O_3, and there are 0.0042 g of H and 0.0329 g of O in the H_2O. Summing up, there is 0.1275 g of P, 0.5719 g of La, 0.0042 g of H, and 0.2964 g of O in the original 1.000 g of hydrated lanthanum phosphate. From these masses, compute the chemical amounts of the four elements by dividing by their molar masses. The outcome is:

$$P_{0.004116}La_{0.004117}H_{0.00417}O_{0.01853} \Longrightarrow P_2La_2H_2O_9 \Longrightarrow \boxed{(LaPO_4)_2 \cdot H_2O}$$

(b) The equation is $\boxed{2\,(LaPO_4)_2 \cdot H_2O \longrightarrow P_4O_{10} + 2\,La_2O_3 + 2\,H_2O}$.

2-108 The balanced equation is $4\,KClO_3 \longrightarrow 3\,KClO_4 + KCl$, which means that 4 mol of $KClO_3$ is equivalent in this system to 3 mol of $KClO_4$ (see second conversion factor below). The theoretical yield of the $KClO_4$ must be computed and then compared to the observed yield of 3.00 g:

$$4.00 \text{ g } KClO_3 \times \left(\frac{1 \text{ mol } KClO_3}{122.549 \text{ g } KClO_3}\right) \times \left(\frac{3 \text{ mol } KClO_4}{4 \text{ mol } KClO_3}\right)$$

$$\times \left(\frac{138.548 \text{ g } KClO_4}{1 \text{ mol } KClO_4}\right) = 3.392 \text{ g } KClO_4$$

The comparison is performed by dividing the actual by the theoretical and multiplying by 100%. The result is $\boxed{88.5\%}$.

2-110 Write the balanced equation

$$4\,KO_2 + 2\,CO_2 \longrightarrow 2\,K_2CO_3 + 3\,O_2$$

Now, if 117 L of O_2 is 98.5% of the theoretical yield, then $117/0.985 = 118.8$ L of O_2 is 100%. According to the law of combining volumes, $2/3 \times 118$ or $\boxed{78.7 \text{ L}}$ of CO_2 must have been used.

2-112 (a) The volume of piperidine is known in U.S. gallons. The density and molar mass of piperidine are needed to compute first the mass and then the chemical amount of piperidine. If the density is in g cm^{-3}, which is a common unit, then the number of cm^3 per U.S. gallon will be needed. According to a standard handbook, the density of piperidine at room conditions is 0.862 g cm^{-3}, and there are 3785 cm^3 per gallon. (In the absence of a reference value, it often works to take density of organic liquids to equal 0.7 to 1 g cm^{-3}.)

(b) It is possible that a nitrogen-containing by-product is evolved in the synthesis of the phencyclidine. If so, *less* than one mole of phencyclidine would be produced for every one mole of piperidine used.[2]

(c) Represent piperidine as "pip" and phencyclidine as "phcyc":

$$5.0 \text{ gall pip} \times \left(\frac{3785 \text{ cm}^3 \text{ pip}}{1 \text{ gall pip}}\right) \times \left(\frac{0.862 \text{ g pip}}{1 \text{ cn}^3 \text{ pip}}\right) \times \left(\frac{1 \text{ mol pip}}{85.15 \text{ g pip}}\right)$$

$$\times \left(\frac{1 \text{ mol phcyc}}{1 \text{ mol phcyc}}\right) \times \left(\frac{243.4 \text{ g phcyc}}{1 \text{ mol phcyc}}\right) \times \left(\frac{1 \text{ kg phcyc}}{1000 \text{ g phcyc}}\right) = \boxed{47 \text{ kg phcyc}}$$

---

[2]Compare to problem 2-54 in which fully half of the sulfur reacted in making CS_2 ends up in a by-product.

2-114 Set up two equations in two unknowns:

$$0.027 = \frac{m_O}{(1.0 + m_{C_5H_{12}O})} \qquad m_O = \frac{16.0}{88.15} m_{C_5H_{12}O}$$

where $m_{C_5H_{12}O}$ is the mass of MTBE that must be added, and m_O is the mass of oxygen that is added by doing so. Note that 88.15 is the relative molecular mass of MTBE. Eliminating m_O between the two equations and solving gives $\boxed{0.17 \text{ kg}}$ of MTBE.

Answers to Even-Numbered Problems, Chapter 3

3-2 Tungsten and molybdenum are in the chromium group of the periodic table, so we predict $\boxed{Mo(CO)_6}$ and $\boxed{W(CO)_6}$ as the formulas of their carbonyls. Based on similar comparisons, the formula of rhenium carbonyl should be $\boxed{Re_2(CO)_{10}}$, and the formula of ruthenium carbonyl $\boxed{Ru(CO)_5}$.

3-4 Count the number of valence electons each atom can gain or lose. Then use the principle of electrical neutrality. (a) MgO   (b) CsCl   (c) AlBr₃

3-6

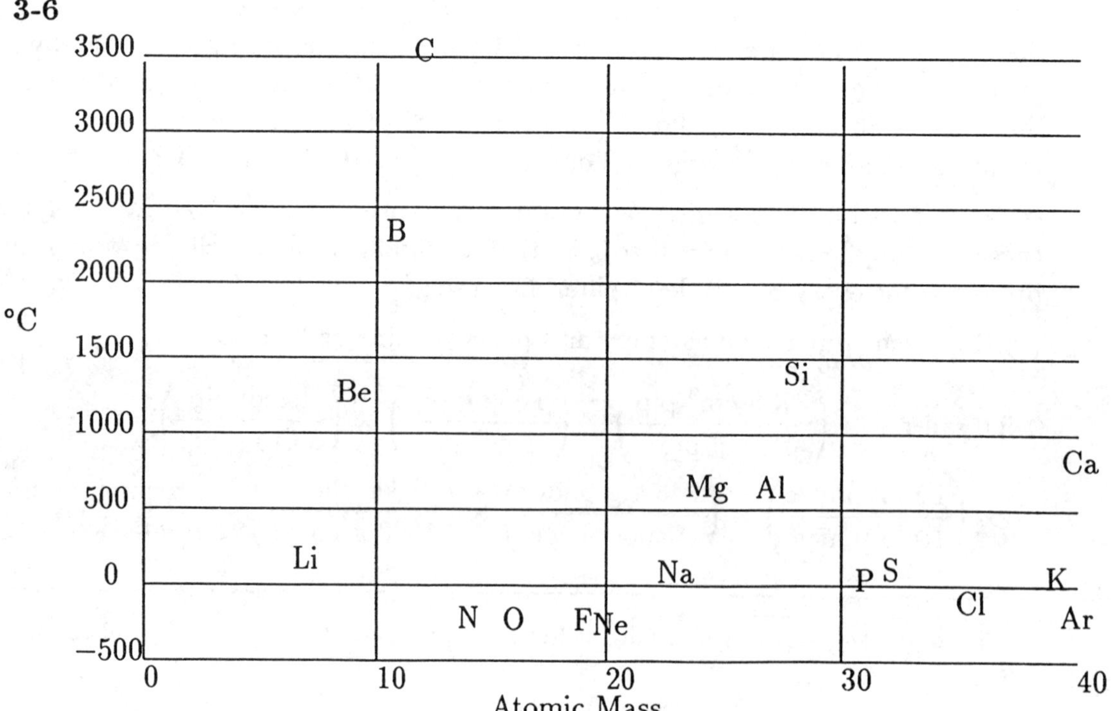

The melting points of the 18 elements from Li to Ca inclusive show a distinctly different pattern when plotted against atomic mass than when plotted against atomic number (problem 3-5). The relative atomic mass of potassium is less than that of argon whereas its atomic number is larger than that of argon.

3-8 (a) The element Q is in group VI of the periodic table.

(b) The possibilities for Q are oxygen, sulfur, selenium, tellurium and polonium. Perhaps the simplest way to get the answer is to compute the percentage of the chalcogen in K_2O, K_2S, K_2Se, K_2Te, and K_2Po. Only K_2S comes close, and it comes very close, to the quoted percentage.

3-10 The only metallic element that is a liquid at 298 K ("room temperature") is mercury. Other elements however come close: cesium melts at 301.55 K, and gallium melts at 302.93 K.

3-12 In the following the melting points (top line, in °C), boiling points (middle line, in °C) and densities (bottom line, in g cm^{-3}) of the four immediate neighbors of technetium are arrayed as in the periodic table.

	Mn mp 1244 bp 1962 d 7.2	
Mo mp 2610 bp 5560 d 10.2	Tc	**Ru** mp 2310 bp 3900 d 12.3
	Re mp 3180 bp 5627 d 20.5	

There is no reason to treat trends across the periodic table as any more or less important than trends down the table. Hence, we just average the four melting points to get a predicted melting point, and do the same with the boiling points and the densities to predict Tc's boiling point and density. The predictions are 2336°C for melting , 4262°C for boiling , and 12.6 g cm^{-3} for d . The experimental values are 2172°C for melting, to which the prediction comes reasonably close, 4877°C for boiling, and 11.50 g cm^{-3} for the density.

3-14 The predicted formulas are GeH_4, HF, H_2Te, BiH_3.

3-16 (a) Xe has 46 core electrons and 8 valence electrons: $:\ddot{\ddot{Xe}}:$

(b) Sn has 46 core elecrons and 4 valence electrons: $\cdot\dot{Sn}\cdot$

(c) B has 2 core electrons and 3 valence electrons: $\cdot\dot{B}\cdot$

(d) Bi has 78 core electrons and 5 valence electrons: $:\dot{\ddot{Bi}}\cdot$

3-18 (a) $(:\ddot{O}\cdot)^-$ (b) $(Ca\cdot)^+$ (c) $(K\cdot)$ (d) $(\cdot Cs\cdot)^-$

3-20

	valence electrons	total electrons	core electrons
O^-	7	9	2
Ca^+	1	19	18
K	1	19	18
Cs^-	2	56	54

3-22 (a) Lithium telluride (Li_2Te) is a compound between lithium and tellurium. It forms by transfer of electrons:

$$2\,Li\cdot \ + \ \cdot\ddot{Te}\cdot \ \longrightarrow \ \left(Li^+\right)\left(:\ddot{Te}:^{2-}\right)\left(Li^+\right)$$

(b) Barium and bromine form $BaBr_2$, barium bromide Each Br atom gains an electron and the Ba atom loses two:

$$\cdot Ba\cdot + 2\left(:\ddot{Br}\cdot\right) \longrightarrow \left(:\ddot{Br}:^-\right)\left(Ba^{2+}\right)\left(:\ddot{Br}:^-\right)$$

(c) Magnesium and astatine form $MgAt_2$, magnesium astatide. Each At atom gains an electron and the Mg atom loses two:

$$\cdot Mg\cdot + 2\left(:\ddot{At}\cdot\right) \longrightarrow \left(:\ddot{At}:^-\right)\left(Mg^{2+}\right)\left(:\ddot{At}:^-\right)$$

(d) Aluminum and chlorine form $AlCl_3$, aluminum chloride. Each chlorine atom gains one electron, and each Al atom loses three:

$$\cdot\dot{Al}\cdot + 3\left(:\ddot{Cl}\cdot\right) \longrightarrow \left(Al^{3+}\right)\left(:\ddot{Cl}:^-\right)_3$$

Note that aluminum chloride actually has substantial covalent character. It consists of dimeric molecules (Al_2Cl_6 molecules) in the low-temperature vapor. In these molecules, two Cl atoms serve as bridges between $AlCl_2$ groups. All of the atoms have valence octets of electrons, but the bridging Cl's have -1 formal charges, and the two Al's have $+1$ formal charges. At high temperature, the dimer dissociates to planar triangular $AlCl_3$ molecules.

(e) Cesium and sulfur form Cs_2S, cesium sulfide. Each S atom gains two electrons and each Cs atom loses one:

$$2\ \text{Cs·} + (\cdot \ddot{\text{S}} \cdot) \longrightarrow (\text{Cs}^+)\ (:\ddot{\text{S}}:{}^{2-})\ (\text{Cs}^+)$$

3-24 (a) potassium nitrite   (b) strontium permanganate   (c) magnesium dichromate   (d) sodium dihydrogen phosphate   (e) barium chloride   (f) sodium chlorate

---

3-26 (a) Cs_2SO_3 (b) $Sr(SCN)_2$ (c) LiH   (d) Na_2O_2 (e) $(NH_4)_2Cr_2O_7$ (f) $RbHSO_4$

3-28 The formula is $\boxed{NH_4H_2PO_4}$, called $\boxed{\text{ammonium dihydrogen phosphate}}$.

3-30 By analogy with the naming of other mixed salts of dianions, the name is $\boxed{\text{sodium hydrogen tartrate}}$.

3-32 (a) The anion "CO_2^-" must have 17 valence electrons, but 18 valence electrons are shown in the Lewis structure.

(b) The $HBCl_3$ molecule must have $1 + 3 + 3(7) = 25$ valence electrons in its Lewis representation. The diagram in the problem shows 26 valence electrons.

(c) A correct Lewis structure of C_2H_2 has only 10 valence electrons. This structure has six too many electrons on each hydrogen.

3-34 (a) The formal charge on the central chlorine atom is $+3$; all four oxygen atoms have formal charges of -1.

(b) The formal charge of the central sulfur in the Lewis diagram for sulfur dioxide is $+1$; the left-hand oxygen (which is single bonded) has a formal charge of -1, and the right-hand oxygen (which is double-bonded) has a formal charge of 0.

(c) The formal charge on the central bromine in this representation of the bonding in the bromite ion is $+1$. The two oxygens both have formal charges of -1.

(d) The formal charge on the central nitrogen in this Lewis diagram for nitrate ion is $+1$. The left-hand oxygen has a formal charge of 0, and the other two oxygen atoms have formal charges of -1.

3-36 In "Cl—Cl—O", the formal charges on the three atoms are, from left to right, 0, $+1$, and -1. In the isomer "Cl—O—Cl", the formal charges on the atoms are all equal to zero. The structure with all formal charges equal to zero is probably favored.

3-38 (a) The "Z" represents a $\boxed{\text{Group V}}$ element, such as N.

(b) The unknown main-group element is a $\boxed{\text{Group VII}}$ element, such as Cl.

(c) The unknown main-group element is a $\boxed{\text{Group VI}}$ element, such as S.

(d) The "Z" is a $\boxed{\text{Group V}}$ element, such as N.

3-40 The Lewis dot[3] structures are:

(a)   (b)   (c)

$$\begin{array}{c} \text{H} \\ | \\ \text{H}-\!\!\underset{|}{\overset{|}{\text{C}}}\!\!-\text{H} \\ | \\ \text{H} \end{array} \qquad\qquad \ddot{\text{O}}=\text{C}=\ddot{\text{O}} \qquad\qquad :\!\ddot{\text{Cl}}-\!\underset{\overset{|}{:\ddot{\text{Cl}}:}}{\text{P}}\!-\ddot{\text{Cl}}\!:$$

3-42 The Lewis structure for the $ClBr_2^-$ ion is $\boxed{(:\!\ddot{\text{Br}}-\ddot{\text{Cl}}-\ddot{\text{Br}}\!:\,)^-}$ with a formal charge on the central Cl of -1.

3-44

$$\text{H}-\!\underset{\overset{|}{\text{H}}}{\overset{\overset{\text{H}}{|}}{\text{C}}}\!-\!\underset{}{\overset{\overset{\ddot{\text{O}}:}{\|}}{\text{C}}}\!-\ddot{\text{O}}-\text{H}$$

The C—H bonds in acetic acid should be about 1.10×10^{-10} m in length. The C=O bond should be about 1.20×10^{-10} m, the C—O about 1.43×10^{-10} m, and the C—C bond about 1.54×10^{-10} m.

---

[3]Shared pairs of dots are represented by lines ("—").

3-46 The octet rule is satisfied for all atoms in these structures. Non-zero formal charges are indicated near the atom, and the non-zero overall charges of molecular ions are shown outside of enclosing brackets:

a) $\ddot{\text{O}} : : \text{C} : : \ddot{\text{N}}$

(b)
$$\left[\; :\ddot{\text{Cl}}-\overset{\overset{\displaystyle :\ddot{\text{O}}:}{|}}{\underset{\underset{\displaystyle :\ddot{\text{O}}:}{|}}{\text{P}}}-\ddot{\text{Cl}}: \; \right]^{-}$$

(c)
$$\left[\; :\overset{-1}{\ddot{\text{O}}}-\overset{\overset{\displaystyle :\ddot{\text{Cl}}:}{|}}{\underset{\underset{\displaystyle :\ddot{\text{Cl}}:}{|}}{\overset{+1}{\text{P}}}}-\overset{-1}{\ddot{\text{O}}}: \; \right]^{-}$$

In Lewis structures, a line equals a pair of dots, but lines are rarely used for lone pairs.

3-48 The carbon-oxygen bonds should be intermediate in length between the length of the C—O bond (1.43×10^{-10} m) and the C=O bond (1.20×10^{-10} m).

$$\left(:\ddot{\text{O}}-\overset{\underset{\displaystyle :\ddot{\text{O}}:}{\|}}{\text{C}}-\ddot{\text{O}}: \quad \longleftrightarrow \quad :\ddot{\text{O}}-\overset{\underset{\displaystyle :\ddot{\text{O}}:}{|}}{\text{C}}=\ddot{\text{O}} \quad \longleftrightarrow \quad \ddot{\text{O}}=\overset{\underset{\displaystyle :\ddot{\text{O}}:}{|}}{\text{C}}-\ddot{\text{O}}: \right)^{2-}$$

3-50

$$:\text{S}\overset{\displaystyle \ddot{\text{S}}-\ddot{\text{S}}:}{} \qquad :\ddot{\text{S}}\overset{\displaystyle \ddot{\text{S}}:}{\underset{\displaystyle \ddot{\text{S}}}{}}$$

In the Lewis structure on the left there is a formal charge of $+1$ on the central S and -1 on the single-bonded terminal S. In the cyclic structure (right) all three sulfur atoms have formal charges of zero. It should be emphasized in discussing this problem that the cyclic and linear structure are *not* resonance forms of each other.

3-52 Each of the four phosphorus atoms has a lone pair of electrons. This accounts for 8 electrons. Each of the six dotted lines is replaced by a pair of electrons. This uses 12 electrons. Thus 20 electrons are used. These 20 electrons are all the valence electrons furnished by the four P atoms in P_4.

3-54

(a)

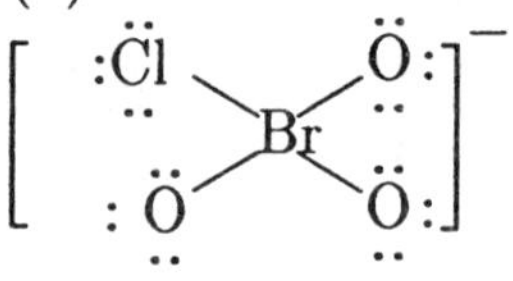

(b)

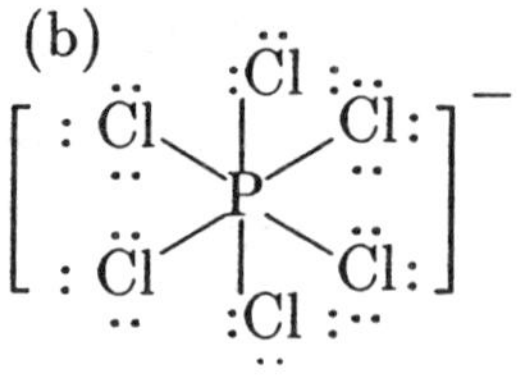

(c)

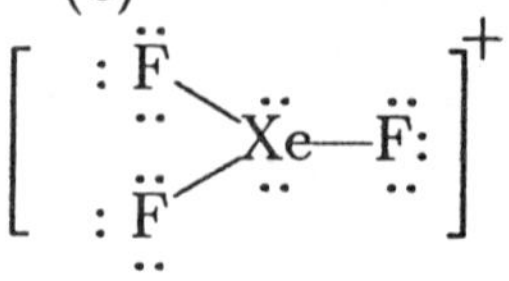

In the above Lewis structure of BrO_4^- ion, each O has a formal charge of -1 and the Br has a formal charge of $+3$. The octet rule is obeyed for every atom. A structure in which all the bonds were double bonds would make the formal charges on the O's all equal zero, expand the valence shell on the central Br to 16, and put a formal charge of -1 on the central Br. Intermediate resonance structures having one, two or three double bonds can also be written. The valence shell on the central Br is expanded to 10, 12 and 14 respectively in these structures.

In the case of PCl_6^- and XeF_3^+ ions, valence shell expansion is not an option but a necessity in drawing a Lewis structure.

3-56 (a) Phosphorus trifluoride PF_3 has a central P with $\boxed{SN\ 4}$.
(b) Sulfuryl chloride SO_2Cl_2 has a central S with $\boxed{SN\ 4}$.
(c) The PF_6^- ion (hexafluorophosphate(V) ion) has a central P with $\boxed{SN\ 6}$.
(d) The ClO_2^- ion (chlorite ion) has a central Cl with $\boxed{SN\ 4}$.
(e) Germane (GeH_4) has a central Ge with $\boxed{SN\ 4}$.

3-58 (a) PF_3, pyramidal (b) SO_2Cl_2 tetrahedral (c) PF_6^- octahedral (d) ClO_2^- bent (e) GeH_4 tetrahedral.

3-60 (a) In TeH_2, the central Te has SN 4. This molecule is bent with the nominal H—Te—H angle distorted to a value *less* than $109.47°$ (the tetrahedral angle).

(b) In AsF_3, the central As has SN 4. The molecule is a trigonal pyramid in which the F—As—F is distorted to a value somewhat less than $109.47°$.

(c) In PCl_4^+, the central P has SN 4. The ion is an undistorted tetrahedron of Cl's about the central P.

(d) In XeF_5^+, the central Xe has SN 6. The ion is a square pyramid with the four angles F_{eq}—Xe—F_{ax} distorted to less than $90°$.

3-62 There are many possible answers for each part. Examples: (a) ClO_4^- (b) CO_2 (c) SbF_6^- (d) ClO_3^-.

3-64 (a) PF_3, Polar (b) SO_2Cl_2 Polar (c) PF_6^- Nonpolar (d) ClO_2^- Polar (e) GeH_4 Nonpolar.

3-66 (a) The $GaCl_4^-$ ion has a central Ga with SN 4; it is predicted to be tetrahedral. The $SbCl_4^-$ ion has a central Sb with SN 5. It is predicted to have a seesaw geometry since one of the five electron pairs is a lone pair.

(b) The $SbCl_2^+$ ion, in which the central Sb has SN 3, is a bent molecular ion, but the $GaCl_2^+$ ion, in which the central Ga has SN 2, is linear. It follows that the formulation $\boxed{(SbCl_2^+)(GaCl_4^-)}$ is more likely correct.

3-68 (a) The observation of a nonzero dipole moment for O_3 rules out a symmetrical linear geometry for the molecule. The molecule of ozone could be linear if the two O-to-O bonds were of different lengths, or if the molecule were bent.

(b) VSEPR theory assigns a steric number of 3 to the central O and predicts that the molecules of ozone are $\boxed{\text{bent}}$.

3-70 (a) H_2Se　(b) N_2F_4　(c) AsH_3　(d) PF_5

3-72 (a) tetraarsenic hexaoxide　(b) hydrogen fluoride　(c) hydrazine (systematic names like dinitrogen tetrahydride or tetrahydrogen dinitride are possible) (d) dichlorine oxide

3-74 The point of the problem is to translate the names into formulas and simultaneously come up with a chemical equation: $Cl_2O_7 \longrightarrow ClO_3 + ClO_4$.

3-76 (a) molybdenum(V) sulfide　(b) nickel(III) fluoride　(c) mercury(I) chloride (Beginners are often troubled by the "I" in this name when both the subscripts are 2's.) (d) niobium(IV) chloride　(e) chromium(III) oxide

3-78 (a) MoI_3　(b) $TlCl$　(c) MnO_2　(d) NbO_2　(e) Cu_2O

3-80 This reduction is represented: $V_2O_5 + 2\,CO \longrightarrow V_2O_3 + 2\,CO_2$.

3-82 Express the percentages of the two elements in the compound in grams and divide each mass by the molar masses given in the problem, including the wrong molar mass of In. The answers are both approximately 1.08 mol, so the molar ratio of the two elements in the compound is 1 to 1. The formula is InO, based on the molar mass of 76 g mol^{-1} for indium.

(b) The true atomic mass of indium is 3/2 of 76, so the experimental percentage of indium in the yellow oxide must have fewer indium atoms than in part (a)—after all, each atom weighs more. We therefore write "$In_{2/3}O$," which translates to In_2O_3 when cleared of the undesirable fractional subscript.

3-84 (a) Element 113 would come directly beneath thallium (Tl) in Group III in the periodic table. It would be expected to be a metal resembling thallium in its properties.

(b) Element 117 would be the next-to-last element in the seventh period. It would come beneath At in Group VII in the periodic table. Element 117 would be even more metallic that At but still probably would form a compound with potassium of formula K(Uus) where Uus is the symbol of element 117.

3-86 The formula of sodium selenate should be $\boxed{Na_2SeO_4}$.

3-88 (a) In the Lewis structure on the left in the problem, the hydrogens and the carbon have formal charges of 0, the $\boxed{\text{S has an fc of } +1}$, and the $\boxed{\text{O has an fc of } -1}$. In the structure on the right, the hydrogens and the oxygen have zero formal charges, the $\boxed{\text{C has fc } -1}$, and $\boxed{\text{S has fc } +1}$.

(b) The only Lewis structure for the molecule of sulfine in which all atoms have formal charges of zero violates the octet rule for the sulfur atom (which has two double bonds and sees a total of ten valence electrons).

$$\begin{array}{c} H \\ \diagdown \\ \diagup C = \ddot{S} = \ddot{O}: \\ H \end{array}$$

3-90 $:N \equiv N:$ $H - \ddot{N} = \ddot{N} - H$

$$\begin{array}{ccc} H & & H \\ \diagdown & \ddot{} & \diagup \\ \ddot{N} & - & \ddot{N} \\ \diagup & & \diagdown \\ H & & H \end{array}$$

The $N \equiv N$ has a length of 1.1×10^{-10} m, the $N = N$ double bond is longer (1.25×10^{-10} m), and the $N - N$ bond is longer yet (1.45×10^{-10} m).

3-92 (a) Ammonia and hydrazoic acid are represented:

$$\begin{array}{c} \ddot{} \\ H - \overset{..}{N} - H \\ | \\ H \end{array}$$

$$H - \ddot{N} - N \equiv \ddot{N} \longleftrightarrow H - \ddot{N} = N = \ddot{N}$$

(b) Both sulfur dioxide and disulfur oxide have 18 valence electrons.

$$: \ddot{O} = \ddot{S} - \ddot{O} : \quad \longleftrightarrow \quad : \ddot{O} - \ddot{S} = \ddot{O} : \qquad : \ddot{S} = \ddot{S} - \ddot{O} : \quad \longleftrightarrow \quad : \ddot{S} - \ddot{S} = \ddot{O} :$$

3-94 The following resonance structures of nitryl chloride together imply equivalent N-to-O bonds:

Nitryl chloride is a reactive gas that decomposes readily to nitrogen dioxide and chlorine. Despite this, it is named as a salt: nitryl ion, a +1 cation, in combination with chloride ion, a −1 anion. The following Lewis structure takes up the suggestion of this name by showing ionic bonding to the Cl. Note that the two N-to-O bonds are still equivalent, but the pair of electrons previously between N and Cl now belongs entirely to the Cl. The octet rule is satisfied for all atoms:

3-96 The steric number of the central Xe in XeF_2 is 5; the molecule is ⌊linear⌋. In XeF_4, the *SN* of the central Xe is 6; the molecule is ⌊square planar⌋. In XeO_3, the *SN* of the central Xe is 4; the molecule is ⌊pyramidal⌋ (like ammonia). In XeO_4, the *SN* of the central Xe is 4; the molecule is ⌊tetrahedral⌋. In H_4XeO_6, the 6 O's are bonded to a central Xe that has *SN* 6. The molecule is ⌊octahedral⌋ with 2 O's and four OH's distributed at the six corners. In $XeOF_4$, the central Xe has *SN* 6. The molecule consists of a ⌊square pyramid⌋ surrounding the Xe with the 4 F's at the corners of the base and the O at the apex (this allows the greatest distance between the long pair and the O).

3-98 The central S has an steric number of 5. Putting the double-bonded O at an equatorial position in the trigonal bipyramidal geometry about the central S minimizes 90° interactions with the four F's.

3-100 The Lewis dot structures are:

(a)   (b)   (c)   (d)   (e)

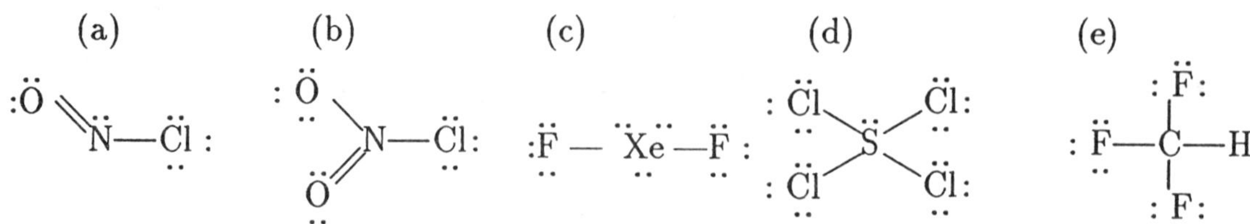

The NOCl molecule is bent and polar; the O_2NCl molecule is (nearly) trigonal about the N and polar; the XeF_2 molecule is linear and non-polar; the SCl_4 molecule has a seesaw geometry and is polar; the CHF_3 molecule is (nearly) tetrahedral and polar.

3-102 a) The molecule of nitrogen dioxide has 17 valence electrons. At least one atom cannot achieve an octet. The following four resonance structures are the candidates for best single Lewis structure:

$$\left[\ :\ddot{O}-\dot{N}=\ddot{O}\ \longleftrightarrow\ \ddot{O}=\dot{N}-\ddot{O}:\ \longleftrightarrow\ \cdot\ddot{O}-\dot{N}=\ddot{O}\ \longleftrightarrow\ :\ddot{O}=\dot{N}-\ddot{O}\cdot\ \right]$$

Other candidate structures break the octet rule on more than one atom or use octet expansion rather than octet deficiency. If the N atom is octet-deficient (left two structures), then formal charges build up as shown. If an O is octet-deficient (right two structures), all atoms have f.c. zero. The best structure puts the odd electron on the ☐ O atoms.

b) Four resonance structure can be drawn varying the relative positions of the double and single bonds:

3-104 When xenon forms covalent bonds, it generally has an expanded octet. XeF^+ ion is an apparent exception to this statement. Note however that it can be regarded as derived from the Xe^+ ion (with seven valence electrons).

(a) $\left(:\overset{+1}{\ddot{Xe}}-\ddot{F}:\right)^+$ (b) $:\ddot{F}-\ddot{Xe}-\ddot{F}:$

3-106 The preparation of chlorine dioxide is represented:

$$2\,NaClO_2 + Cl_2 \longrightarrow 2\,NaCl + 2\,ClO_2$$

3-108 The formation of dinitrogen tetrafluoride is represented:

$$2\,NF_3 + 2\,Cu \longrightarrow N_2F_4 + 2\,CuF$$

3-110 The "double salt" $CuCl_2 \cdot 2KCl$ is $\boxed{K_2[CuCl_4]}$ in which the four chloride ions serve as ligands. Similarly, $CoF_3 \cdot 3NaF$ is the complex $\boxed{Na_3[CoF_6]}$.

Answers to Even-Numbered Problems, Chapter 4

4-2 The dissolution of ethanol in water is written $\boxed{C_2H_5OH(l) \longrightarrow C_2H_5OH(aq)}$. No dissociation is shown, and water is not explicitly indicated as a reactant. Compare however to the equation in problem 4-90a.

4-4 If tin(II) fluoride is a strong electrolyte, the dissolution is best represented $SnF_2(s) \longrightarrow Sn^{2+}(aq) + 2\,F^-(aq)$. If it is a weak electrolyte or non-electrolyte, the equation $SnF_2(s) \longrightarrow SnF_2(aq)$ is better. Notice that water is not indicated explicitly as a reactant.

4-6 (a) Formic acid is a organic acid of low molar mass; it should be soluble in water: $HCOOH(l) \longrightarrow HCOOH(aq)$. In fact, it is completely $\boxed{\text{miscible with water}}$.
(b) Heptane has almost completely non-polar molecules and should be poorly soluble in water. In fact, it is quite $\boxed{\text{immiscible with water}}$. After being shaken in contact with water, it quickly separates as a separate layer on top of the aqueous phase (it is less dense than water).
(c) Potassium rubidium sulfate, a mixed salt should behave like potassium sulfate and rubidium sulfate: it should be $\boxed{\text{soluble in water}}$.

4-8 At the start, $KClO_4$ and KCl are dissolved in the some solution. As water evaporates from the solution, the solubility of $KClO_4$ is exceeded first (see text Table 4-1); it is only slightly soluble, and KCl is soluble.

4-10 Water that is in contact with the air dissolves quantities of both oxygen and nitrogen. Gases are generally less soluble in hot water than in cold. This suggests $\boxed{\text{boiling the water}}$ to drive off the dissolved oxygen. Dissolved nitrogen is required at the same time. Boiling is in fact a standard method of purging dissolved gases from water and other solvents.

4-12 Net ionic equations omit spectator ions and molecules. These are the species that are not actually reacting:

(a) $Ba^{2+}(aq) + SO_4^{2-}(aq) \longrightarrow BaSO_4(s)$

(b) $3\,Cl_2(g) + 6\,OH^-(aq) \longrightarrow ClO_3^-(aq) + 5\,Cl^-(aq) + 3\,H_2O(l)$

(c) $Hg_2^{2+}(aq) + 2\,I^-(aq) \longrightarrow Hg_2I_2(s)$

(d) $3\,OCl^-(aq) + I^-(aq) \longrightarrow IO_3^-(aq) + 3\,Cl^-(aq)$

4-14 The reaction is the precipitation of $NiCO_3$: $Na_2CO_3(aq) + Ni(NO_3)_2(aq) \longrightarrow NiCO_3(s) + 2\,NaNO_3(aq)$. Two net ionic equations may be offered as answers:

$$H_2CO_3(aq) + Ni^{2+}(aq) \longrightarrow NiCO_3(s) + 2\,H^+(aq)$$

$$CO_3^{2-}(aq) + Ni^{2+}(aq) \longrightarrow NiCO_3(s)$$

The first is preferable to the second because the concentration of carbonate ions would be quite low under "controlled acidic conditions." Note that the solution must be acidic. If it is not, then the precipitate is a basic nickel carbonate. Also, the nickel(II) carbonate in fact precipitates as the hexahydrate, $NiCO_3 \cdot 6H_2O$.

4-16 Although NaCl is quite soluble in water, it is evidently less soluble in a solution of hydrochloric acid. Addition of $HCl(g)$ to aqueous NaCl changes the solvent to $HCl(aq)$.

4-18 Predicting the products requires a knowledge of solubility rules (Table 4-1, text page 130).

(a) $Ca^{2+}(aq) + 2\,F^-(aq) \longrightarrow CaF_2(s)$

(b) $Hg_2^{2+}(aq) + SO_4^{2-}(aq) \longrightarrow Hg_2SO_4(s)$

(c) $NaBr(aq) + Mg(CH_3COO)_2(aq) \longrightarrow NR$

(d) $2\,I^-(aq) + Pb^{2+}ClO_3(s) \longrightarrow PbI_2(s) + 2\,ClO_3^-$

4-20 (a) $3\,Ca^{2+}(aq) + 2\,PO_4^{3-}(aq) \longrightarrow Ca_3(PO_4)_2(s)$

(b) NR (Both $CuSO_4$ nor KCl are soluble in water.)

(c) NR (But $BaCO_3$ precipitates unless air (a source of CO_2) is excluded. Also,

the solubility of $Ba(OH)_2 \cdot 8H_2O$ is relative low (5.6 g per 100 mL of water) near room temperature.)
(d) $Pb^{2+}(aq) + 2\,I^-(aq) \longrightarrow PbI_2(s)$

4-22 The two simultaneous precipitation reactions are:

$$Ag^+(aq) + Cl^-(aq) \longrightarrow AgCl(s) \qquad Ca^{2+}(aq) + 2\,F^-(aq) \longrightarrow CaF_2(s)$$

4-24 The student could $\boxed{\text{add } NaCl(aq)}$. The right amount will replace all dissolved $Ag^+(aq)$ with $Na^+(aq)$. The progress of the removal could in principle be followed by observing the formation of the precipitate of $AgCl(s)$.

4-26 (a) hypoiodous acid   (b) acetic acid   (c) nitrous acid   (d) hydroselenic acid

---

4-28 The way the formulas are written often reveals whether the compound is an acid or a base. In acids, acidic hydrogens are very commonly segregated to the left side of the formula.
(a) The anhydride of this acid is As_2O_5, diarsenic pentaoxide, which is also sometimes called arsenic(V) oxide.
(b) The anhydride of this acid is molybdenum(VI) oxide, MoO_3.
(c) The anhydride of this base is Rb_2O, rubidium oxide.
(d) The anhydride of this acid is SO_2, sulfur dioxide.

4-30 (a) N_2O_3 is the acid anhydride of nitrous acid (HNO_2).
(b) Li_2O is the base anhydride of the lithium hydroxide (LiOH).
(c) P_4O_{10} is the acid anhydride of phosphoric acid (H_3PO_4).
(d) BaO is the base anhydride of the barium hydroxide ($Ba(OH)_2$).

4-32 These reactions are all characteristic reactions of acids (see text page 143).
(a) The products are aqueous $\boxed{\text{lithium nitrate}}$ and $\boxed{\text{water}}$. (An acid plus a base gives water and a salt.)
(b) The products are aqueous $\boxed{\text{nickel(II) chloride}}$ and gaseous $\boxed{\text{hydrogen}}$. (An acid plus (many, not all) metals gives a salt plus gaseous hydrogen.)
(c) The products are aqueous $\boxed{\text{ammonium iodide}}$, gaseous $\boxed{\text{carbon dioxide}}$, and $\boxed{\text{water}}$. (An acid acts on a carbonate to liberate CO_2 and produce a salt and water.)

4-34 These reactions are all characteristic reactions of bases (see text page 144).
(a) The products are aqueous $\boxed{\text{rubidium chloride}}$, $\boxed{\text{ammonia}}$, in good part as

a gas, and [water]. (An ammonium salt plus a base gives gaseous ammonia, water and a salt.)

(b) The products are aqueous [potassium selenate] and [water]. (A base reacts with oxides of nonmetals to produce a salt and water.)

(c) The products are aqueous [cesium fluoride] and [water]. (A base neutralizes an acid to produce water and a salt. Cesium fluoride is expected to be soluble in water based on the solubility rules.

4-36 (a) $C_6H_5COOH + NH_3 \longrightarrow NH_4C_6H_5COO$

(b) $2\,HClO_4 + Mg(OH)_2 \longrightarrow Mg(ClO_4)_2 + 2\,H_2O$

(c) $HNO_3 + CsOH \longrightarrow CsNO_3 + H_2O$

4-38 (a) $2\,NaOH + H_2SO_3 \longrightarrow 2\,H_2O + Na_2SO_3$.

The reaction is the neutralization of the base sodium hydroxide by the acid sulfurous acid. The salt is sodium sulfite.

(b) $Ca(OH)_2 + 2\,C_6H_5COOH \longrightarrow 2\,H_2O + Ca(C_6H_5COO)_2$

The reaction is the neutralization of the base calcium hydroxide by the acid benzoic acid. The salt is calcium benzoate.

(c) $PbO + H_2SO_4 \longrightarrow H_2O + PbSO_4$.

This reaction is the neutralization of the base lead(II) oxide by sulfuric acid. The salt is lead(II) sulfate.

(d) $Cu(OH)_2 + 2\,HCl \longrightarrow 2\,H_2O + CuCl_2$.

This reaction is the neutralization of the base copper(II) hydroxide by hydrochloric acid. The salt is copper(II) chloride.

4-40 (a) The four compounds are perchloric acid, tetraphosphorus decaoxide, dichlorine heptachloride, and phosphoric acid.

(b) $12\,HClO_4 + P_4O_{10} \longrightarrow 6\,Cl_2O_7 + 4\,H_3PO_4$

4-42 The salt comes from sodium hydroxide and sulfuric acid, so it is sodium sulfate (Na_2SO_4). Sodium hydrogen sulfate $(NaHSO_4)$ might form but the context suggests the complete neutralization of the sulfuric acid.

4-44 The metal oxide dissolves in the acid

$$\boxed{MnO(s) + 2\,HCl(aq) \longrightarrow MnCl_2(aq) + H_2O(l)}$$

The products are $\boxed{\text{manganese(II) chloride}}$ and $\boxed{\text{water}}$.

4-46 (a) In C_2H_6, the oxidation numbers are $\boxed{C, -3}$; $\boxed{H, +1}$.

(b) In $C_2H_4O_2$, the oxidation numbers are $\boxed{C, 0}$; $\boxed{H, +1}$; and $\boxed{O, -2}$.

(c) In NH_4NO_3, the oxidation numbers are $\boxed{O, -2}$ and $\boxed{H, +1}$. If the formula were written $H_4N_2O_3$, then the nitrogen would have an oxidation state of $+1$. The separation of the N's in the formula suggests there are two kinds of nitrogen: the ammonium nitrogen with $\boxed{N, -3}$, and the nitrate nitrogen with $\boxed{N, +5}$. This is the way all chemists would assign the oxidation states. Mentioning the alternative is an excellent reminder of the arbitrary nature of oxidation numbers.

(d) In KO_2, we have $\boxed{K, +1}$; $\boxed{O, -1/2}$.

4-48 The balanced equation is $2\,CaSO_4 + C \longrightarrow 2CaO + 2\,SO_2 + CO_2$. The carbon increases its oxidation number from 0 to $+4$; two sulfate sulfur atoms lower their oxidation numbers from $+6$ to $+4$. Four electrons are given up in the first change; four electrons are gained in the second.

4-50 (a) Nitrogen changes oxidation state from $+4$ in N_2O_4 to $+3$ in NOCl and $+5$ in KNO_3.

(b) Sulfur changes oxidation state from -2 in H_2S to $+6$ in SF_6. Oxygen changes oxidation state from $+1$ in O_2F_2 to 0 in O_2.

(c) Phosphorus changes oxidation state from $+5$ in $POBr_3$ to $+2$ in PO. Magnesium changes from 0 in Mg to $+2$ in $MgBr_2$.

(d) Chlorine changes oxidation state from -1 in BCl_3 to 0 in Cl_2. Sulfur changes from $+4$ in SF_4 to $+2$ in SCl_2.

4-52 This question refers to the equations in problem 4-50.

(a) $\boxed{N_2O_4 \text{ is both oxidized and reduced}}$. Half of the nitrogen gains 1 electron per atom to give the NOCl, and half loses 1 electrons per atom to give the KNO_3.

(b) The $\boxed{\text{S is oxidized}}$ from -2 to $+6$ and $\boxed{\text{O is reduced}}$. It takes four moles of O_2F_2 to absorb the electrons from one mole of H_2S because the change in the oxidation number of each of the eight oxygen atoms is only from $+1$ to zero.

(c) The $\boxed{POBr_3}$ is reduced. Each P goes from the $+5$ to the $+2$ state. There are two P's in the balanced equation, so six electrons are gained. $\boxed{\text{Mg is reduced}}$

to furnish six electrons (two from each of three Mg atoms oxidized from the 0 to the +2 state).

(d) Some $\boxed{\text{chloride ion is oxidized}}$ from the -1 to the 0 oxidation state. This oxidation releases six electrons. These are absorbed by the $\boxed{\text{reduction of } SF_4}$. The S's in three SF_4 molecules are reduced from +4 to +2. Not all of the Cl is oxidized; some appears in the -1 state on the product side of the equation.

4-54 The iodine is in the +5 oxidation state on the left side of the equation and in the 0 oxidation state on the right side. Carbon is in the +2 oxidation state on the left side and in the +4 oxidation state on the right. All oxygen stays in the -2 oxidation state throughout the reaction. This is an oxidation-reduction even in the most literal sense—one of the reactants, the I_2O_5, is reduced in its oxygen content and the other, the CO, is increased in its oxygen content.

4-56 $2\,PbSO_4(s) + 2\,H_2O(aq) \longrightarrow PbO_2(aq) + Pb(s) + 2\,H_2SO_4(aq)$.

In this reaction, half of the lead(II) sulfate is oxidized to lead(IV) oxide (PbO_2), and half is reduced to metallic lead.

4-58 (a) The likely product of the reaction is HBr(g):

$$Br_2(g) + H_2(aq) \longrightarrow 2HBr(g)$$

This is the reduction of bromine with hydrogen (or the oxidation of hydrogen with bromine). It is also a hydrogenation reaction (the hydrogenation of bromine) and a bromination reaction (the bromination of hydrogen). It is a $\boxed{\text{combination reaction}}$ between two elements.

(b) The likely product of the reaction is ammonia $NH_3(g)$

$$NH_2NH_2(g) + H_2(g) \longrightarrow 2\,NH_3(g)$$

This is the reduction of hydrazine with hydrogen or the hydrogenation of hydrazine. It is a $\boxed{\text{redox combination}}$ between an element and a compound.

(c) The likely products of this reaction are CO_2 and H_2O:

$$2\,CH_4(g) + 3\,O_2(g) \longrightarrow 2\,CO_2(g) + 2\,H_2O(g)$$

This reaction is the oxidation of methane, or the $\boxed{\text{combustion of methane}}$. It is also the reduction of oxygen. Incomplete oxygenation to give CO(g) along with H_2O is also possible.

(d) The likely product of this reaction is calcium bromide:

$$Ca(s) + Br_2(g) \longrightarrow CaBr_2(s)$$

The reaction is a reduction of bromine and an oxidation of calcium. It is a boxed{redox combination} reaction.

4-60 The reaction is the boxed{disproportionation} of HNO_2:

$$3\,HNO_2(aq) \longrightarrow HNO_3(aq) + 2\,NO(g) + H_2O(l)$$

Note that only one element changes oxidation state. Some nitrogen is oxidized from the +3 state to the +5 state, and some nitrogen is reduced from the +3 state to the +2 state.

4-62 Because X contains no lead, the coefficients for $Pb(NO_3)_2$ and PbO in the balanced equation for this reaction must be equal. Also, because neither lead(II) oxide nor oxygen contains any nitrogen, the gaseous product X must contain all of the nitrogen from the lead(II) nitrate. It follows that the coefficient for product X must be twice the coefficient for $Pb(NO_3)_2$. By Avogadro's hypothesis (see text Section 1-4), the coefficient for product X equals the coefficient for O_2 multiplied by 4. Writing a chemical equation to fit these facts is possible only as follows:

$$2\,Pb(NO_3)_2(l) \longrightarrow 2\,PbO(s) + O_2(g) + 4\,NO_2(g)$$

The product X is boxed{nitrogen dioxide}.

In this reaction, nitrogen in the nitrate ion is reduced from +5 to +4, and some of the oxygen in the nitrate ion is oxidized from −2 to 0. The reaction boxed{is a redox reaction}, a redox decomposition. It is not a disproportionation because two elements change oxidation state (one up and one down), not just one.

4-64 The formation of hydrogen bromide by the combination of its elements is a redox reaction: $H_2 + Br_2 \longrightarrow 2\,HBr$. The HBr, a gas at room conditions, reacts with water by dissolving. It also reacts with potassium carbonate in a neutralization reaction, and with magnesium in a redox reaction. The balanced equations are:

$$HBr(g) \longrightarrow HBr(aq) \quad \text{or} \quad HBr(g) \longrightarrow H^+(aq) + Br^-(aq)$$

$$2\,HBr(g) + K_2CO_3(s) \longrightarrow 2\,KBr(aq) + H_2O(l) + CO_2(g)$$

$$2\,HBr(g) + Mg(s) \longrightarrow H_2(g) + MgBr_2(s)$$

4-66 (a) The balanced equation is

$$2\,HBr(aq) + ZnO(s) \longrightarrow ZnBr_2(aq) + H_2O(l)$$

This is an $\boxed{\text{acid-base}}$ reaction. Note that ZnO is the base anhydride of $Zn(OH)_2$.

(b) As is stated on text page 152, cesium forms the superoxide when it reacts with oxygen. The balanced equation is

$$Cs(s) + O_2)(g) \longrightarrow CsO_2(g)$$

This is a $\boxed{\text{redox}}$ reaction.

(c) In strict terms, this reaction does not fit any of the categories, and it is difficult to predict the products until it is noticed that the reactants are a base anhydride and an acid anhydride. Since the combination of a base and an acid gives a salt plus water it seems likely that the combination of a base anhydride and an acid anhydride gives a salt *without* the water. The balanced equation is

$$MgO(s) + SO_3(g) \longrightarrow MgSO_4(s)$$

Based on the thinking just described, it is also reasonable to say the reaction is an $\boxed{\text{acid-base}}$ reaction.

(d) Concurrent $\boxed{\text{redox}}$ reactions occur. Students usually want to represent everything that occurs in a single equation, which is an error. The sodium hydride will be oxidized to H_2 by the water and also oxidized to H_2 by the copper(II) ion in solution:

$$2NaH(s) + Cu^{2+}(aq) \longrightarrow Cu(s) + 2\,Na^+(aq) + H_2(g)$$

$$2NaH(s) + 2\,H_2O(l) \longrightarrow 2\,OH^-(aq) + 2\,Na^+(aq) + 2\,H_2(g)$$

Note that the $OH^-(aq)$ ion and $Cu^{2+}(aq)$ ion formed by these reactions will react further to give a precipitate of $Cu(OH)_2(s)$.

4-68 The reaction is $Ag^+(aq) + Cl^-(aq) \longrightarrow AgCl(s)$. If the silver ion is the limiting reactant, the theoretical yield is

$$25.0 \text{ mL AgNO}_3 \text{ solution} \times \left(\frac{0.150 \text{ mol AgNO}_3}{1000 \text{ mL AgNO3 solution}} \right) \times \left(\frac{1 \text{ mol Ag}^+}{1 \text{ mol AgNO}_3} \right)$$

$$\times \left(\frac{1 \text{ mol AgCl}}{1 \text{ mol Ag}^+} \right) \times \left(\frac{143.321 \text{ g AgCl}}{1 \text{ mol AgCl}} \right) = 0.537 \text{ g AgCl}$$

If chloride ion is the limiting reactant, the theoretical yield is

$$15.0 \text{ mL CaCl}_2 \text{ solution} \times \left(\frac{0.100 \text{ mol ClCl}_2}{1000 \text{ mL CaCl}_2 \text{ solution}} \right) \times \left(\frac{2 \text{ mol Cl}^-}{1 \text{ mol CaCl}_2} \right)$$

$$\times \left(\frac{1 \text{ mol AgCl}}{1 \text{ mol Cl}^-} \right) \times \left(\frac{143.321 \text{ g AgCl}}{1 \text{ mol AgCl}} \right) = 0.430 \text{ g AgCl}$$

The answer is the smaller of these or $\boxed{0.430 \text{ g AgCl}}$.

4-70 Assume that two solutions mix to give a combined volume equal to the sum of the individual volumes. The first solution contains 300 mmol of fructose and the second contains 120 mmol of fructose. The total amount of fructose is 420 mmol. If the final volume of solution is 800.0 mL, then the concentration of the fructose is

$$c_{\text{fructose}} = \frac{420 \text{ mmol}}{800.0 \text{ mL}} = \boxed{0.525 \text{ mol L}^{-1}}$$

4-72 (a) A balanced equation for the reaction between potassium iodate and potassium iodide is

$$\boxed{5 \text{ KI}(aq) + \text{KIO}_3(aq) + 6 \text{ HCl}(aq) \longrightarrow 3 \text{ I}_2(aq) + 6 \text{ KCl}(aq) + 3 \text{ H}_2\text{O}(l)}$$

The net ionic equation is

$$5 \text{ I}^-(aq) + \text{IO}_3^-(aq) + 6 \text{ H}^+(aq) \longrightarrow 3 \text{ I}_2(aq) + 6 \text{ K}^+(aq) + 3 \text{ H}_2\text{O}(l)$$

(b) The 50.00 mL of 0.01000 molar KIO_3 contains 0.5000 mmol of IO_3^- ion; the 1.00 g of KI contains 6.024 mmol of I_ ion; the 10.00 mL of 3 molar HCl contains 30 mmol of H^+ ion. The balanced chemical equations show that iodide ion is consumed at 5 times the rate of iodate ion and that H^+ ion is consumed at 6 times the rate of iodate ion. However, the amount (in moles) of iodide ion is about 12 times the amount of iodate and the amount of H^+ ion is about 60 times the amount of iodate ion. Therefore $\boxed{\text{potassium iodate}}$ is the limiting reactant.

4-74

$$2.13 \text{ g N}_2\text{O}_3 \times \left(\frac{1 \text{ mol N}_2\text{O}_3}{76.01 \text{ g N}_2\text{O}_3} \right) \times \left(\frac{2 \text{ mol NO}_2^-}{1 \text{ mol N}_2\text{O}_3} \right)$$

$$\times \left(\frac{1 \text{ mol NaNO}_2}{1 \text{ mol NO}_2^-} \right) = 0.0560 \text{ mol NaNO}_2$$

This chemical amount of $NaNO_2$, when present in 856 mL (0.856 L) of solution, gives a solution with the concentration:

$$c_{NaNO_2} = \frac{0.0560 \text{ mol}}{0.856 \text{ L}} = \boxed{0.0655 \text{ mol L}^{-1}}$$

4-76

$$2200 \times 10^3 \text{ g Ca}_5(PO_4)_3F \times \left(\frac{1 \text{ mol Ca}_5(PO_4)_3F}{504.3 \text{ g Ca}_5(PO_4)_3F}\right) \times \left(\frac{3 \text{ mol H}_3PO_4}{1 \text{ mol Ca}_5(PO_4)_3F}\right)$$

$$\times \left(\frac{1 \text{ L solution}}{6.3 \text{ mol H}_3PO_4}\right) = \boxed{2.1 \times 10^3 \text{ L solution}}$$

4-78 The volume of 0.646 M solution required is

$$6.07 \text{ g NO} \times \left(\frac{1 \text{ mol NO}}{30.01 \text{ g NO}}\right) \times \left(\frac{6 \text{ mol NaNO}_2}{4 \text{ mol NO}}\right)$$

$$\times \left(\frac{1 \text{ L solution}}{0.646 \text{ mol NaNO}_2}\right) = \boxed{0.470 \text{ L solution}}$$

4-80 (a) $PCl_5 + 4\,H_2O \longrightarrow H_3PO_4 + 5\,HCl$

(b) The 0.0293 mol of PCl_5 generates 0.0293 mol of H_3PO_4 and exactly fives times this amount of HCl. Both acids are collected in enough water so that the final volume of the mixed solution is 697 mL. The concentration of the H_3PO_4 is 0.0293 mol/0.697 L = $\boxed{0.0420 \text{ mol L}^{-1}}$, and the concentration of the HCl is five times this, or $\boxed{0.210 \text{ mol L}^{-1}}$.

4-82 The titration of the potassium acid phthalate is represented

$$NaOH(aq) + KHC_8H_4O_4(aq) \longrightarrow NaKC_8H_4O_4(aq) + H_2O(aq)$$

and the titration of the sulfuric acid is

$$2\,NaOH(aq) + H_2SO_4(aq) \longrightarrow Na_2SO_4(aq) + 2\,H_2O(aq)$$

The potassium acid phthalate and sodium hydroxide react in a 1 : 1 molar ratio. Therefore, the chemical amount of sodium hydroxide is

$$0.6753 \text{ g KHC}_4H_8O_4 \times \left(\frac{1 \text{ mol KHC}_8H_4O_4}{204.22 \text{ g KHC}_8H_4O_4}\right) \times \left(\frac{1 \text{ mol NaOH}}{1 \text{ mol KHC}_8H_4O_4}\right)$$

$$= 0.003307 \text{ mol NaOH}$$

The molarity of the NaOH solution equals this chemical amount divided by the volume of solution

$$c_{\text{NaOH}} = \frac{0.003307 \text{ mol NaOH}}{0.03498 \text{ L}} = \boxed{0.09453 \text{ mol L}^{-1}}$$

The chemical amount of sulfuric acid consumed in the second reaction is

$$21.49 \text{ mL NaOH} \times \left(\frac{0.09453 \text{ mol NaOH}}{1000 \text{ mL NaOH}}\right) \times \left(\frac{2 \text{ mol H}_2\text{SO}_4}{1 \text{ mol NaOH}}\right)$$
$$= 0.0040629 \text{ mol H}_2\text{SO}_4$$

The concentration of the sulfuric acid equals this chemical amount divided by the volume of sulfuric acid used. It equals

$$c_{\text{H}_2\text{SO}_4} = \frac{0.0040629 \text{ mol}}{0.02000 \text{ L}} = \boxed{0.2031 \text{ mol L}^{-1}}$$

4-84 The 0.217 g sample of As_2O_3 (molar mass 197.84 g mol^{-1}) contains 1.097 mmol of As_2O_3. Upon dissolution it gives 2.194 mmol of H_3AsO_3 because there are two As atoms per As_2O_3 formula unit. The titration consumes 4.388 mmol (twice 2.194 mmol) of Ce^{4+} because Ce^{4+} and H_3AsO_3 react in a 2 : 1 molar ratio according to the balanced equation. The 4.388 mmol of Ce^{4+} ion is delivered by the addition of 21.47 mL of solution in the course of the titration. The concentration of the Ce^{4+} ion in the titrating solution is

$$c_{\text{Ce}^{4+}} = \frac{n_{\text{Ce}^{4+}}}{V} = \frac{4.388 \text{ mmol}}{21.47 \text{ mL}} = \boxed{0.204 \text{ mol L}^{-1}}$$

4-86 (a) The molar mass of LiClO_3 is 90.39 g mol^{-1}. Hence, 315 g of LiClO_3 equals 3.48 mol. If this much LiClO_3 dissociates completely in 100 mL of water it generates 6.96 mol of ions. The 100 g of water contains 5.55 mol of H_2O. (the mass divided by the molar mass, which is 18.01 g mol^{-1}). The solution contains $\boxed{5.55/6.96 = 0.80 \text{ mol of of water per mol of ions}}$, assuming complete dissociation.

(b) $\boxed{\text{No}}$, the LiClO_3 is not completely dissociated. The solution contains insufficient water molecules to furnish even one H_2O per ion. Clearly, very numerous ion-ion interactions must exist in a saturated solution of LiClO_3.

4-88 Table 4-1 shows that neither mercury(I) chloride nor barium sulfate is soluble in water. A $\boxed{\text{(white) precipitate}}$ will be observed when the two solutions are mixed. The precipitate is a mixture of two solids formed by the concurrent reactions:

$$Hg_2^{2+}(aq) + 2\,Cl^-(aq) \longrightarrow Hg_2Cl_2(s)$$

$$Ba^{2+}(aq) + SO_4^{2-}(aq) \longrightarrow BaSO_4(s)$$

4-90 (a) The ore bauxite is clearly a base. If it is neutralized with sulfuric acid, a salt with the proper ratio of aluminum and sulfate results:

$$2\,Al(OH)_3(s) + 3\,H_2SO_4(aq) + 15\,H_2O(l) \longrightarrow Al_2(SO_4)_3 \cdot 18H_2O(s)$$

Neutralization actually gives a solution of aluminum sulfate. Reducing the volume of this solution causes the alum to precipitate. This particular solid forms because it is less soluble than other hydrates of aluminum sulfate or anhydrous $Al_2(SO_4)_3$ itself.

(b) $Al^{3+}(aq) + 3\,OH^-(aq) \longrightarrow Al(OH)_3(s)$.

4-92 The sulfate ion from the sulfuric acid removed poisonous Pb^{2+} ion from solution by reacting with it to form insoluble $PbSO_4$. This protected the workers. Even if they drank suspended solid $PbSO_4(s)$, it was too insoluble to give up significant amounts of lead in its passage through the body. The sour taste of the water (and tooth damage?) was a small price to pay.

4-94 (a) Assign oxidation numbers to the elements in the reactants and the products according to the rules on text page 147. If any oxidation numbers change, the reaction is a redox reaction. Neither hydrogen nor fluorine changes oxidation number, but the oxygen in OF_2 on the reactant side is in the $+2$ oxidation state, and the oxygen in the H_2O is in the -2 oxidation state and all oxygen on the product side is in the 0 oxidation state. Hence OF_2 is reduced, water is oxidized, and this $\boxed{\text{is a redox reaction}}$.

(b) The oxoacid that would form through the hypothetical reaction of OF_2 as an acid anhydride is $\boxed{\text{HOF, hypofluorous acid}}$.

4-96 In the ion Bi_5^{3+} the oxidation number of the bismuth is $+3/5$. Because the oxidation number of F is conventionally -1, the oxidation number of the As is $+5$. The elevated dot in the formula means that SO_2 is loosely associated with this salt in its solid state. The oxidation number of S in SO_2 is $+4$, and the oxidation number of O is -2.

4-98 (a) The oxidation number of lead in PbO is $+2$; in PbO_2 it is $+4$, in Pb_2O_3 it is $+3$, and in Pb_3O_4 it is $+8/3$.

(b) Note that the formula "Pb_2O_3" is the sum of PbO and PbO_2. Perhaps in Pb_2O_3 half of the lead is Pb(II) and half is Pb(IV). Similarly, the lead in Pb_3O_4 may be 2/3 Pb(II) and 1/3 Pb(IV).

4-100 Hydrogen forms a $+1$ ion in many of its compounds, just like the metals of Group I. The $+1$ oxidation state in fact predominates in its chemistry. On the other hand, it is a gas, not a metal, at room conditions. Also, it does in some situations form a -1 ion like the elements of Group VII.

4-102 (a) Dehydration refers to the $\boxed{\text{removal of water}}$ from a compound, that is, to the reverse of hydration. An example is $H_2SO_4 \longrightarrow H_2O + SO_3$

(b) Deoxygenation refers to the $\boxed{\text{removal of oxygen}}$ from a compound. It is the reverse of oxygenation. An example is $2\,CO \longrightarrow O_2 + 2\,C$.

(c) Reproportionation is the reverse of disproportionation.

(d) Fluoridation is the addition of or displacement by the fluoride ion. It is not the same as fluorination, which is taking up of fluorine by treatment with some fluorinating agent. Fluoridation of the drinking water leads to fluoridation of apatite in the enamel of the tooth gives fluoroapatite $Ca_5(PO_4)_3)OH(s) + F^-(aq) \longrightarrow Ca_5(PO_4)_3)F(s) + OH^-(aq)$

4-104 (a) The acid anhydride of H_3AsO_3 is As_2O_3, arsenic(III) oxide; the acid anhydride of H_3AsO_4 is As_2O_5, arsenic(V) oxide. Because arsenic is a semimetal, these two compounds are sometimes named diarsenic trioxide and diarsenic pentaoxide. The names arsenic trioxide and arsenic pentaoxide are also often encountered. Finally, As_2O_3 has been shown to consist of As_4O_6 molecules in the gas phase. The name would then be tetraarsenic hexaoxide.

(b) The two oxoacids are arsenious acid, in which the stem "arseni-" is combined with the "-ous" ending, and arsenic acid in which the same stem is combined with the "-ic" ending (and the double i is shortened to a single i).

(c) We expect the acid in which the As is in the lower oxidation state to be the one that is amphoteric. This is H_3AsO_3.

(d) The chemical formulas of the oxoacids of arsenic differ from these of nitrogen in the same oxidation state by having an additional molecule of water: H_3AsO_3 $(= HAsO_2 \cdot H_2O)$ is analogous to HNO_2 plus water; H_3AsO_4 $(= HAsO_3 \cdot H_2O)$ is analogous to HNO_3 plus water.

(e) In the Lewis structure of the As(V) oxoacid the central arsenic is surrounded by four O's and has a single bond to each. In the analogous oxoacid of N(V), the central nitrogen is surrounded by only three O's, with which it forms two single bonds and one double bond. The greater double-bond-forming ability of the second-period element (N) in effect allows it (or causes it) to have one less molecule of water for the sake of a double bond. Similar considerations apply in the comparison of the formulas of the As(III) and N(III) oxoacids.

4-106 (a) Acid-base: $H_3PO_4(aq) + NaOH(aq) \longrightarrow NaH_2PO_4(aq) + H_2O(l)$.

(b) Redox: $Fe_3O_4(s) + 4\,C(s) \longrightarrow 3\,Fe(s) + 4\,CO(g)$.

(c) Precipitation:
$3\,Ca(NO_3)_2(aq) + 2\,K_3PO_4(aq) \longrightarrow Ca_3(PO_4)_2(s) + 6\,KNO_3(aq)$.

(d) Redox: $K_2Cr_2O_7(aq) + 14\,HI(aq) \longrightarrow 2\,CrI_3(s) + 3\,I_2(s) + 2\,KI(aq) + 7\,H_2O(l)$.

4-108 (a) The molar volumes are the volumes of solution that contain one mole of solute. They are simply the reciprocals of the concentrations multiplied by 1000 to convert from liters per mole to milliliters per mole.

(b) Put 1.34 L of 12.1 M HCl in a flask and add enough water to bring the final volume to 2.50 L.

(c) It is the $\boxed{\text{acetic acid}}$. It can easily be shown from the definitions that for a pure substance the molarity multiplied by the molar mass equals the density times 1000. This is true only for acetic acid in this table.

Answers to Even-Numbered Problems, Chapter 5

5-2 The gaseous product from the reaction of $NaOH(aq)$ with $NH_4HS(s)$ is $\boxed{NH_3}$. Addition of a nonvolatile aqueous base drives off a basic gas. This is a characteristic reaction of bases (see Section 4-3 of the text).

5-4 When trying to predict the products of a reaction, look for small stable molecules (like those of water, carbon dioxide or ammonia) that might be split off by larger molecules (see text page 151). The products of this reaction are easily predicted in this context. The balanced equation is:

$$\boxed{NH_4CO_2NH_2(s) \longrightarrow 2\,NH_3(g) + CO_2(g)}$$

5-6 The acidification of aqueous solutions of KCN drives poisonous $HCN(g)$ out of the solution: $CN^-(aq) + H^+(aq) \longrightarrow HCN(g)$.

5-8 The identity of the second gas among the products must be inferred. The named products contain the chromium, hydrogen and oxygen from the original compound. Because gaseous nitrogen (N_2) is a small stable molecule it is a natural (and correct) surmise as the nitrogen-containing product:

$$\boxed{(NH_4)_2Cr_2O_7(s) \longrightarrow Cr_2O_3(s) + N_2(g) + 4\,H_2O(g)}$$

5-10 Two good methods of solution exist. The first is direct substitution into the formula $P = gdh$. When the density of Ga and the height of the column of Ga are expressed in base SI units, the pressure comes out as a combination of base SI units:

$$P = gdh = 9.807\,\mathrm{m\,s}^{-2}\left(6.114 \times 10^3\,\mathrm{kg\,m}^{-3}\right) 1.68\,\mathrm{m} = 1.007 \times 10^5\,\mathrm{kg\,m}^{-1}\mathrm{s}^{-2}$$

This combination of base SI units is the pascal (Pa). Division by 1.01325×10^5 Pa atm^{-1} converts the answer to atm. It is $\boxed{0.994\ \text{atm}}$.

The second method is to set up a ratio. The density of the Ga (6.114 g cm^{-3}) is less than the density of Hg (13.596 g cm^{-3}). A column of mercury that is shorter in proportion to the ratio of these densities suffices to balance the given pressure. Once the pressure in cm of Hg is obtained, it needs to be converted to standard atmospheres. The combined steps are:

$$P = 168 \text{ m Ga} \times \left(\frac{6.114 \text{ cm Hg}}{13.596 \text{ cm Ga}}\right) \times \left(\frac{1 \text{ atm}}{76.000 \text{ cm Hg}}\right) = \boxed{0.994\ \text{atm}}$$

5-12 Convert the pressure from atmospheres to pascals (Pa) and then use the barometric formula $P = dgh$ to compute the depth (h) of water that exerts such a pressure. In this computation, g is the acceleration of the earth's gravity (9.807 m s^{-2}) and d is the density of the water. Take the density of the seawater as 1.0×10^3 kg m^{-3}. The density of water is not too much affected by amounts of dissolved salts in the sea or by the compression of overlying layers.

$$P = \left(\frac{1.01325 \times 10^5 \text{ Pa}}{1 \text{ atm}}\right) \times 414 \text{ atm} = 4.19 \times 10^7 \text{ Pa}$$

$$h = \frac{P}{dg} = \frac{4.19 \times 10^7\,\mathrm{kg\,m}^{-1}\,\mathrm{s}^{-2}}{(1.0 \times 10^3\,\mathrm{kg\,m}^{-3})\,9.807\,\mathrm{m\,s}^{-2}} = \boxed{4.3 \times 10^3 \text{ m}}$$

One pascal equals one a kg m^{-1}s^{-2}, as given in Table 5-2 and discussed in Appendix B.

5-14 A 76-cm column of mercury exerts the same pressure at its base that the atmosphere does on the earth at sea level. The density of the mercury in such a column is 13.6 g cm^{-3}, which equals 13.6×10^3 g L^{-1}, and is uniform over the length of the column. If air with a uniform density of a mere 1.3 g L^{-1} replaces the mercury, its column must be substantially longer to exert the same pressure. It is longer in proportion to the ratio of the densities. The depth (or height) of the suppositious atmosphere is thus

$$h = 76 \text{ cm} \times \left(\frac{13.6 \times 10^3 \text{ g L}^{-1}}{1.3 \text{ g L}^{-1}} \right) = 8.0 \times 10^5 \text{ cm} = \boxed{8.0 \text{ km}}$$

The real atmosphere becomes attenuated with increasing altitude and so is much deeper than 8 km. See text Section 18-5.

5-16 The conversions can be carried out using the information in text Table 5-2 (see text page 185):

$$5 \times 10^{-10} \text{ torr} \times \left(\frac{1 \text{ atm}}{760 \text{ torr}} \right) = \boxed{7 \times 10^{-13} \text{ atm}}$$

$$5 \times 10^{-10} \text{ torr} \times \left(\frac{1 \text{ atm}}{760 \text{ torr}} \right) \times \left(\frac{1.01325 \times 10^5 \text{ Pa}}{1 \text{ atm}} \right) = \boxed{7 \times 10^{-8} \text{ Pa}}$$

5-18 Assume that the CO_2 behaves ideally and use Boyle's law in the form $P_1 V_1 = P_2 V_2$. This is justified because the chemical amount of CO_2 in the sample does not change and the temperature, from the context of the problem, is also constant. Realizing this latter point is often hard for students. The computation is:

$$P_2 = \left(\frac{V_1}{V_2} \right) P_1 = \left(\frac{12.07 \text{ cm}^3}{11.51 \text{ cm}^3} \right) 1.106 \text{ atm} = \boxed{1.160 \text{ atm}}$$

Any unit of volume works in this equation as long as the two volumes are given in the *same* units.

5-20 Although the entire Stirling cycle is described, only the relationship used in problem 5-18 is necessary for the actual question. Students find this problem more difficult than 5-18 if they are not accustomed to sifting out extra information. The computation is:

$$P_2 = \left(\frac{V_1}{V_2} \right) P_1 = \left(\frac{0.350 \text{ L}}{1.31 \text{ L}} \right) 1.23 \text{ atm} = \boxed{0.329 \text{ atm}}$$

5-22 All of the problems can be solved by use of Boyle's law starting from the form $P_2V_2 = P_1V_1$.

(a) $\quad V_2 = V_1\left(\dfrac{P_1}{P_2}\right) = 485\,\text{L}\left(\dfrac{512\,\text{torr}\left(\frac{1\,\text{atm}}{760\,\text{torr}}\right)\left(\frac{1.01325\times10^2\,\text{kPa}}{1\,\text{atm}}\right)}{50.0\,\text{kPa}}\right) = \boxed{662\,\text{L}}$

(b) $\quad V_2 = V_1\left(\dfrac{P_1}{P_2}\right) = 781\,\text{mL}\left(\dfrac{1.50\,\text{atm}\left(\frac{1.01325\times10^2\,\text{kPa}}{1\,\text{atm}}\right)}{110.0\,\text{kPa}}\right) = \boxed{1079\,\text{mL}}$

(c) $\quad P_2 = P_1\left(\dfrac{V_1}{V_2}\right) = 995\,\text{torr}\left(\dfrac{1\,\text{atm}}{760\,\text{torr}}\right)\left(\dfrac{60.0\,\text{cm}^3\left(\frac{1\,\text{mL}}{1\,\text{cm}^3}\right)}{90.0\,\text{mL}}\right) = \boxed{0.873\,\text{atm}}$

(d) $\quad P_2 = 2.00\,\text{atm}\left(\dfrac{1.01325\times10^2\,\text{kPa}}{1\,\text{atm}}\right)\left(\dfrac{y\,\text{mL}}{3.25y\,\text{mL}}\right) = \boxed{62.4\,\text{kPa}}$

5-24 At all temperatures, $t_\text{F} = 1.8\,t_\text{C} + 32$. At the special temperature in question, $t_\text{F} = t_\text{C}$. Combining these two equations and solving gives an answer of

$$t_\text{F} = t_\text{C} = \boxed{-40°}$$

5-26 The problem emphasizes the difference between the direct dependence of ideal-gas volume on the absolute temperature and the linear dependence of ideal-gas volume on the Celsius temperature. Only in terms of an absolute temperature is it correct to write

$$V_2 = \left(\frac{T_2}{T_1}\right)V_1$$

Converting the Celsius temperatures given in the problem to absolute temperatures and substituting gives

$$V_2 = \left(\frac{313.15\,\text{K}}{293.15\,\text{K}}\right)4.00\,\text{L} = \boxed{4.27\,\text{L}}$$

5-28 The problem illustrates the operation of a gas thermometer. The temperature of a sample of an ideal gas is directly proportional to the absolute temperature as long as the pressure and the amount of gas in the sample do not change. Hence

$$T_2 = \left(\frac{V_2}{V_1}\right)T_1$$

If the temperature is converted from the Celsius scale to the Kelvin scale and the volumes are substituted:

$$T_2 = \left(\frac{V_2}{V_1}\right) T_1 = \left(\frac{5.26 \text{ L}}{5.40 \text{ L}}\right) 299.65 \text{ K} = 291.88 \text{ K}$$

Converting back to the Celsius scale gives $\boxed{18.7°\text{C}}$.

5-30 Use Charles's law in the form:

$$V_2 = \left(\frac{T_2}{T_1}\right) V_1$$

T_2 is 503.15 K and T_1 is 293.15 K. Substituting gives

$$V_2 = \left(\frac{503.15 \text{ K}}{293.15 \text{ K}}\right) 3.41 \text{ L} = \boxed{5.85 \text{ L}}$$

5-32 The first three parts can be solved by use of Charles's law in the form $V_2 = V_1(T_2/T_1)$. All temperatures must be converted to kelvins.

(a) $\quad V_2 = V_1\left(\dfrac{T_2}{T_1}\right) = 24.2\,\text{L}\left(\dfrac{(273.15 + 0.0)\text{ K}}{(273.15 + 25.0)\text{ K}}\right) = \boxed{22.2 \text{ L}}$

(b) $\quad V_2 = V_1\left(\dfrac{T_2}{T_1}\right) = 781\,\text{mL}\left(\dfrac{450\text{ K}}{150\text{ K}}\right) = \boxed{2.34 \times 10^3 \text{ mL}}$

(c) $\quad V_2 = V_1\left(\dfrac{T_2}{T_1}\right) = 781\,\text{mL}\left(\dfrac{(150 + 273.15)\text{ K}}{150\text{ K}}\right) = \boxed{2.20 \times 10^3 \text{ mL}}$

(d) Rewrite Charles's law as $T_2 = T_1(V_2/V_1)$. Then

$$T_2 = T_1\left(\frac{V_2}{V_1}\right) = 200\text{ K}\left(\frac{3.25y \text{ mL}}{y \text{ mL}}\right) = \boxed{650 \text{ K}}$$

5-34 The pressure of an ideal gas changes in direct proportion to its absolute temperature, if the amount and volume of the gas are kept constant:

$$P_2 = P_1\left(\frac{T_2}{T_1}\right) \quad n \text{ and } V \text{ constant}$$

In this case T_2 equals 273.15 K, T_1 equals 293.15 K (20.0°C) and P_1 equals 20.0 atm. It is assumed that the volume of the steel tank itself does not change when the tank is chilled and that the helium behaves ideally. Then:

$$P_2 = P_1\left(\frac{T_2}{T_1}\right) = 20.0\text{ atm}\left(\frac{273.15 \text{ K}}{293.15 \text{ K}}\right) = \boxed{18.6 \text{ atm}}$$

5-36 The gaseous IF_7 is synthesized at 300°C and a pressure that is substantially less than room pressure. Heating it to 400°C tends to increase its volume, but compressing it tends to decrease its volume. The effects compete. The amount of gas n does not change when T and P change because no decomposition or other reactions take place. Therefore,

$$\frac{P_1 V_1}{T_1} = \frac{P_2 V_2}{T_2}$$

Solving for V_2 gives

$$V_2 = \left(\frac{P_1}{P_2}\right)\left(\frac{T_2}{T_1}\right) V_1 = \left(\frac{0.459 \text{ atm}}{0.980 \text{ atm}}\right)\left(\frac{673.15 \text{ K}}{573.15 \text{ K}}\right) 63.6 \text{ L} = \boxed{35.0 \text{ L}}$$

5-38 To prevent dangerous spurting when the container is opened, the pressure inside must be brought below 0.96 atm. The pressure at 20°C (293.15 K) is 1.47 atm, and the pressure of an ideal gas at constant volume is directly proportional to the absolute temperature. Because $0.96/1.47 = 0.653$, the required temperature is $0.653 \times 293.15 = 191.4$ K. Subtracting 273.15 K from this Kelvin temperature and multiplying by 1°C/1 K gives $\boxed{-82°C}$.

5-40

$$m_{\text{Ne}} = \mathcal{M}_{\text{Ne}} n_{\text{Ne}} = \mathcal{M}\left(\frac{PV}{RT}\right)_{\text{Ne}}$$

$$= 20.18 \text{ g mol}^{-1} \frac{(2.00 \text{ atm})(35.0 \text{ L})}{(0.08206 \text{ L atm mol}^{-1} \text{ K}^{-1})(273.15 + 25.0) \text{ K}} = \boxed{57.7 \text{ g}}$$

5-42 First calculate the chemical amount of O_2 present in 0.30 kg of O_2. It is 9.375 mol O_2.

(a) Assuming that O_2 is an ideal gas, the pressure in the scuba tank is:

$$P = \frac{nRT}{V} = \frac{9.375 \text{ mol } (0.08206 \text{ L atm mol}^{-1} \text{ K}^{-1})278.15 \text{ K}}{2.32 \text{ L}} = \boxed{92 \text{ atm}}$$

$$= 92 \text{ atm}\left(\frac{14.7 \text{ psi}}{1 \text{ atm}}\right) = \boxed{1.4 \times 10^3 \text{ psi}}$$

(b) Convert the Celsius temperature to kelvins. Then:

$$V = \frac{nRT}{P} = \frac{9.375 \text{ mol } (0.08206 \text{ L atm mol}^{-1} \text{ K}^{-1})303.15 \text{ K}}{0.98 \text{ atm}} = \boxed{2.4 \times 10^2 \text{ L}}$$

5-44 Assume that the average temperature and pressure of the specified portion of the atmosphere are 273 K and 1 atm. Under these conditions a mole of gas occupies 22.4 liters. A mole of gas contains N_0 molecules. Hence, to within a power of ten:

$$\frac{1.657 \text{ L CH}_4}{10^6 \text{ L air}} \left(\frac{1 \text{ mol CH}_4}{22.4 \text{ L CH}_4}\right) \left(6.02 \times 10^{23} \frac{\text{molec CH}_4}{\text{mol CH}_4}\right) \approx \boxed{10^{16} \frac{\text{molec CH}_4}{\text{L air}}}$$

5-46 Assume ideal-gas behavior by the hydrazine:

$$d_{\text{NH}_2\text{NH}_2} = \frac{P}{RT} \mathcal{M} = \left(\frac{1.47 \text{ atm}}{(0.08206 \text{ L atm mol}^{-1} \text{ K}^{-1})623.15 \text{ K}}\right) 32.05 \text{ g mol}^{-1}$$

$$= \boxed{0.921 \text{ g L}^{-1}}$$

5-48 At STP, 1.00 mol of an ideal gas occupies 22.4 L. Assuming that the gaseous compound and the $\text{Cl}_2(g)$ are ideal, then:

$$n_{\text{cmpd}} = \frac{0.153 \text{ L}}{22.4 \text{ L mol}^{-1}} = 0.006830 \text{ mol}$$

$$n_{\text{Cl}_2} = \frac{0.688 \text{ L}}{22.4 \text{ L mol}^{-1}} = 0.03071 \text{ mol}$$

From the second result, it follows that the chemical amount of Cl is 0.06142 mol. Compute the mass of chlorine:

$$m_{\text{Cl}_2} = 0.03071 \text{ mol} \times (70.906 \text{ g mol}^{-1}) = 2.178 \text{ g Cl}_2$$

By difference, the mass of boron in the original sample equals 0.6642 g. Convert this to moles:

$$n_{\text{B}} = \frac{0.6642 \text{ g B}}{10.811 \text{ g mol}^{-1}} = 0.06143 \text{ mol B}$$

The original gaseous sample contains 0.006830 mol of compound. It furnishes chemical amounts of B and Cl that are both almost exactly 9 times this. Therefore, its molecular formula is $\boxed{\text{B}_9\text{Cl}_9}$.

5-50 (a) The volume of one mole of an ideal gas at a pressure of 1.00 atm and a temperature of 25°C (298.15 K) is obtained from the ideal gas law:

$$V = \frac{nRT}{P} = \frac{(1 \text{ mol})(0.08206 \text{ L atm mol}^{-1} \text{ K}^{-1})(298.15 \text{ K})}{1.00 \text{ atm}} = 24.46 \text{ L}$$

Assume that the O_2 is ideal and use this molar volume as a conversion factor:

$$5.00\,\text{g BaO}_2 \times \left(\frac{1\,\text{mol BaO}_2}{169.33\,\text{g BaO}_2}\right) \times \left(\frac{1\,\text{mol O}_2}{2\,\text{mol BaO}_2}\right) \times \left(\frac{24.46\,\text{L O}_2}{1\,\text{mol O}_2}\right) = \boxed{0.361\,\text{L O}_2}$$

(b) The mass of the BaO is computed similarly:

$$5.00\,\text{g BaO}_2 \times \left(\frac{1\,\text{mol BaO}_2}{169.33\,\text{g BaO}_2}\right) \times \left(\frac{1\,\text{mol BaO}}{1\,\text{mol BaO}_2}\right) \times \left(\frac{153.33\,\text{g BaO}}{1\,\text{mol BaO}}\right) = \boxed{4.53\,\text{g BaO}}$$

5-52 (a) The balanced equation is: $\boxed{\text{Fe}(s) + \text{H}_2\text{SO}_4(aq) \longrightarrow \text{H}_2(g) + \text{FeSO}_4(aq)}$.

(b) The 300×10^3 g mass of sulfuric acid is converted to chemical amount by dividing by its molar mass (98.08 g mol^{-1}). The equation states that 1 mol of H_2 forms for every 1 mol of H_2SO_4, so the reaction gives 3.059×10^3 mol of H_2. Substituting this value as n_{H_2} in the ideal gas equation with $T = 300$ K and $P = 1.0$ atm gives $V_{\text{H}_2} = \boxed{7.5 \times 10^4\,\text{L}}$.

(c) The formula $V = 4/3\pi r^3$ relates the volume of a sphere to its radius. A liter is equal to a cubic decimeter, so inserting the volume of gas in liters into this formula and solving for r gives the radius of the spherical balloon in decimeters. It is 26 dm, which is 2.6 m.

5-54 (a) The chemical equation is

$$\boxed{2\,\text{Al}(s) + 6\,\text{HCl}(aq) \longrightarrow 3\,\text{H}_2(g) + 2\,\text{AlCl}_3(aq)}$$

(b) Assume that the hydrogen behaves ideally and substitute the volume, pressure, absolute temperature into the ideal gas equation along with R in appropriate units:

$$n_{\text{H}_2} = \frac{PV}{RT} = \frac{(0.750\,\text{atm})\,(10.0\,\text{L})}{(0.08206\,\text{L atm mol}^{-1}\,\text{K}^{-1})\,(303.15\,\text{K})} = 0.302\,\text{mol}$$

According to the balanced equation, 2/3 of this chemical amount of aluminum reacts. Therefore:

$$m_{\text{Al}} = 0.3015\,\text{mol H}_2 \times \left(\frac{2\,\text{mol Al}}{3\,\text{mol H}_2}\right) \times \left(\frac{26.98\,\text{g Al}}{\text{mol Al}}\right) = \boxed{5.42\,\text{g Al}}$$

5-56 First compute the minimum required chemical amount of ozone:

$$n_{O_3} = 4.69\,\text{g}\,KO_3 \times \left(\frac{1\,\text{mol}\,KO_3}{87.10\,\text{g}\,KO_3}\right) \times \left(\frac{5\,\text{mol}\,O_3}{2\,\text{mol}\,KO_3}\right) = 0.135\,\text{mol}\,O_3$$

Next, use the ideal gas law to determine the volume of this amount of gaseous ozone under the stated conditions:

$$V_{O_3} = \frac{nRT}{P} = \frac{(0.1346\,\text{mol})(0.08206\,\text{L atm mol}^{-1}\,\text{K}^{-1})(258.15\,\text{K})}{0.134\,\text{atm}} = \boxed{21.3\,\text{L}}$$

5-58 Assume that the gaseous hydrocarbon behaves ideally. Then, its chemical amount in the original sample is

$$n = \frac{PV}{RT} = \frac{(3.40\,\text{atm})(25.4\,\text{L})}{(0.08206\,\text{L atm mol}^{-1}\,\text{K}^{-1})(400\,\text{K})} = \boxed{2.631\,\text{mol}}$$

Burning the hydrocarbon takes off all of its hydrogen in the form of water and all of its carbon in the form of carbon dioxide. The 47.4 g of H_2O is 2.631 mol H_2O; the 231.6 g of CO_2 is 5.262 mol of CO_2. The original sample must contain 5.26 mol of H and 5.26 mol of C. For every 1 mol of hydrocarbon, 2 mol of C and 2 mol of H are present. The molecular formula is therefore $\boxed{C_2H_2}$.

5-60 In problem 5-59, the partial pressure of O_2 was 1.02 atm. It is reduced to 0.51 atm by cutting the amount of O_2 in half with no changes in P and T, as long as the O_2 behaves as an ideal gas. The partial pressure of the other gas remains equal to 1.41 atm. Hence the mole fraction of the oxygen is

$$X_{O_2} = \frac{0.51}{0.51 + 1.41} = \boxed{0.27}$$

5-62 Convert the amounts of ammonia, nitrogen and hydrogen, which are given in grams, to moles. There are 8.749 mol of NH_3, 27.308 mol of N_2 and 81.35 mol of H_2. The total chemical amount of gas in the reaction mixture is $8.749 + 27.31 + 81.35 = 117.41$ mol. The ammonia contributes 8.749 mol so its mole fraction is $8.749/117.41 = 0.0745$. The partial pressure of the ammonia equals its mole fraction multiplied by the total pressure of 210 atm as long as the mixture behaves according to Dalton's law. Then $P_{NH_3} = \boxed{15.6\,\text{atm}}$.

5-64 Assume that the Venusian atmosphere follows the ideal gas law and Dalton's law of partial pressures. The chemical amounts of the gases in the Venusian atmosphere are then in direct proportion to their relative volumes in the mixture.

That is, the mole fractions of the gases equal their volume fractions: 0.965 for the CO_2 and 0.035 for the N_2. The partial pressures of the components of the atmosphere equal the total pressure multiplied by these mole fractions:

$$P_{CO_2} = 0.965(90.8\,\text{atm}) = \boxed{87.6\,\text{atm}}$$
$$P_{N_2} = 0.035(90.8\,\text{atm}) = \boxed{3.18\,\text{atm}}$$

5-66 The way the mole fraction of a substance in a reaction mixture changes over the course of the reaction is not intuitively obvious, as this problem shows.

(a) Before the reaction, the mole fraction of the Br_2 is $4.5/(4.5 + 33.1) = 0.12$.

(b) The balanced equation shows that formation of 2.2 mol of BrF_5 consumes 1.1 mol of Br_2 and 5.5 mol of F_2. Hence, at the point in the reaction reaction at which 2.2 mol of BrF_5 is present 3.4 mol of Br_2 and 27.6 mol of F_2 are left. At that point, 33.2 mol of substances of all kinds are present. The mole fraction of Br_2 is $3.4/33.2 = \boxed{0.10}$. Despite the fact that about a quarter of the Br_2 has been consumed, the mole fraction of Br_2 has dropped by only a sixth.

5-68 The molar mass of sodium is 0.02299 kg mol^{-1}. Substitute this value, the given T, and a value of R in the proper units into the formula for root-mean-square speed:

$$u_{\text{rms}}(\text{Na}) = \sqrt{\frac{3RT}{M}} = \sqrt{\frac{3(8.315 \text{ J K}^{-1} \text{ mol}^{-1})(0.00024 \text{ K})}{0.02299 \text{ kg mol}^{-1}}} = \boxed{0.51\,\text{m s}^{-1}}$$

It is assumed that 500 atoms is a large enough sample to have a meaningful temperature.

5-70 To calculate the rms speeds of helium, argon, and xenon atoms at 2000 K substitute the molar masses in kg mol^{-1} of the three gases successively into the expression

$$u_{\text{rms}} = \sqrt{\frac{3RT}{M}}$$

along with $T = 2000$ K and $R = 8.315$ J K^{-1} mol^{-1}. The answers are 3.53 km s^{-1} for helium, 1.12 km s^{-1} for argon and 0.616 km s^{-1} for xenon. These values are respectively $\boxed{31.5\%,\ 10.0\%\text{ and }5.50\%}$ of 11.2 km s^{-1}, the given escape velocity. Clearly, a greater fraction of helium will be able to escape.

5-72 The mass of a ClO_2 molecule is less than the mass of a Cl_2 molecule. At the same temperature the average kinetic energies of the two species are equal. It follows that the average speed of ClO_2 molecules exceeds that of the Cl_2 molecules at every temperature. Accordingly, the percentage of chlorine molecules with speeds in excess of 400 m s^{-1} is $\boxed{\text{less than}}$ the percentage of chlorine dioxide molecules with speeds in excess of 400 m s^{-1}.

5-74 The problem is a straightforward application of Graham's law of effusion. The relative rates of effusion are given by

$$\frac{\text{rate}_{F_2}}{\text{rate}_{BrF_5}} = \sqrt{\frac{\mathcal{M}_{BrF_5}}{\mathcal{M}_{F_2}}}$$

Substituting the molar masses gives $\boxed{2.145}$ as the ratio of the rates of effusion. As effusion continues the remaining mixture becomes enriched in the heavier gas, and this ratio drops.

5-76 Solve the van der Waals equation for the pressure:

$$P = \frac{nRT}{V - nb} - a\left(\frac{n^2}{V^2}\right)$$

Look up the values of a and b for water in text Table 5-3 (text page 212). They equal 5.464 atm L^2 mol^{-2} and 0.03049 L mol^{-1}. Substitute them, together with $V = 2500 \times 10^3$ L, $R = 0.08206$ L atm mol^{-1} K^{-1}, $n = 140 \times 10^6/18.0153$ mol, and $T = 813.15$ K to obtain $\boxed{P = 176 \text{ atm}}$. This pressure equals 2.59×10^3 psi.

5-78 The van der Waals equation works better than the ideal gas equation to represent the behavior of real gases, but this does not make it exactly correct. In the problem, experimental values of P, V, and T are given for a sample of CO_2 of known n and a value of a is computed assuming that b is 0.04267 L mol^{-1}. Solving the equation for a gives

$$a = \frac{V^2}{n^2}\left(\frac{nRT}{V - nb} - P\right)$$

Substitution of the various quantities in units involving K, mol, L and atm, gives $a = \boxed{4.600 \text{ atm L}^2\text{mol}^{-2}}$. It is also instructive to compute what b would be if a were equal to the value given in Table 5-3 (3.592 atm L^{2}mol^{-2}).

5-80 Mercury expands with increasing temperature. On a hot day the column that the atmosphere maintains is therefore longer than if the barometer were put in a freezer. The answer to the question is consequently expected to be less than 760.0 torr = 1 atm. From the equation $P = dgh$ the height of the column of liquid in a barometer is inversely proportional to the density of the liquid that it contains. The height that this column would have at 0.0°C is

$$h_{0°} = h_{35°} \times \left(\frac{13.5094 \, \text{g cm}^{-3}}{13.5955 \, \text{g cm}^{-3}}\right) = 760.0 \, \text{mm} \times 0.993667 = 755.2 \, \text{mm}$$

This height in a mercury barometer at 0.0°C means a pressure of $\boxed{0.9937 \text{ atm}}$.

5-82 The $CO_2(g)$ generated by the reaction expands the bag. The expanded bag displaces more air. The bag and its contents are buoyed up by a larger force than before expansion. A similar but larger change in the buoyant force occurs when a balloon immersed in water is expanded. This is the reason the reading on the digital balance drops. If the bag were in a vacuum, there would be no change on the readout. $\boxed{\text{The law of conservation of mass is not violated}}$.

5-84 The volume of a ideal gas goes to zero as the gas is cooled toward absolute zero. Therefore, set $V = 0$ in the given equation and solve for t_F:

$$V = 0 = 209.4 \, \text{L} + \left(0.456\frac{\text{L}}{°\text{F}}\right) \times t_F$$

$$t_F = \frac{-209.4 \, \text{L}}{0.456 \, \text{L} \, (°\text{F})^{-1}} = \boxed{-459°\text{F}}$$

5-86 Reasoning by analogy to the textbook statements of Charles's law and Boyle's law, one concludes that Amontons's law is that at constant volume, the pressure of a sample of a gas is directly proportional to its absolute temperature.

5-88 A hydrocarbon contains only the elements carbon and hydrogen. Let the molecular formula of the hydrocarbon be C_xH_y. The hydrocarbon and oxygen combine in a 2 to 9 ratio by volume. It follows that they combine in a 2 to 9 molar ratio as well, because the chemical amounts (of ideal gases) are in direct proportion to their volumes when T and P are the same. The equation for the combustion reaction is

$$2\,C_xH_y(g) + 9\,O_2(g) \longrightarrow 2x\,CO_2(g) + y\,H_2O(g)$$

Balancing oxygen between the two sides of this equation requires the relationship $18 = 4x + y$. The chemical amount of hydrogen in the sample of the hydrocarbon is

$$n_H = 0.135 \, \text{g H}_2\text{O} \times \left(\frac{1 \, \text{mol H}_2\text{O}}{18.0153 \, \text{g H}_2\text{O}}\right) \times \left(\frac{2 \, \text{mol H}}{1 \, \text{mol H}_2\text{O}}\right) = 0.01499 \, \text{mol H}$$

and the chemical amount of carbon is

$$n_C = 0.330 \text{ g CO}_2 \times \left(\frac{1 \text{ mol CO}_2}{44.0098 \text{ g CO}_2}\right) \times \left(\frac{1 \text{ mol C}}{1 \text{ mol CO}_2}\right) = 0.00750 \text{ mol C}$$

Clearly, the chemical amount of H is twice that of C, so regardless of the size of the sample, $y = 2x$. There are now two equations relating the unknowns x and y. Solving gives $x = 3$ and $y = 6$. The hydrocarbon is $\boxed{C_3H_6}$.

5-90 As established in problem 5-50, the molar volume of an ideal gas at 25°C and 1.00 atm pressure equals 24.46 L. From the first balanced equation in "Chemistry in Your Life: How an Air-Bag Works" (text page 181) 3 mol of N_2 is generated from every 2 mol pf NaN_3. Use these facts as follows:

$$70.0 \text{ L N}_2 \times \left(\frac{1 \text{ mol N}_2}{24.46 \text{ L N}_2}\right) \times \left(\frac{2 \text{ mol NaN}_3}{3 \text{ mol N}_2}\right) \times \left(\frac{64.91 \text{ g NaN}_3}{1 \text{ mol NaN}_3}\right) = \boxed{124 \text{ g NaN}_3}$$

5-92 Let the number of molecules in the container prior to the chemical change equal x. $x/2$ of these molecules are unaffected by the change, but $x/2$ of them react. These $x/2$ are consumed, but produce $2/3(x/2)$ new molecules. After the change, the number of molecules is

$$N_{\text{after}} = x - \frac{x}{2} + \left(\frac{2}{3} \times \frac{x}{2}\right) = x\left(1 - \frac{1}{2} + \frac{1}{3}\right) = \frac{5}{6}x$$

The pressure is directly proportional to the number of molecules of the container, as long as the volume and temperature do not change. Hence, the pressure after the chemical change is 5/6 of 0.740 atm, or $\boxed{0.617 \text{ atm}}$.

5-94 (a) Assume that air is an ideal gas. It can be easily shown that its density is then

$$d = \mathcal{M}\left(\frac{P}{RT}\right)$$

A temperature of 95°F is 35.0°C or 308.15 K, and the P and $\mathcal{M}$ are given explicitly in the problem. Substitute these values in the preceding:

$$d = 29.0 \text{ g mol}^{-1}\left(\frac{1.00 \text{ atm}}{(0.08206 \text{ L atm mol}^{-1} \text{ K}^{-1})(308.15 \text{ K})}\right) = \boxed{1.15 \text{ g L}^{-1}}$$

(b) A temperature of 50°F is 10.0°C or 283.15 K. Use this T:

$$d = 29.0 \text{ g mol}^{-1}\left(\frac{1.00 \text{ atm}}{(0.08206 \text{ L atm mol}^{-1} \text{ K}^{-1})(283.15 \text{ K})}\right) = \boxed{1.25 \text{ g L}^{-1}}$$

(c) Saturating dry air with water vapor means adding a component that has a molar mass less that the molar masses of N_2 and O_2, the main components of dry air. The molar mass of moist air is therefore less than 29.0 g mol^{-1}. The density of a gas is directly proportional to its molar mass It follows that moist air is less dense than dry air at any given T and P. If a batted ball truly carries better in low-density air, then $\boxed{\text{yes}}$, high humidity favors the home run.

5-96 (a) Convert the volume of the trap to liters (using 1 L $= 10^{-3}$ m^3) and the amount of its contents to moles. Then apply the ideal gas law:

$$P = n\left(\frac{RT}{V}\right) = \frac{500}{6.022 \times 10^{23} \text{ mol}^{-1}} \left(\frac{(0.08206 \text{ L atm mol}^{-1} \text{ K}^{-1})(0.00024 \text{ K})}{1.0 \times 10^{-12} \text{ L}}\right)$$

$$= \boxed{1.6 \times 10^{-14} \text{ atm}}$$

(b) The mean free path of the sodium atoms in the optical trap (3.9 m) is much longer than that of gaseous sodium atoms at room conditions (about 10^{-7} m).

5-98 (a) Convert the number density to density in moles per liter:

$$\frac{n}{V} = \frac{10 \text{ atoms}}{1 \text{ cm}^3} \times \frac{1 \text{ mol}}{6.022 \times 10^{23} \text{ atoms}} \times \frac{1000 \text{ cm}^3}{1 \text{ L}} = 1.66 \times 10^{-20} \text{ mol L}^{-1}$$

Substitution of this value into $P = (n/V)\,RT$ gives $P = \boxed{1.4 \times 10^{-19} \text{ atm}}$.

(b) The root-mean-square speed is calculated by substituting $T = 100$ K and $\mathcal{M} = 1.008 \times 10^{-3}$ kg mol^{-1} into

$$u_{\text{rms}} = \sqrt{\frac{3RT}{\mathcal{M}}}$$

being sure to use R in the proper units (J K^{-1}mol^{-1}). The answer is 1.56 $\times$ 10^3 m s^{-1}. Multiplying this rms speed by the time between collisions gives the average distance between collisions. It is $\boxed{1.6 \times 10^{12} \text{ m}}$, about 10.5 times the distance between the earth and the sun.

Answers to Even-Numbered Problems, Chapter 6

6-2 The statement is $\boxed{\text{false}}$. The lengths of covalent bonds between large atoms can easily exceed the distance between the centers of small atoms that are interacting by means of nonbonded attractions.

6-4 (a) dispersion forces   (b) dipole-dipole forces (predominant) and dispersion forces (c) dispersion forces   (d) ion-ion forces (predominant) and dispersion forces   (e) hydrogen bonds (predominant) and dispersion forces

---

6-6 An atom of argon should be most strongly attracted by an atom of $\boxed{\text{krypton}}$. An atom of krypton has more electrons and is more polarizable than an atom of argon or neon. The strength of dispersion forces depends on the polarizability of the interacting species.

6-8 The structure has chains of H—O—F molecules, which are bent, linked by O$\cdots$H hydrogen bonds. The fluorine atoms stick out to the sides of these chains.

6-10 The kinetic theory predicts an increase in molecular speed in all three states of matter with increasing temperature. Since diffusion depends on the random motions of the molecule of substances, which become more rapid at higher temperature, the theory predicts that the diffusion constants should $\boxed{\text{increase}}$ with temperature in all states of matter.

6-12 The surface tension of liquid NaCl should be $\boxed{\text{higher}}$ than that of CCl_4 because the intermolecular forces in NaCl (ion-ion forces) are intrinsically stronger than the forces in CCl_4 (dispersion forces).

6-14 Apply the ideal-gas equation to the helium vapors. First, rearrange the equation so that the molar volume is explicitly on the left. Then substitute the values from the problem:

$$\frac{V}{n} = \frac{RT}{P} = \frac{(0.08206 \text{ L atm mol}^{-1} \text{ K}^{-1})2.20 \text{ K}}{0.05256 \text{ atm}} = \boxed{3.43 \text{ L mol}^{-1}}$$

Even at these extreme conditions, the molar volume is still large, a characteristic of a gas. It is, however, smaller than that at STP, 22.4 L mol^{-1}.

6-16 Use mole ratios from the balanced equation to compute the chemical amount of the ammonia that is produced:

$$3.68 \text{ g NH}_4\text{Cl} \times \left(\frac{1 \text{ mol NH}_4\text{Cl}}{53.491 \text{ g NH}_4\text{Cl}}\right) \times \left(\frac{1 \text{ mol NH}_3}{1 \text{ mol NH}_4\text{Cl}}\right) = 0.06880 \text{ mol NH}_3$$

This amount of ammonia exerts a partial pressure of 0.9465 atm in the mixture; the water vapor exerts the other 0.0419 atm. According to Dalton's law, the water vapor and the ammonia vapor do not interact with each other in the container but instead effectively interpenetrate. The volume of the wet ammonia therefore equals the volume that dry ammonia would occupy if present by itself. This volume is calculated using the chemical amount of ammonia and the partial pressure of ammonia in the ideal gas equation:

$$V_{NH_3} = \frac{n_{NH_3}RT}{P} = \frac{(0.06880\ \text{mol})(0.08206\ \text{L atm mol}^{-1}\ \text{K}^{-1})(303.15\ \text{K})}{(0.9465\ \text{atm})}$$

$$= \boxed{1.81\ \text{L}}$$

6-18 Antimony has a higher molar mass than arsenic. Since dispersion forces are the sole intermolecular forces operating in SbF_5 and AsF_5, the boiling point of the former should be higher than that of the latter. Among fluorides of the same metal the one with the metal in the higher oxidation state should have the lower boiling point; these compounds have weaker intermolecular forces. The intermolecular forces in fluorine should be the weakest of all because they are dispersion forces in a low-molar-mass compound. The order of boiling points therefore is

$$\boxed{F_2 \quad < AsF_5 \quad < SbF_5 \quad < SbF_3}$$

6-20 Reading from the diagram (text page 240) across on the $P = 4$ atm line to the liquid/vapor equilibrium line and then down to the temperature axis gives an estimate of 430 K or $\boxed{157°C}$ for the boiling point of water in the pressure cooker.

6-22 The "intermolecular" (interatomic) forces in aluminum are stronger because it melts and boils higher than thallium. Hence, we expect $\boxed{\text{thallium}}$ to be more volatile (have a higher vapor pressure than aluminum at room temperature.

6-24 Gray tin is favored over white tin by lower temperature, but white tin is favored by higher pressure (because it is more dense than gray tin). Suppose that the two forms of tin are present at equilibrium at 1 atm and 13.2°C. Raising the pressure to 2 atm (eventually) converts all the tin to white tin. To restore the gray tin the temperature must be adjusted in the direction that favors gray tin, that is, the temperature should be $\boxed{\text{lower}}$ than 13.2°C.

6-26 (a) The phase diagram of H_2O (text Figure 6-20c) confirms that at equilibrium at room conditions (the starting conditions in the process described in the problem), water is a $\boxed{\text{liquid}}$.

(b) At 400 K and 1 atm, H_2O is a $\boxed{\text{gas}}$, according to the phase diagram.

(c) Although the water starts out as a liquid and ends as a gas in the process described, no phase transitions occur . The water is taken into the supercritical region by the changes described. The change from liquid to supercritical fluid is smooth and gradual. The subsequent change from supercritical fluid to gas is also continuous. There is no abrupt change in density or other physical properties and therefore no phase transition.

6-28 (a) The vapor pressure of solid hydrogen, and liquid hydrogen *both* equal the external pressure on the system at the triple point of hydrogen. The answer is 0.069 atm .

(b) The pressure is maintained at a value below the triple-point pressure of hydrogen. Hence the solid hydrogen converts directly to the gas. The phase transition is a sublimation .

6-30 The tube must contain 0.235 g of ammonia for every cm^3 of volume. If it contains more ammonia than this, the ammonia will be liquid and will become supercritical above 132.23°C without anything to be seen from the outside. If it contains less ammonia, the ammonia will be gaseous below the critical temperature and become supercritical above that temperature, again with nothing to be seen. The interior radius of the tube is $5.0 - 4.20 = 0.80$ mm $= 0.080$ cm. The height of the tube is 15.5 cm so that its volume is

$$V = \pi r^2 h = \pi \, (0.080)^2 \ cm^2 \times 15.5 \ cm = 0.3116 \ cm^3$$

The required mass of ammonia is

$$m_{NH_3} = dV = (0.235 \ g \ cm^{-3}) \times (0.3116 \ cm^3) = \boxed{0.0732 \ g}$$

6-32 The percentage by mass of acetic acid is

$$\left(\frac{m_{CH_3COOH}}{m_{CH_3COOH} + m_{H_2O}} \right) \times 100\% = \left(\frac{29.3 \ g}{29.3 \ g + 575 \ g} \right) \times 100\% = \boxed{4.85\%}$$

To obtain the mole fraction of acetic acid, first compute the molar masses of acetic acid and water.[4] Use them to convert from masses to chemical amounts:

$$\frac{29.3 \ g \ CH_3COOH}{60.053 \ g \ mol^{-1}} = 0.488 \ mol \ CH_3COOH; \qquad \frac{575 \ g \ H_2O}{18.015 \ g \ mol^{-1}} = 31.92 \ mol \ H_2O$$

---

[4]It is necessary to know or to look up the molecular formulas of acetic acid (see text page 137) and water to compute their molar masses.

Then:

$$X_{CH_3COOH} = \frac{0.488 \text{ mol}}{0.488 \text{ mol} + 31.92 \text{ mol}} = \boxed{0.0151}$$

The molality equals the chemical amount in moles of acetic acid divided by the mass in kilograms of the solvent

$$m_{CH_3COOH} = \frac{0.488 \text{ mol}}{0.575 \text{ kg}} = \boxed{0.849 \text{ kg mol}^{-1}}$$

6-34 The volume of 1.000 kg of the solution is

$$1.000 \text{ kg} \times \left(\frac{1000 \text{ g}}{1 \text{ kg}}\right) \times \left(\frac{1 \text{ cm}^3}{1.0250 \text{ g}}\right) \times \left(\frac{1 \text{ L}}{1000 \text{ cm}^3}\right) = 0.9756 \text{ L}$$

1.000 kg of this solution contains 200.0 g of acetic acid, which equals 3.3304 mol of acetic acid (using a molar mass of 60.063 g mol^{-1}). The concentration therefore is

$$c_{CH_3COOH} = \frac{n_{CH_3COOH}}{V} = \frac{3.3304 \text{ mol}}{0.9756 \text{ L}} = \boxed{3.414 \text{ mol L}^{-1}}$$

The molality equals the chemical amount in moles of acetic acid divided by the mass in kilograms of the solvent

$$m_{CH_3COOH} = \frac{3.304 \text{ mol}}{0.8000 \text{ kg}} = \boxed{4.130 \text{ mol kg}^{-1}}$$

6-36 Consider a sample of the 9.20 mol L^{-1} aqueous perchloric acid that has a volume of exactly 1 L. It contains 9.20 mol of perchloric acid. This amount of HClO$_4$ (molar mass 100.46 g mol^{-1}) has a mass equal to 0.924 kg. Meanwhile, the mass of the 1 L of *solution* is 1.54 kg, because the density of the solution is 1.54 kg L^{-1}. Therefore the molality of the perchloric acid is:

$$m_{HClO_4} = \frac{n_{HClO_4}}{\text{kg of water}} = \frac{9.20 \text{ mol}}{(1.54 - 0.924) \text{ kg}} = \boxed{14.9 \text{ mol kg}^{-1}}$$

6-38 The molar mass of maleic acid is 116.07 g mol^{-1}, and the molar mass of diethyl ether is 74.12 g mol^{-1}. Applying the definition of mole fraction to this solution of the two gives:

$$X_{C_4H_4O_4} = \left(\frac{1.800/116.07 \text{ mol}}{(1.800/116.07) \text{ mol} + (100.0/74.12) \text{ mol}}\right) = 0.01136$$

The change in the vapor pressure of the volatile solvent diethyl ether is related to the mole fraction of the nonvolatile solute maleic acid:

$$X_{C_4H_4O_4} = \frac{-\Delta P}{P^o_{ether}} \quad \text{that is,} \quad 0.01136 = \frac{-\Delta P}{0.8517 \text{ atm}}$$

Solving gives ΔP equal to -0.009679 atm. The vapor pressure of the solution equals the vapor pressure of the pure solvent plus this change. The result is $\boxed{0.8420 \text{ atm}}$.

6-40 The vapor pressure of the benzene at 25°C is reduced from 0.1252 atm to 0.1199 atm) by the presence of the nonvolatile solute. The mole fraction of the solute is:

$$X_{solute} = \frac{\Delta P}{-P^o_{benzene}} = \frac{(0.1199 - 0.1252) \text{ atm}}{-0.1252 \text{ atm}} = \boxed{0.042}$$

6-42 The dissolution of the nonvolatile nondissociating compound of indium and chlorine in tin(IV) chloride raises the boiling point of the solvent in proportion to its molality. Compute the molality:

$$m_{cmpd} = \frac{\Delta T}{K_b} = \frac{(116.3 - 114.1)°C}{9.43°C \text{ kg mol}^{-1}} = 0.233 \text{ kg mol}^{-1}$$

When 2.60 g of the compound is dissolved in 50.0 g of solvent, the molality of the resulting solution is the same as if 52.0 g had been dissolved in 1000 g of solvent (scaling up by a factor of 20). Thus 52.0 g of the compound equals 0.233 mol and

$$\mathcal{M} = \frac{52.0 \text{ g}}{0.233 \text{ mol}} = \boxed{2.2 \times 10^2 \text{ g mol}^{-1}}$$

The molar mass of In is approximately 115 g mol^{-1}. This allows room in the molecular formula for only one indium. The remainder of the molar mass must be contributed by chlorine. Three Cl's furnish almost exactly the required molar mass, so the formula is $\boxed{InCl_3}$.

6-44 The ranking of the three nonvolatile solutes must be done on equal masses. Assume 1000 g of each solute. This converts to 17.11 mol of NaCl, 9.01 mol of CaCl$_2$ and 16.67 mol of H$_2$NCONH$_2$. The freezing point depression depends on the number of particles released upon dissolution. The 1000 g of NaCl (for which $\nu = 2$) would, if dissolved and fully dissociated into ions, release 34.22 mol of particles, The 1000 g of CaCl$_2$, for which $\nu = 3$, would release 27.03 mol of particles if fully dissociated, and the 1000 g of H$_2$NCONH$_2$, for which

$\nu = 1$) would release 16.67 mol of particles if fully dissociated. Hence, the order of effectiveness is

$$\boxed{NaCl > CaCl_2 > NH_2CONH_2}$$

This answer assumes that the entire 1000 g of each *can* be dissolved and that the dissociation of the NaCl and CaCl₂ is complete. The relative rates at which the solutes dissolve would matter in the practical selection of a substance for melting ice on the streets. Also, the environmental impact of the substances would be important.

6-46 The chilling solution contains eights parts of water by mass for every one part of salt. If there is 1.000 kg of water, there must be 125 g of salt, according to this ratio. But 125 g of NaCl (molar mass 58.44 g mol^{-1}) is 2.14 mol of NaCl. The molality of the chilling solution is accordingly 2.14 mol kg^{-1}. Substitute this value in the equation for freezing-point depression along with a van't Hoff i of 2 (because NaCl dissociates into two particles per mole in aqueous solution):

$$\Delta T = -iK_f m = -2 \left(\frac{1.86°C \ kg}{mol} \right) (2.14 \ mol \ kg^{-1}) = -8.0°C$$

To get the freezing point of the solution, add this ΔT to the freezing point of pure water. The answer is $\boxed{-8.0°C}$, which is equivalent to 18°F. A solution of NaCl that has this composition actually freezes at $-7.44°C$.

6-48 The observed change in freezing point ΔT is $-0.396°C$, and the K_f of the solvent is 1.86 K kg mol^{-1}. Also, the molality of the solution, computed by dividing the chemical amount of NaHCO₃ (2.02 g/84.0 g mol^{-1}) by the mass of water (0.200 kg) is 0.120 mol kg^{-1}. Solve the freezing-point depression equation for i and substitute:

$$i = \frac{\Delta T_f}{-K_f m} = \frac{-0.396 \ K}{-(1.86 \ K \ kg \ mol^{-1})(0.120 \ mol \ kg^{-1})} = \boxed{1.77}$$

The dissolution must give at least two moles of products per mole of sodium hydrogen carbonate, that is, $\nu \geq 2$. A suitable equation is

$$\boxed{NaHCO_3 \longrightarrow Na^+ + HCO_3^-}$$

6-50 The solution of ethanol in sulfuric acid has a molality of 0.050 mol kg^{-1} because 2.3 g of ethanol (molar mass 46 g mol^{-1}) equals 0.050 mol of ethanol. Compute i for this solution:

$$i = \frac{\Delta T_f}{-K_f m} = \frac{-0.92 \ K}{-(6.12 \ K \ kg \ mol^{-1})(0.050 \ mol \ kg^{-1})} = 3.0$$

The ethanol must react to give $\boxed{\text{three particles}}$ per molecule upon dissolution in sulfuric acid. The reaction has been studied. It is not exactly a dissociation:

$$C_2H_5OH + 2\,H_2SO_4 \longrightarrow C_2H_5HSO_4 + HSO_4^- + H_3O^+$$

6-52 Assume that the density of the ethanolic solution equals the density of the solvent and substitute in the barometric formula to compute the osmotic pressure:

$$\Pi = gdh = 9.807\,\mathrm{m\,s^{-2}}\left(0.789 \times 10^3\,\mathrm{kg\,m^{-3}}\right) 0.263\,\mathrm{m} = 2035\,\mathrm{Pa}$$

This osmotic pressure equals 0.0201 atm. Now, compute the molar mass of the protein from the osmotic pressure and the other quantities:

$$\mathcal{M} = \frac{mRT}{\Pi V} = \frac{1.25\ \mathrm{g}(0.08206\ \mathrm{L\ atm\ mol^{-1}\ K^{-1}})(293.15\ \mathrm{K})}{(0.0201\ \mathrm{atm})(0.0100\ \mathrm{L})}$$

$$= \boxed{1.50 \times 10^5\ \mathrm{g\ mol^{-1}}}$$

6-54 Use the relationship

$$\alpha = \frac{i-1}{\nu-1}$$

given on text page 259 with $\alpha = 0.76$ and $\nu = 3$ to compute i. It equals 2.52. Then:

$$\Delta T_f = -iK_f m = 2.52(1.86\ \mathrm{K\ kg\ mol^{-1}})(0.15\ \mathrm{mol\ kg^{-1}}) = -0.70\ \mathrm{K}$$

The freezing point of the solution is therefore $\boxed{-0.70^\circ\mathrm{C}}$.

6-56 (a) From Table 5-1, the partial pressures of N_2 and O_2 in air are 0.781 atm and 0.210 atm respectively. Henry's law gives the mole fractions in solution:

$$X_{N_2} = \frac{0.781\ \mathrm{atm}}{8.57 \times 10^4\ \mathrm{atm}} = 9.11 \times 10^{-6} \qquad X_{O_2} = \frac{0.210\ \mathrm{atm}}{4.347 \times 10^4\ \mathrm{atm}} = 4.84 \times 10^{-6}$$

These solutions are dilute. They contain 55.5 mol of water per liter of solution, and the chemical amount of dissolved gas can be neglected compared to 55.5 mol. The n's of the two gases in 1.00-L samples of solution are then:

$$\frac{n_{N_2}}{55.5\ \mathrm{mol}} = 9.11 \times 10^{-6} \qquad n_{N_2} = \boxed{5.06 \times 10^{-4}\ \mathrm{mol}}$$

$$\frac{n_{O_2}}{55.5\ \mathrm{mol}} = 4.84 \times 10^{-6} \qquad n_{O_2} = \boxed{2.69 \times 10^{-4}\ \mathrm{mol}}$$

(b) A given partial pressure causes less helium to dissolve in the diver's blood than nitrogen because $\boxed{\text{helium has a lower solubility}}$ (larger Henry's law constant) in water than nitrogen. If the gases formed ideal solutions with water then they would have the same molar solubility in water, and the Henry's law constant would equal the vapor pressure of pure water at the temperature in question. Using helium in place of nitrogen in breathing mixtures moderates the risk of "bends." because a smaller amount of gas dissolves in the blood. Some gas must be mixed with oxygen to dilute it for breathing. Breathing pure oxygen under pressures exceeding 1 or 2 atm so thoroughly saturates the blood with dissolved oxygen that deadly oxygen poisoning results.

6-58 The volume of methane that is expelled by boiling the methane-in-benzene solution is related to its chemical amount by the ideal gas law. At STP, 0.510 L of gas amounts to 0.0228 mol of gas. This amount of methane was present in 1.00 kg of benzene which is 12.80 mol of benzene (molar mass 78.114 g mol^{-1}). The mole fraction of dissolved methane therefore must have been

$$\frac{0.0228}{12.80 + 0.0228} = 1.78 \times 10^{-3}$$

This fraction of dissolved methane was forced into the benzene by a pressure of 1.00 atm of methane. Rearranging Henry's law for this case gives

$$k_H = \frac{P_{\text{methane}}}{X_{\text{methane}}} = \frac{1.00 \text{ atm}}{1.78 \times 10^{-3}} = \boxed{562 \text{ atm}}$$

6-60 Raoult's law gives the vapor pressure of a volatile component above an ideal solution. The vapor pressure is simply the vapor pressure that the component would have, if it were pure, multiplied by its mole fraction in the solution. Therefore, the vapor pressure of the toluene above this solution is

$$P_{\text{toluene}} = X_{\text{toluene}} P^{\circ}_{\text{toluene}} = \left(\frac{0.900}{0.400 + 0.900}\right) 0.534 \text{ atm} = 0.370 \text{ atm}$$

at this temperature. The vapor pressure of the benzene is 0.412 atm, by a similar calculation. The total pressure of the vapors above the solution is 0.782 atm, the sum of the partial pressures of the two volatile components. The fraction of this total pressure that is due to toluene is

$$f_{\text{toluene}} = \frac{0.370 \text{ atm}}{0.782 \text{ atm}} = \boxed{0.473}$$

This same number is also the mole fraction of the toluene in the vapors because molecules of toluene in the vapor exert pressure in proportion to the number present as long as the mixture of gases follows Dalton's law.

6-62 This problem resembles 6-60 with the difference that the chemical amounts of the two components in the solution must be calculated from their masses.

(a) The chemical amount of benzene (molar mass 78.11 g mol^{-1}) in 50.0 g is 0.640 mol. The chemical amount of hexane (molar mass 86.18 g mol^{-1}) is 0.580 mol. The mole fraction of benzene in the solution is:

$$X_{\text{benzene}} = \frac{0.640}{0.640 + 0.580} = \boxed{0.525}$$

(b) The total vapor pressure over the solution is the vapor pressure of the benzene plus the vapor pressure of the hexane. These two vapor pressures are computed using Raoult's law: They are

$$P_{\text{benzene}} = X_{\text{benzene}} P^{\circ}_{\text{benzene}} = 0.525(0.1355 \text{ atm}) = 0.0711 \text{ atm}$$
$$P_{\text{hexane}} = X_{\text{hexane}} P^{\circ}_{\text{hexane}} = 0.475(0.2128 \text{ atm}) = 0.1011 \text{ atm}$$

The total vapor pressure is therefore $\boxed{0.172 \text{ atm}}$.

(c) The mole fraction of benzene in the vapor is $0.0711/0.172 = \boxed{0.413}$.

6-64 The rate of coagulation of the colloid will $\boxed{\text{decrease}}$ because the presence of more iron(III) ions on the surfaces of the colloidal particles increases the electrostatic repulsions that keep the particles apart.

6-66 The "floc" settles out of the water as a loose low-density sediment, It carries colloidal impurities particles with it and so removes them.

6-68 Compute the ratio a/b for the four substances

$$\frac{a}{b} = \frac{1.390}{0.03913} = 35.52 \text{ L atm mol}^{-1} \quad \text{for N}_2$$
$$\frac{a}{b} = \frac{0.2444}{0.02661} = 9.185 \text{ L atm mol}^{-1} \quad \text{for H}_2$$
$$\frac{a}{b} = \frac{6.714}{0.05636} = 119.1 \text{ L atm mol}^{-1} \quad \text{for SO}_2$$
$$\frac{a}{b} = \frac{3.667}{0.04081} = 89.86 \text{ L atm mol}^{-1} \quad \text{for HCl}$$

The strength of the attractive forces goes as the magnitude of a/b so the ranking is

$$\boxed{\text{SO}_2 > \text{HCl} > \text{N}_2 > \text{H}_2}$$

6-70 The chemical amount of air that was present in the 6.00-L portion of air mixed with the vapors of the unknown can be computed because its physical state after purification is fully described

$$n_{air} = \frac{PV}{RT} = \frac{(1.000 \text{ atm})(3.75 \text{ L})}{(0.08206 \text{ L atm mol}^{-1} \text{ K}^{-1})(223.15 \text{ K})} = 0.2048 \text{ mol}$$

Now, compute the pressure that this chemical amount of air exerted as part of the 6.00-L mixture

$$P_{air} = \frac{n_{air} RT}{V}$$

$$P_{air} = \frac{(0.2048 \text{ mol})(0.08206 \text{ L atm mol}^{-1} \text{ K}^{-1})(298.15 \text{ K})}{6.00 \text{ L}} = 0.835 \text{ atm}$$

But the total pressure above the unknown was 0.980 atm. By Dalton's law

$$P_{unknown} = 0.980 - 0.835 = \boxed{0.145 \text{ atm}}$$

6-72

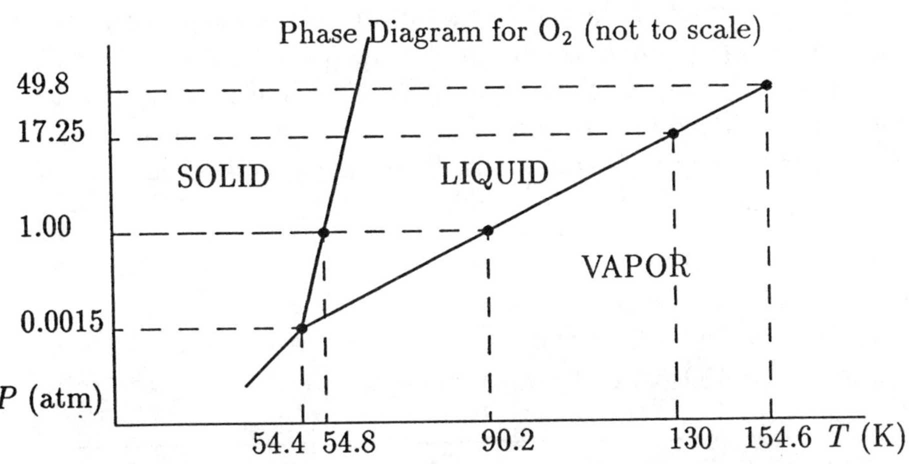

6-74 (a) A 400 atm pressure would lower the melting point of ice to about $\boxed{-4°C}$.

(b) The pressure exerted by the blade of the skate is

$$P = \frac{F}{A} = \frac{ma}{A} = \frac{(75 \text{ kg})(9.8 \text{ m s}^{-2})}{8.0 \times 10^{-5} \text{ m}^2} = 9.2 \times 10^6 \text{ Pa}$$

This pressure equals about 90 atm. It lowers the melting point of the ice by $\boxed{0.9°C}$. Therefore, ice at $-5°C$ does not melt under the pressure of the blade.

6-76 For the sake of concreteness, suppose that 1.00 kg of Na_2CO_3 is dissolved in 5.00 kg of water. The molar mass of Na_2CO_3 is 106.0 g mol^{-1}, so the 5.00 kg of water contains 9.43 mol of Na_2CO_3 for a molality of $\boxed{1.89 \text{ mol kg}^{-1}}$. Now suppose that the recipe is followed and 2.70 kg of $Na_2CO_3 \cdot 10H_2O$ is dissolved in 5.00 kg of water. The molar mass of the decahydrate is 286.15 g mol^{-1}, so that the solution contains 9.44 mol of $Na_2CO_3 \cdot 10H_2O$ per 5.00 kg of water. This gives a molality of the decahydrate equal to 1.89 mol kg^{-1}. This result is the same number obtained using the first option in the recipe. It is the molality of the decahydrate. The molality of the second solution in terms of "plain" (anhydrous) Na_2CO_3 is different! The second solution contains 9.44 mol of Na_2CO_3, because one formula unit of Na_2CO_3 comes from every formula unit of the decahydrate that dissolves. But this amount of solute is mixed with $5.00 + 1.70$ kg of water. The extra 1.70 kg of water comes from the (10×9.44) mol of water that 9.44 mol of $Na_2CO_3 \cdot 10H_2O$ contributes as it dissolves. In terms of anhydrous Na_2CO_3 as a solute, the molality of the second solution is $\boxed{1.41 \text{ mol kg}^{-1}}$.

6-78 (a) The element iodine is distributed in the solution in different forms, but the total amount put into the solution, which is mixed from pure compounds, is always known. The total mass of the iodine in 100 mL of Donovan's solution equals the fraction by mass of iodine in AsI_3 (molar mass 455.64 g mol^{-1}) multiplied by 1.00 g of AsI_3 plus the fraction by mass of iodine in HgI_2 (molar mass 454.4 g mol^{-1}) multiplied by 1.00 g of HgI_2. The necessary fractions are computed from the formulas of the compounds. The mass of iodine per liter of solution is 10 times this answer or $\boxed{13.9 \text{ g L}^{-1}}$.

(b) The 0.100 M AsI_3 solution furnishes 4.556 g of AsI_3 per 100 mL. To make 3.50 L of Donovan's solution, 35.0 g of AsI_3 is needed. Therefore, measure out $(35.0/4.556) \times 100$ mL $= 768$ mL of the AsI_3 solution. Add to it 35.0 g of $HgI_2(s)$ and 31.5 g of $NaHCO_3(s)$ along with enough water to bring the total volume to 3.50 L.

6-80 The change in the vapor pressure ΔP is $0.3868 - 0.3914 = -0.0046$ atm. The mole fraction of the sulfur present in the sulfur-CS_2 system is therefore

$$X_{\text{sulfur}} = \frac{-\Delta P}{P^{\circ}_{CS_2}} = \frac{-(-0.0046 \text{ atm})}{0.3914 \text{ atm}} = 0.01175$$

The 1.00 kg of CS_2 equals 13.13 mol of CS_2. Hence

$$X_{\text{sulfur}} = 0.01175 = \frac{n_{\text{sulfur}}}{n_{\text{sulfur}} + n_{CS_2}} = \frac{n_{\text{sulfur}}}{n_{\text{sulfur}} + 13.13}$$

Solving for the chemical amount of sulfur gives 0.156 mol. This amount of sulfur is simultaneously 40.0 g of sulfur. Hence, the molar mass of sulfur in this solution is 40.0 g/0.156 mol = 256 g mol^{-1}. The molecular formula of the sulfur in the solution must be $\boxed{S_8}$.

6-82 Consider a portion of solution that contains 1.000 mol of substances. This sample consists of 0.390 mol of NaI and 0.610 mol of H_2O. Compute the mass of water in the solution and the molality of the NaI:[5]

$$\text{mass}_{H_2O} = m_{H_2O} = 0.610 \text{ mol } H_2O \times \frac{18.0153 \text{ g}}{\text{mol } H_2O} = 10.99 \text{ g} = 0.01099 \text{ kg}$$

$$\text{molality}_{NaI} = m_{NaI} = \frac{0.390 \text{ mol NaI}}{0.01099 \text{ kg solvent}} = 35.49 \text{ mol kg}^{-1}$$

The boiling-point elevation according to the expression in the chapter is:

$$\Delta T_b = imK_b = i(35.49 \text{ mol kg}^{-1})(0.52 \text{ K kg mol}^{-1}) = i(18.45)°C$$

The i for NaI dissolved in water has an apparent maximum of 2, corresponding to complete dissociation to Na^+ and I^- ions. Putting $i = 2$ into the preceding equation predicts a maximum boiling-point elevation of $\boxed{36.9°C}$. This is less than the observed elevation (44°C). Reversing the calculation gives an i of 2.38 for this solution. Observation of a large i in a concentrated solution counters the trend displayed in text Table 6-2 (text page 259) in which larger i's go with more dilute solutions.

6-84 The molality of the HgCl is

$$m_{HgCl} = \frac{1.36 \text{ g HgCl}/(236.04 \text{ g mol}^{-1} \text{ HgCl})}{0.100 \text{ kg solvent}} = 0.0576 \text{ mol kg}^{-1}$$

Compute the van't Hoff i for the solution:

$$i = \frac{\Delta T}{-mK_f} = \frac{-0.99 \text{ K}}{-(0.0576 \text{ mol kg}^{-1})(34.3 \text{ K kg mol}^{-1})} = 0.501$$

An i that is less than 1 implies association. Instead of breaking apart to give more particles, the "HgCl" units in this solution cluster together. Clustering in dimers, molecules having the doubled formula, is consistent with $i = 0.5$. Dimers have molecular formula $\boxed{Hg_2Cl_2}$ and molar mass $\boxed{472.09 \text{ g mol}^{-1}}$. Incomplete clustering into trimers or higher polymers cannot be ruled out on the basis of the data given in the problem.

---

[5]Note that m indeed stands for both mass and molality.

6-86 Ethylene glycol does not dissociate when it dissolves (see text page 127). Assume that it also does not associate in aqueous solution (as solutes do in problems 6-81 and 6-84). Then, $i = 1$ and the molality of the required solution is:

$$m_{CH_2OHCH_2OH} = \frac{\Delta T}{-iK_f} = -\frac{-5.0°\ C}{-1(1.86°C\ kg\ mol^{-1})} = 2.69\ mol\ kg^{-1}$$

This solution contains 2.69 mol of CH_2OHCH_2OH (molar mass 62.07 g mol^{-1}) per 1000 g of water. This equals 166.9 g of ethylene glycol per 1000 g of water, which is the same as 166.9 g of ethylene glycol for every 1166.9 g of solution. The percentage by mass of the ethylene glycol in the solution is $166.9/1166.9 \times 100\% = \boxed{14.3\%}$.

6-88 Many students are surprised that adding salt to ice in a bucket lowers the temperature in the bucket. Some of the salt dissolves in the liquid water that is present. This lowers the freezing point of the liquid. Some ice melts to chill the solution to its new, lower freezing point. The remaining ice, which is in thermal contact with the solution, is also chilled.

6-90 The vapor pressure above the beaker containing the more concentrated solution of NaCl will be less than the vapor pressure above the beaker with the more dilute solution because the presence of a nonvolatile solute lowers the vapor pressure of a solvent in proportion to the amount of the solute (Raoult's law). Water will migrate from the dilute solution to the concentrated solution. If the amount of water vapor in the space under the (tight-fitting) bell jar is negligible, then the amount of water that will eventually transfer can be calculated based on the fact that the concentrations of NaCl in the two beakers must ultimately become equal to each other. Say that x g of water transfers. Then:

$$\frac{5\ g\ NaCl}{(95 - x)\ g\ H_2O} = \frac{10\ g\ NaCl}{(90 + x)\ g\ H_2O}$$

Solving this equation for x reveals that the transfer of 33.3 g of H_2O from the beaker with the more dilute solution into the beaker with the more concentrated solution makes the concentrations of NaCl in the two beakers become equal.

6-92 (a) In pickling, a cucumber is immersed in a solution having a high osmotic pressures. Fluids depart the cucumber to dilute the surrounding solution. The cucumber becomes $\boxed{smaller}$.

(b) Foodstuffs contain microorganisms that grow rapidly under ordinary conditions and assist the decay of the food. Pickling kills these organisms by removing fluids from their cells.

6-94 The mole fraction of benzene in the aqueous solution of benzene defined in the problem is

$$X_{C_6H_6} = \frac{n_{C_6H_6}}{n_{C_6H_6} + n_{H_2O}} = \frac{(2.0/78.11)}{(2.0/78.11) + (1.0 \times 10^6/18.015)} = 4.6 \times 10^{-7}$$

Write Henry's law for a solution of benzene in water:

$$P_{C_6H_6} = k_H X_{C_6H_6} = (301 \text{ atm})(4.6 \times 10^{-7}) = \boxed{1.4 \times 10^{-4} \text{ atm}}$$

Insert this pressure, and T in kelvins into the rearranged ideal gas equation:

$$\frac{n}{V} = \frac{P}{RT} = \frac{1.4 \times 10^{-4} \text{ atm}}{(0.08206 \text{ L atm K}^{-1}\text{mol}^{-1})(298.15 \text{ K})} = 5.7 \times 10^{-6} \text{ mol L}^{-1}$$

The answer is the equilibrium concentration of benzene in the vapor above the solution. This concentration of vaporous benzene is equivalent to

$$\frac{5.7 \times 10^{-6} \text{ mol}}{L} \times \left(\frac{6.022 \times 10^{23} \text{ molecule}}{mol}\right) = 3.4 \times 10^{18} \text{ molecule L}^{-1}$$

This means there are $\boxed{3.4 \times 10^{15}}$ molecules per cubic centimeter.

6-96 According to the data in problem 6-60, the vapor pressure of toluene is 0.534 atm at 90°C, and the vapor pressure of benzene is 1.34 atm. In order for the solution to boil, its total vapor pressure must equal 1.00 atm. This total vapor pressure is the sum of the vapor pressures of the two components, each of which is given by Raoult's law. If the mole fraction of the toluene equals X, then the mole fraction of the benzene is $1 - X$, and

$$0.534X + 1.34(1 - X) = 1.00$$

Solving this equation for X gives 0.422 as the mole fraction of the toluene.

6-98 Consider a 2.8 g sample of the gold-ruby glass. This sample has a volume of 1.0 cm^3. It contains $0.000080 \times 2.8 \text{ g} = 2.24 \times 10^{-4}$ g of gold. Taking the density of gold as 19.3 g cm^{-3} allows computation of the volume of this gold

$$V = \frac{m}{d} = \frac{2.24 \times 10^{-4} \text{ g}}{19.3 \text{ g cm}^{-3}} = 1.16 \times 10^{-5} \text{ cm}^3$$

But each particle of colloidal gold is very small and has a volume of only

$$V = \frac{4}{3}\pi r^3 = \frac{4}{3}\pi(3.0 \times 10^{-7} \text{ cm})^3 = 1.131 \times 10^{-19} \text{ cm}^3/\text{ particle}$$

The number of particles of gold in this quantity of the gold-ruby glass is

$$\frac{1.16 \times 10^{-5} \text{ cm}^3}{1.131 \times 10^{-19} \text{ cm}^3 \, / \text{ particle}} = 1.03 \times 10^{14} \text{ particles}$$

Because the selected sample of glass has a volume of 1.0 cm^3, the answer is $\boxed{1.0 \times 10^{14} \text{ particles cm}^{-3}}$.

Answers to Even-Numbered Problems, Chapter 7

7-2 The form of an equilibrium expression depends on how the equation is written. The following expressions correspond to the equations as they are written in the problem. If all the coefficients in one of the equations are multiplied by a common factor, the exponents in the K expression are multiplied by that same factor.

(a) $\boxed{\dfrac{P_{\text{Cl}_2\text{O}}^2}{P_{\text{Cl}_2}^2 P_{\text{O}_2}} = K}$ (b) $\boxed{\dfrac{P_{\text{NOBr}}^2}{P_{\text{N}_2} P_{\text{O}_2} P_{\text{Br}_2}} = K}$ (c) $\boxed{\dfrac{P_{\text{H}_2\text{O}}^4 P_{\text{CO}_2}^3}{P_{\text{C}_3\text{H}_8} P_{\text{O}_2}^5} = K}$

7-4 One chemical equation and its corresponding equilibrium expression are:

$$\boxed{4\,\text{NH}_3(g) + 5\,\text{O}_2(g) \rightleftharpoons 6\,\text{H}_2\text{O}(g) + 4\,\text{NO}(g) \qquad \frac{P_{\text{H}_2\text{O}}^6 P_{\text{NO}}^4}{P_{\text{NH}_3}^4 P_{\text{O}_2}^5} = K}$$

Other equations involve the same ratio of chemical amounts of reactants and products, for example the coefficients 8, 10, 12, 8 could replace 4, 5, 6, 4. Each different equation has a different equilibrium expression. It is customary to refer to the equation with the smallest integral coefficients as "the" equation for a reaction.

7-6 (a) The equation with the smallest whole-number coefficients is

$$\boxed{\text{F}_3\text{SSF}(g) \rightleftharpoons 2\,\text{SF}_2(g)}$$

(b) Write the equilibrium expression for this equation and insert the partial pressures listed in the problem. Before substitution, each partial pressure must be divided by a reference pressure of 1 atm. This amounts to omitting their units. Thus:

$$K = \frac{P_{\text{SF}_2}^2}{P_{\text{F}_3\text{SSF}}} = \frac{(1.1 \times 10^{-4} \text{ atm})^2}{(0.0484 \text{ atm})} \implies \frac{(1.1 \times 10^{-4})^2}{(0.0484)} = \boxed{2.5 \times 10^{-7}}$$

7-8 This problem gives a chemical equation explicitly. Consequently, the question of doubled or tripled coefficients does not arise. Computing K requires the partial pressures of all three gases at equilibrium. Set up a three-line table as in Example 7-2 on text page 287:

$$SbCl_5(g) \rightleftharpoons SbCl_3(g) + Cl_2(g)$$

	$SbCl_5$	$SbCl_3$	Cl_2
Initial partial pressure (atm)	x	0.0	0.0
Change in partial pressure(atm)	$-0.718x$	$+0.718x$	$+0.718x$
Equilibrium partial pressure (atm)	$0.282x$	$0.718x$	$0.718x$

The change in the partial pressure of $SbCl_5(g)$ equals -0.718 times its original value because 718 out of every 1000 molecules of $SbCl_5$ dissociates. The original pressure is not known at this stage and is represented by x. The total pressure at equilibrium is the sum of the partial pressures of the three gases and equals 1.000 atm:

$$1.000 \text{ atm} = P_{SbCl_5} + P_{SbCl_3} + P_{Cl_2} = 0.282x + 0.718x + 0.718x$$

Solving gives $x = 0.582$ atm. The equilibrium partial pressures of the three gases are easily computed:

$$P_{SbCl_5} = 0.164 \text{ atm} \qquad P_{SbCl_3} = P_{Cl_2} = 0.418 \text{ atm}$$

Substitution in the equilibrium expression gives

$$K = \frac{P_{SbCl_3} P_{Cl_2}}{P_{SbCl_5}} = \frac{(0.418)(0.418)}{(0.164)} = \boxed{1.07}$$

7-10 The table of changes for attaining equilibrium in the H_2/I_2 system is:

$$H_2(g) + I_2(g) \rightleftharpoons 2HI(g)$$

	H_2	I_2	HI
Initial partial pressure (atm)	2.75	1.50	0
Change in partial pressure(atm)	$-x$	$-x$	$+2.79$
Equilibrium partial pressure (atm)	$2.75 - x$	$1.50 - x$	2.79

According to the balanced equation, the change x in the partial pressures of the H_2 and I_2 must equal half the change in the partial pressure of the HI and have the opposite sign. Hence $x = +1.395$ atm, and the equilibrium partial pressures of the H_2 and I_2 are 1.355 and 0.105 atm respectively. The equilibrium constant is calculated by inserting the three equilibrium partial pressures into the proper expression:

$$K = \frac{P_{HI}^2}{P_{H_2} P_{I_2}} = \frac{(2.79)^2}{(1.355)(0.105)} = \boxed{54.7}$$

This is the quite close to the answer to problem 9. Problems 7-9 and 7-10 as a pair confirm that the same K governs distinct starting mixtures.

7-12 (a) As the hint points out, the vapor density of the contents of the flask cannot change—the reaction changes neither the contents of the container nor the volume of the container. Before any $NOBr(g)$ has a chance to react:

$$P_{NOBr} = \left(\frac{n}{V}\right) RT = \left(\frac{2.219 \text{ g NOBr}}{L} \times \frac{1 \text{ mol NOBr}}{109.91 \text{ g NOBr}}\right) RT$$

$$= (0.0202 \text{ mol L}^{-1})(0.08206 \text{ L atm mol}^{-1} \text{ K}^{-1})(350 \text{ K} = 0.5798 \text{ atm}$$

Set up a table of changes:

	$NOBr(g) \rightleftharpoons$	$NO(g) +$	$1/2\,Br_2(g)$
Initial partial pressure (atm)	0.5798	0.0	0.0
Change in partial pressure(atm)	$-x$	$+x$	$+1/2x$
Equilibrium partial pressure (atm)	$0.5798 - x$	x	$1/2x$

The final pressure in the flask, the sum of the three partial pressures, is 0.675 atm. It follows that

$$P_{NOBr} + P_{NO} + P_{Br_2} = (0.5798 - x) + x + 1/2x = 0.675$$

Solving gives $x = 0.190$. The three partial pressures are (from left to right) $\boxed{0.389 \text{ atm}}$, $\boxed{0.190 \text{ atm}}$, and $\boxed{0.095 \text{ atm}}$ at equilibrium.

(b) To calculate the equilibrium constant, substitute the equilibrium partial pressures into the proper expression:

$$K = \frac{P_{Br_2}^{\frac{1}{2}} P_{NO}}{P_{NOBr}} = \frac{(0.095)^{\frac{1}{2}} (0.190)}{0.389} = \boxed{0.151}$$

7-14 The second equation is the first equation reversed with every coefficient divided by 6. The equilibrium constant for the second equation is accordingly the reciprocal of that of the first raised to the 1/6 power:

$$K_2 = \left(\frac{1}{32.6}\right)^{\frac{1}{6}} = \sqrt[6]{\frac{1}{32.6}} = \boxed{0.559}$$

7-16 The third equation written in the problem is equal to the first added to the reverse of the second. The equilibrium constant for the third reaction therefore equals the equilibrium constant of the first multiplied by the reciprocal of the equilibrium constant of the second:

$$K_3 = K_1(K_2)^{-1} = \frac{K_1}{K_2} = \frac{7.0 \times 10^3}{38 \times 10^3} = \boxed{0.18}$$

7-18 The K for this reaction equals 0.0200, and the reaction quotient has the form

$$Q = \frac{P_{SbCl_3} P_{ClF}^5}{P_{SbF_5} P_{Cl_2}^4}$$

(a) Putting the given partial pressures into the preceding expression gives

$$Q = \frac{(0.655)(0.410)^5}{(1.20)(0.200)^4} = \boxed{3.95}$$

Since Q exceeds K the reaction proceeds to the $\boxed{\text{left}}$ to come to equilibrium.

(b) For this set of initial conditions

$$Q = \frac{(0.125)(1.20)^5}{(0.400)(10.0)^4} = \boxed{7.78 \times 10^{-5}}$$

which is less than K. The reaction proceeds to the $\boxed{\text{right}}$.

(c) For this set of initial conditions

$$Q = \frac{(0.800)(0.800)^5}{(0.800)(0.800)^4} = \boxed{0.800}$$

which exceeds K. The reaction proceeds to the $\boxed{\text{left}}$.

7-20 (a) The reaction quotient Q for all reactions, including $F_3SSF(g) \rightleftharpoons 2\,SF_2(g)$, tends toward K as the system tends toward equilibrium. In this case, the system is not at equilibrium at the start because Q differs from K:

$$Q = \frac{P_{SF_2}^2}{P_{F_3SSF}} = \frac{(2.3 \times 10^{-4})^2}{0.0484} = \boxed{1.09 \times 10^{-6}}$$

(b) The value of Q exceeds K, which equals 2.5×10^{-7} (problem 7-6). The reaction will tend to proceed to the $\boxed{\text{left}}$, generating F_3SSF and consuming SF_2, until Q becomes equal to K.

7-22 Insert partial pressures from the given list into the reaction quotient expression for this reaction to obtain Q at the moment the four compounds are mixed:

$$Q = \frac{P_{NO}P_{CO_2}}{P_{NO_2}P_{CO}} = \frac{(1.4)(1.4)}{(3.4)(3.4)} = 0.17$$

The equilibrium constant for this reaction must $\boxed{\text{exceed } 0.17}$ based on the observation that brown NO_2 is consumed as the reaction goes toward equilibrium (removal of NO_2 increases Q).

7-24 The equilibrium law for the reaction at 1107 K is:

$$\frac{P_{HI}^2}{P_{H_2}P_{I_2}} = K = 38.6$$

The partial pressure of $I_2(g)$ is 0.0252 atm, and the partial pressure of $HI(g)$ is 0.427 atm. Substitution of these values into the equation, and solution for the partial pressure of $H_2(g)$ gives $P_{H_2} = \boxed{0.187 \text{ atm}}$.

7-26 (a) The reaction quotient has the same mathematical form as the equilibrium law but uses partial pressures as they are at any moment, rather than at equilibrium. At the moment of mixing the $Br_2(g)$ and $Cl_2(g)$, the partial pressure of the product $BrCl(g)$ is 0. Therefore Q is $\boxed{0}$ at that moment.

(b) As the reaction proceeds, $BrCl(g)$ is produced and $Br_2(g)$ and $Cl_2(g)$ are consumed until Q becomes equal to K:

	$Br_2(g)$ +	$Cl_2(g)$ $\rightleftharpoons$	$2\,BrCl(g)$
Init. pressure (atm)	0.100	0.100	0
Change in pressure (atm)	$-x$	$-x$	$+2x$
Equil. pressure (atm)	$0.100 - x$	$0.100 - x$	$2x$

Substitute the equilibrium pressures into the equilibrium law and solve by taking the square root of both sides and rearranging:

$$\frac{(2x)^2}{(0.100 - x)(0.100 - x)} = 58.0 \quad \text{from which} \quad x = 0.0792.$$

The partial pressure of $BrCl(g)$ is $2x$, which is $\boxed{0.158 \text{ atm}}$.

7-28 (a) The molar mass of the isopropyl alcohol is 60.096 g mol^{-1}. The 10.00 g sample is 0.1664 mol of the alcohol. The alcohol exerts a pressure of 0.6174 atm in the 10.00 L container at 452.15 K (179°C), assuming it behaves as an ideal gas. This is *before* the reaction has a chance to occur. The reaction generates acetone and hydrogen at the expense of isopropyl alcohol. Let x represent the decrease in the partial pressure of the isopropyl alcohol as the reaction comes to equilibrium. Then:

	$(CH_3)_2CHOH(g)$	$\rightleftharpoons$	$(CH_3)_2CO(g)$	$+$	$H_2(g)$
Init. pressure (atm)	0.6174		0		0
Change in pressure (atm)	$-x$		$+x$		$+x$
Equil. pressure (atm)	$0.6174 - x$		x		x

Write the equilibrium constant expression and substitute the equilibrium partial pressures:

$$K = \frac{P_{\text{acetone}} P_{H_2}}{P_{\text{alcohol}}} = \frac{x \cdot x}{(0.6174 - x)} = 0.444$$

Solution for x give 0.3467, so $P_{\text{acetone}} = \boxed{0.3467 \text{ atm}}$ at equilibrium.

(b) If there were no reaction, the partial pressure of isopropyl alcohol would be 0.6174 atm. In fact however, 0.3467 atm of the isopropyl alcohol is dissociated. This means that the fraction of the isopropyl alcohol dissociated is $0.3467/0.6174 = \boxed{0.562}$.

7-30 The P_{HI} is unknown at the moment that the glass vessel is filled at 1107 K. Let it be y. The partial pressures of I_2 and H_2 (the products) are both zero at this moment. Attainment of equilibrium clearly reduces P_{HI} by some amount. Call it $2x$. It concurrently increases the partial pressures of both H_2 and I_2 by x:

	$2\,HI(g)$	$\rightleftharpoons$	$I_2(g)$	$+$	$H_2(g)$
Init. pressure (atm)	y		0		0
Change in pressure (atm)	$-2x$		$+x$		$+x$
Equil. pressure (atm)	$y - 2x$		x		x

Substitution of the equilibrium partial pressures into the expression for K gives

$$K = \frac{P_{I_2} P_{H_2}}{P_{HI}^2} = \frac{x \cdot x}{(y - 2x)^2} = 0.0259$$

The total pressure at equilibrium, which is given in the problem, equals the sum of the equilibrium partial pressures of the three components of the mixture: $(y - 2x) + x + x = 6.45$ atm. Clearly, $y = 6.45$ atm, and

$$\frac{x^2}{(6.45 - 2x)^2} = 0.0259 \quad \text{so that} \quad \frac{x}{(6.45 - 2x)} = \sqrt{0.0259} = 0.1609$$

Solving the second equation gives $x = 0.785$. Hence, $\boxed{P_{H_2} = P_{I_2} = 0.79 \text{ atm}}$, and $\boxed{P_{HI} = 4.88 \text{ atm}}$ at equilibrium in this vessel.

7-32 Consider the *reverse* of the reaction given in the problem and write the initial pressures, changes and final partial pressures in the usual way:

$$OF_2(g) \rightleftharpoons F_2(g) + 1/2\,O_2(g)$$

	OF_2	F_2	$1/2\,O_2$
Init. pressure (atm)	y	0	0
Change in pressure (atm)	$-x$	$+x$	$+1/2x$
Equil. pressure (atm)	1.00	x	$1/2x$

From the law of mass action:

$$K = \frac{1}{40.1} = \frac{(1/2x)^{\frac{1}{2}}x}{1.00}$$

The equilibrium constant is $1/40.1$ because the set-up is for the reverse of the reaction written in the problem. Squaring both sides gives

$$\left(\frac{1}{40.1}\right)^2 = (0.500x)x^2$$

from which $x = 0.1075$. The equilibrium partial pressure of the $F_2(g)$ is $\boxed{0.107 \text{ atm}}$, and that of the $O_2(g)$ is half as much, $\boxed{0.054 \text{ atm}}$. It is also possible to compute y, the initial pressure of the $OF_2(g)$, but this is not requested.

7-34 The equilibrium constant is given and the reaction starts with a single pure reactant. Set up a table of changes for the attainment of equilibrium between that reactant and a single product

$$3\,S_8(g) \rightleftharpoons 4\,S_6(g)$$

	$3\,S_8$	$4\,S_6$
Initial partial pressure (atm)	0.15	0
Change in partial pressure(atm)	$-3x$	$+4x$
Equilibrium partial pressure (atm)	$0.15 - 3x$	$4x$

Insert the equilibrium partial pressures into the equilibrium expression:

$$K = 1.6 \times 10^{-4} = \frac{P_{S_6}^4}{P_{S_8}^3} = \frac{(4x)^4}{(0.15 - 3x)^3} \quad \text{or} \quad 1.6 \times 10^{-4} = \frac{(4x)^4}{(0.15 - 3x)^3}$$

The equation is difficult to solve analytically. However, x must be positive and less than 0.05. Also, x is likely to be quite small because K is small. Neglecting x relative to 0.15 makes the preceding equation easy to solve and gives $x = 0.0068$. Adjusting this first "guess" slightly by successive trials in the preceding equation gives $x = 0.00615$. It follows that:

$$P_{S_8} = 0.15 - 3(0.00615) = \boxed{0.13 \text{ atm}}$$

7-36 The concentration of each component gas in the container depends on its partial pressure:

$$c_{\text{component}} = \left(\frac{n}{V}\right)_{\text{component}} = \left(\frac{P_{\text{component}}}{RT}\right)$$

At 300°C (573.15 K), the quantity RT equals 47.031 L atm mol^{-1}. Substitution of the partial pressures given in problem 7-18 produces these answers:

	$[SbF_5]$	$[Cl_2]$	$[SbCl_3]$	$[ClF]$
(a)	0.0255 mol L^{-1}	0.00425 mol L^{-1}	0.0139 mol L^{-1}	0.00872 mol L^{-1}
(b)	0.00850	0.213	0.00266	0.0255
(c)	0.0170	0.0170	0.0170	0.0170

7-38 According to problem 7-25, we have

$$SO_2Cl_2(g) \rightleftharpoons SO_2(g) + Cl_2(g) \qquad K_{373.15} = 2.40$$

Data concerning two of the three components of the equilibrium are given in terms of concentrations. The partial pressure of a component such as SO_2 in a gaseous mixture is related to its concentration by

$$P_{SO_2} = \left(\frac{n_{SO_2}}{V}\right) RT = [SO_2]RT$$

as long as the mixture obeys Dalton's law and the ideal-gas law. Substitute such concentration terms in the equilibrium expression:

$$K = \frac{P_{Cl_2} P_{SO_2}}{P_{SO_2Cl_2}} = \frac{[Cl_2]RT[SO_2]RT}{[SO_2Cl_2]RT} = \frac{[Cl_2][SO_2]}{[SO_2Cl_2]}RT$$

The reference pressure for the numerical K (2.40) quoted in problem 7-25 equals 1 atm. Therefore, using R in L atm mol^{-1} K^{-1} assures that all concentrations come out in mol L^{-1}. The computation is:

$$2.40 = \frac{(6.9 \times 10^{-3} \text{ mol L}^{-1})[SO_2]}{(3.6 \times 10^{-4} \text{ mol L}^{-1})}(0.08206 \text{ L atm mol}^{-1} \text{ K}^{-1})(373.15 \text{ K})$$

$$[SO_2] = \boxed{4.1 \times 10^{-3} \text{ mol L}^{-1}}$$

7-40 The table of changes for the approach to equilibrium in this system is:

$$H_2(g) + I_2(g) \rightleftharpoons 2\,HI(g)$$

	$H_2(g)$	$I_2(g)$	$2\,HI(g)$
Initial partial pressure (atm)	5.73	5.73	0
Change in partial pressure (atm)	$-x$	$-x$	$+2x$
Equilibrium partial pressure (atm)	$5.73 - x$	$5.73 - x$	$2x$

Substitute the equilibrium values in the equilibrium expression:

$$K = \frac{P_{\mathrm{HI}}^2}{P_{\mathrm{H}_2} P_{\mathrm{I}_2}} = \frac{(2x)^2}{(5.73 - x)^2} = 53.7$$

Taking the square root of both sides of this equation gives

$$\frac{(2x)}{(5.73 - x)} = \sqrt{53.7} = 7.328$$

From this, $x = 4.501$. Based on this value, the equilibrium partial pressures for H_2 and I_2 equal 1.229 atm and an equilibrium partial pressure of 9.002 atm for HI.

Now the system is perturbed by removing 1.00 atm of H_2. This sets a new "initial" P_{H_2} of 0.229 atm:

	$H_2(g)$ +	$I_2(g)$ $\rightleftharpoons$	$2\,HI(g)$
Initial partial pressure (atm)	$1.229 - 1.00$	1.229	9.002
Change in partial pressure(atm)	$+y$	$+y$	$-2y$
Equilibrium partial pressure (atm)	$0.229 + y$	$1.229 + y$	$9.002 - 2y$

Substitute the new final partial pressures into the equilibrium expression:

$$K = 53.7 = \frac{P_{\mathrm{HI}}^2}{P_{\mathrm{H}_2} P_{\mathrm{I}_2}} = \frac{(9.002 - 2y)^2}{(0.228 + y)(1.228 + y)}$$

Solving this equation by approximation gives $y = 0.479$. This corresponds to an equilibrium P_{HI} of $9.002 - 2(0.479) = \boxed{8.04\ \mathrm{atm}}$.

7-42 (a) If $O_2(g)$ is added to the equilibrium $SO_3(g) \rightleftharpoons SO_2(g) + 1/2\,O_2(g)$ at constant V and T, the equilibrium $\boxed{\text{shifts left}}$.

(b) If the mixture is compressed at constant T, the equilibrium $\boxed{\text{shifts left}}$.

(c) Cooling the mixture causes the equilibrium to $\boxed{\text{shift left}}$; the equilibrium constant of an endothermic reaction decreases with decreasing temperature.

(d) Pumping a non-reactive gas into the equilibrium mixture at constant T and P must expand the container. Increasing the volume favors the products. The equilibrium $\boxed{\text{shifts right}}$.

(e) If an inert gas is pumped in at constant volume, the total pressure in the container rises, but the partial pressures of the reactants and products are unchanged; the equilibrium is $\boxed{\text{unaffected}}$.

7-44 Use LeChatelier's principle. $\boxed{\text{Increasing } P}$ and $\boxed{\text{lowering } T}$ shift the equilibrium toward the right (to give more methanol) because the right side has fewer moles of gas and because the reaction is exothermic.

7-46 (a) Since K increases with increasing temperature, the reaction is $\boxed{\text{endothermic}}$.

(b) The equilibrium is shifted to the left (the reactants) when the non-reactive gas neon is admitted in such a way that the volume increases. Hence the reactants take up more volume than the products and are favored by a system of larger volume. This means the total number of gas molecules $\boxed{\text{decreases}}$ from as products form in this reaction.

7-48

(a) $\boxed{\dfrac{1}{P_{C_2H_2}^3 P_{H_2}^3} = K}$ (b) $\boxed{\dfrac{P_{CO}^2}{P_{CO_2}} = K}$ (c) $\boxed{\dfrac{P_{HF}^4 P_{CO_2}}{P_{CF_4}} = K}$ (d) $\boxed{P_{F_2} = K}$

7-50

(a) $\boxed{\dfrac{[OH^-]^8[I_2]^3}{[I^-]^6[MnO_4^-]^2} = K}$ (b) $\boxed{\dfrac{[I_2]}{[Cu^{2+}]^2[I^-]^4} = K}$ (c) $\boxed{\dfrac{[H_3O^+]^2}{P_{O_2}^{1/2}[Sn^{2+}]} = K}$

7-52 (a) When $P_{N_2O_4}$ is graphed against $P_{NO_2}^2$, the result is close to a straight line:

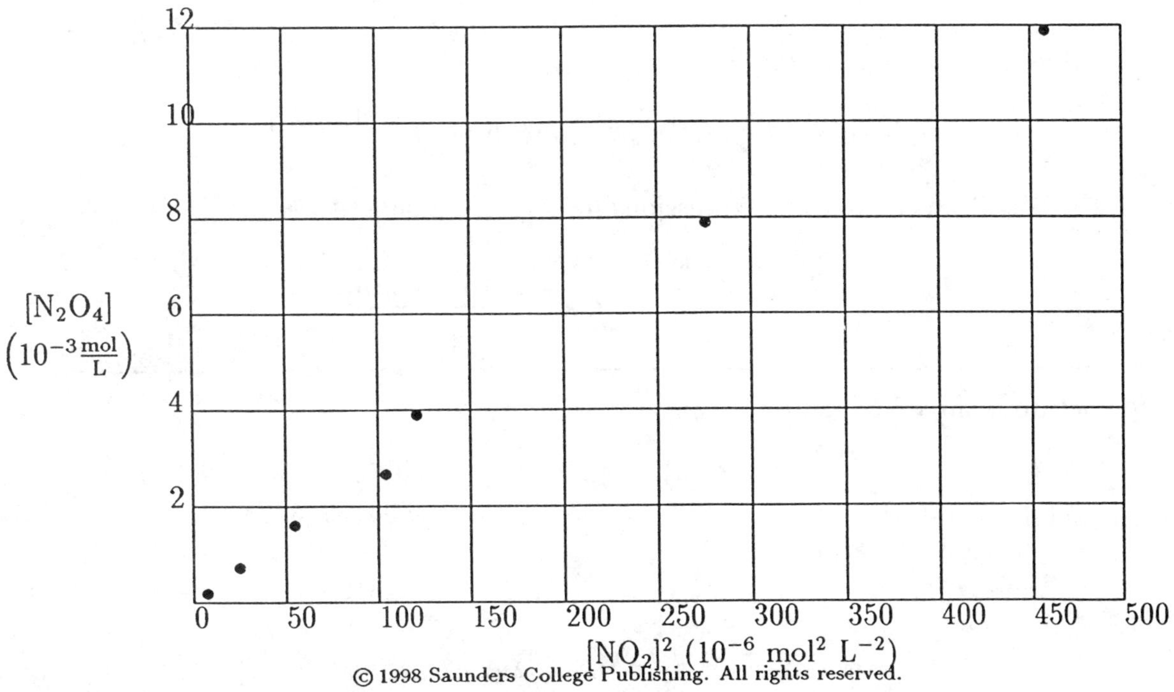

The slope of this line (with units discarded) is the numerical value of K for the reaction $2\,NO_2 \rightleftharpoons N_2O_4$. The fifth of the determinations, which come from an authentic, published experimental report may be in error.

(b) The seven pairs of equilibrium concentrations give seven different values of K. Their mean is $\boxed{28.4}$. The graph can also be fitted by linear least-squares to give the relationship

$$[N_2O_4] = 26.26[NO_2]^2 + 0.002152$$

from which K equals 26.3.

7-54 (a) Putting the values given in the problem into the reaction quotient expression generates

$$Q = \frac{P_{HI}^2}{P_{H_2S}P_{I_2}} = \frac{0^2}{(0.050)(0.461)} = 0$$

Only reactants are present; some $\boxed{\text{S is produced}}$ to reach equilibrium.

(b) For this set of initial conditions

$$Q = \frac{9.0^2}{(0.050)(0.461)} = 3.5 \times 10^3$$

Thus Q exceeds K. $\boxed{\text{S is consumed}}$ as the reaction comes to equilibrium.

7-56 (a) The decomposition of $NaHCO_3$ at 110°C produces a solid and two gases. The equilibrium partial pressures of the two gases (H_2O and CO_2) equal each other and add up to 1.648 atm. Clearly both equal 0.824 atm. Inserting these values into the equilibrium expression gives an equilibrium constant:

$$K_{110°C} = P_{H_2O}P_{CO_2} = (0.824)(0.824) = \boxed{0.679}$$

(b) Solve the equilibrium expressions for P_{H_2O} and substitute:

$$P_{H_2O} = \frac{K}{P_{CO_2}} = \frac{0.679}{0.800} = \boxed{0.849\ \text{atm}}$$

7-58 Set up a three-line table of changes

	$2\,NH_3(g)\ +$	$CO_2(g) \rightleftharpoons$	$H_2O(g)\ +$	$CO(NH_2)_2(s)$
Init. P (atm)	0	0	0.03126	---
Change in P (atm)	$+2x$	$+x$	---	---
Equil. P (atm)	$2x$	x	0.03126	---

The partial pressure of water equals 0.03126 atm both before and after the reaction comes to equilibrium, presumably by the action of a second equilibrium. The equilibrium expression is

$$\frac{P_{H_2O}}{P_{NH_3}^2 P_{CO_2}} = K \qquad \text{from which} \qquad \frac{0.03126}{(2x)^2 \cdot x} = 0.615$$

The only real root is $x = 0.233$. The partial pressure of ammonia at equilibrium is twice this or $\boxed{0.466 \text{ atm}}$, and the partial pressure of carbon dioxide is $\boxed{0.233 \text{ atm}}$.

7-60 The equilibrium concentration of iodine in the aqueous layer is given as 4.16×10^{-5} mol L^{-1}; the volume of this layer is not given. Suppose that it is V L. Then the chemical amount of I_2 in the aqueous layer at equilibrium is $(4.16 \times 10^{-5}V)$ mol. Before equilibrium (before the mixture was shaken), the amount of I_2 in the aqueous layer was much more, namely, $(2.50 \times 10^{-2}V)$ mol. The amount of I_2 that transfers to the CS_2 layer in attaining the partition equilibrium is

$$\left((2.50 \times 10^{-2}) - (4.16 \times 10^{-5})\right) \text{ mol } L^{-1} \times V \text{ L} \approx (2.50 \times 10^{-2})V \text{ mol}$$

The final concentration of the iodine in the CS_2 is this chemical amount divided by the volume of the CS_2, which equals V L. It is therefore 2.50×10^{-2} mol L^{-1}. The partition equilibrium constant is

$$K = \frac{[I_2]_{(CS_2)}}{[I_2]_{(aq)}} = \frac{2.50 \times 10^{-2}}{4.16 \times 10^{-5}} = \boxed{600}$$

7-62 (a) The equilibrium constant for the dissolution of the citric acid according to the equation given in the problem is, in simple theory, just the concentration of the dissolved citric acid. The molar mass of citric acid equals 192.1 g mol^{-1}, so that a saturated aqueous solution of citric acid contains 3.75 mol L^{-1} and $K = \boxed{3.75}$. This assumes no ionization of the citric acid in water and an ideal solution. Neither assumption is very defensible. The K for the dissolution in ether is, by similar reasoning, $\boxed{0.11}$.

(b) The transfer of citric acid from water into ether equals the reverse of the dissolution of citric acid in water added to the dissolution of citric acid in ether. The partition coefficient is therefore $0.1145/3.75 = \boxed{0.031}$.

7-64 The synthetic reaction is:

$$CO(g) + 2\,H_2(g) \rightleftharpoons CH_3OH(g) \qquad \frac{P_{CH_3OH}}{P_{CO}P_{H_2}^2} = 6.08 \times 10^{-3} = K_{225°C}$$

Let the equilibrium partial pressure of the H_2 be $2x$. Then, $P_{CO} = x$. The equilibrium partial pressure of the methanol equals 0.500 atm. Substitution gives:

$$\frac{0.500}{x(2x)^2} = 6.08 \times 10^{-3}$$

Solving for x gives 2.74. At equilibrium

$$P_{CO} = \boxed{2.74 \text{ atm}} \quad \text{and} \quad P_{H_2} = \boxed{5.48 \text{ atm}}$$

7-66 (a) Calculation of the degree of conversion of the *tert*-butanol requires knowledge of its partial pressure at equilibrium. Set up a three-line table:

	$(CH_3)_3COH(g)$ $\rightleftharpoons$	$(CH_3)_2CCH_2(g)$ +	$H_2O(g)$
Init. pressure (atm)	0.100	0.0	0.0
Change in pressure (atm)	$-x$	$+x$	$+x$
Equil. pressure (atm)	$0.100 - x$	x	x

Putting the equilibrium values into the equilibrium expression gives

$$K = \frac{x^2}{(0.100 - x)} = 2.42$$

from which x (using the quadratic formula) is 0.0962 atm. The fraction of *tert*-butanol that is converted at equilibrium is $\boxed{0.962}$.

(b) A similar calculation gives x equal to 2.473 atm so that the fraction converted is $\boxed{0.495}$. Higher pressure favors the reactants, the side with fewer moles of gas, at equilibrium.

7-68 The reaction vessel contains 1.46 g L^{-1} at equilibrium. Its contents must have had this same density before the reaction started because a fixed-volume container does not change its V, and matter neither enters nor leaves the container. The initial partial pressure of PCl_5 therefore was:

$$P_{PCl_5} = \left(\frac{n}{V}\right)RT = \left(\frac{1.46 \text{ g}/208.24 \text{ g mol}^{-1}}{L}\right)RT$$

$$= (0.00701 \text{ mol L}^{-1})(0.08206 \text{ L atm mol}^{-1} \text{ K}^{-1})(523 \text{ K}) = 0.3009 \text{ atm}$$

At equilibrium in this container:

$$P_{PCl_3} = P_{Cl_2} \quad \text{and} \quad P_{PCl_3} + P_{Cl_2} + P_{PCl_5} = 0.569 \text{ atm}$$

It follows that:

$$P_{PCl_3} = P_{Cl_2} = 0.2681 \text{ atm}$$

$$K = \frac{P_{PCl_3} P_{Cl_2}}{P_{PCl_5}} = \frac{(0.2681)(0.2681)}{0.3009} = \boxed{0.239}$$

7-70 (a) Acetic acid is present as a dimer or as a monomer. According to the problem, the sum of the partial pressures of the two is 0.725 atm. Also, the two partial pressures are related by an equilibrium expression. This leads to the simultaneous equations:

$$P_{monomer} + P_{dimer} = 0.725 \quad \text{and} \quad K = \frac{P_{dimer}}{P_{monomer}^2} = 3.72$$

Eliminating $P_{monomer}$ and solving (by means of the quadratic formula) gives $P_{dimer} = \boxed{0.398 \text{ atm}}$.

(b) Solving for $P_{monomer}$ in the previous part gave 0.327 atm. If none of the acetic acid were dimerized, the partial pressure of the monomer would be $0.327 + (2 \times 0.398) = 1.123$ atm. The pressure of monomer that is actually present is 0.327 atm. This is 29.1% of 1.123 atm; it follows that $\boxed{70.9\%}$ of the acetic acid is present as the dimer.

7-72 This is a three-phase heterogeneous reaction. The K-expression is:

$$\boxed{\frac{[Au(CN)_2^-]^4[OH^-]^4}{[CN^-]^8 P_{O_2}} = K}$$

7-74 The second chemical equation in the problem is the first reversed and multiplied through by 2. Hence, $\boxed{K_2 = 1/K_1^2}$ where the subscripts refer to the equations in the order given in the problem.

7-76 Write the equilibrium law for the reaction of $KOH(s)$ with $CO_2(g)$ to give $KHCO_3(s)$. Two of the three compounds in the equilibrium are solids and do not appear in the expression, which is consequently quite simple:

$$\frac{1}{P_{CO_2}} = K$$

The K is given as 6×10^{15} so $P_{CO_2} = \boxed{2 \times 10^{-16} \text{ atm}}$. The amounts of the two solids are immaterial except to assure that neither is consumed completely before equilibrium is reached. If the container were exceedingly large, then the decomposition of more $KHCO_3$ than 9.41 g, which is all that is available, might be required to fill the container with $CO_2(g)$ at the equilibrium pressure. The volume of the container would have to exceed 10^{16} L for this to happen.

7-78 (a) According to the balanced equation, one mole of fructose forms for every one mole of glucose that reacts. At equilibrium in the first experiment, [fructose] = 0.1175 mol L^{-1} and [glucose] = 0.2564 − 0.1175 = 0.1389 mol L^{-1}. Hence:

$$\frac{[\text{fructose}]}{[\text{glucose}]} = K = \frac{0.1175}{0.1389} = 0.8459$$

In the second experiment, equilibrium is approached from the other direction. At equilibrium [fructose] = 0.2666 − 0.1415 = 0.1251 mol L^{-1} and [glucose] = 0.1415 mol L^{-1}. The K is 0.8841, which is about 5% higher. (These are genuine experimental data.) The average K is $\boxed{0.8650}$.

(b) Let x stand for the fraction of glucose converted to fructose at equilibrium and let c_0 stand for the original concentration of glucose. The concentrations of the two substances at equilibrium are

$$[\text{glucose}] = (1 - x)c_0 \quad \text{and} \quad [\text{fructose}] = xc_0$$

Substitute the equilibrium concentrations in the K expression along with average equilibrium constant just calculated:

$$\frac{[\text{fructose}]}{[\text{glucose}]} = \frac{xc_0}{(1 - x)c_0} = 0.865$$

The c_0's cancel out. Solution for x gives 0.464; $\boxed{46.4\%}$ of the glucose is converted.

7-80 Write the equation for the reaction of iron and steam and construct a table of changes:

	$3\,Fe(s)\ +$	$4\,H_2O(g)$	$\rightleftharpoons$	$Fe_3O_4(s)\ +$	$4\,H_2(g)$
Init. P (atm)	— —	22.2		— —	0.0
Change in P (atm)	— —	$-x$		— —	$+x$
Equil. P (atm)	— —	$22.2 - x$		— —	x

The empty slots in the table reflect the fact that the two pure solids do not appear in the equilibrium constant expression for this reaction, which is

$$\frac{P_{H_2}^4}{P_{H_2O}^4} = K = 12.3 = \frac{x^4}{(22.2 - x)^4}$$

Solving for x (by taking the fourth root of both sides of the equation and then rearranging) gives $x = 14.47$; the equilibrium partial pressure of the hydrogen is thus $\boxed{14.5 \text{ atm}}$. Since superheated steam is often handled in steel or iron vessels, this reaction is a matter of potential concern for chemical engineers.

7-82 From problems 7-61 and 7-62:

$$K_1 = \frac{[\text{benzoic acid}]_{ether}}{[\text{benzoic acid}]_{aq}} = 330 \quad \text{and} \quad K_2 = \frac{[\text{citric acid}]_{ether}}{[\text{citric acid}]_{aq}} = 0.030$$

Ether and water are immiscible. When the 50:50 mixture of citric acid and benzoic acid is treated with the mixture of ether and water, most of the benzoic acid ends up in the ether layer, and most of the citric acid ends up in the water layer.

The sample consists of 1.00 g of benzoic acid and 1.00 g of citric acid. Let x equal the mass in grams of the benzoic acid present in the water at equilibrium. Then $1.00 - x$ g is the mass of benzoic acid present in the ether. Let y be the mass in grams of the citric acid present in the water at equilibrium. Then $1.00 - y$ g is the mass of citric acid in the ether at equilibrium. The equilibrium laws for the partition of the two organic acids involve concentrations, not amounts. In this problem, the volumes of the two solvents are equal, so the amounts of the solutes in the two layers have the same ratios as their concentrations. Use this fact to write:

$$\frac{1.00 - x}{x} = 330 \quad \text{and} \quad \frac{1.00 - y}{y} = 0.030$$

Solving the equation in x reveals that there is 3.0×10^{-3} g of benzoic acid in the water at equilibrium and 0.9970 g of benzoic acid in the ether. From the equation in y, there is 0.9708 g of citric acid in the water at equilibrium and 0.0291 g of citric acid in the ether. The total mass of substances dissolved in ether is 1.0261 g, of which 0.9970 g is benzoic acid. The benzoic acid is therefore $\boxed{97.2\%}$ of the solids recovered from the ether. The total mass of substances dissolved in the water is 0.9738 g, of which 0.9708 g is citric acid. The citric acid is $\boxed{99.7\%}$ of the solids recovered from the water layer.

Answers to Even-Numbered Problems, Chapter 8

8-2 $\boxed{\text{All}}$ of the species listed can act as Brønsted-Lowry bases. The formulas of the conjugate acids are HF, HSO_4^-, OH^-, H_2O, and H_3O^+.

8-4 (a) The solution contains carbonate ion from the ionization of the sodium carbonate into sodium ions and carbonate ions. The following reaction makes the solution basic:

$$CO_3^{2-}(aq) + H_2O(l) \longrightarrow HCO_3^-(aq) + OH^-(aq)$$

(b) $\qquad$ $RbOH(aq) \longrightarrow Rb^+(aq) + OH^-$

(c) $\qquad$ $HPO_4^{2-}(aq) + H_2O(l) \longrightarrow H_2PO_4^-(aq) + OH^-(aq)$

8-6 The pOH is the negative logarithm of the concentration of OH^-. It equals $\boxed{1.44}$. Calculate the concentration of H_3O^+ in the bleach using the relationship $K_w = [OH^-][H_3O^+]$ with $K_w = 1.01 \times 10^{-14}$. The pH is the negative logarithm of the $[H_3O^+]$; $pH = \boxed{12.55}$. In both cases, it is correct to round off to two figures past the decimal point.

8-8 For H_3O^+ ion, the range is 3.5×10^{-8} to 4.5×10^{-8} mol L^{-1}. For OH^- ion the range is 2.2×10^{-7} to 2.8×10^{-7} mol L^{-1}.

8-10 The concentration of hydroxide in the original solution is

$$[OH^-] = 10^{-(14.00 - 11.65)} = \boxed{4.467 \times 10^{-3} \text{ M}}$$

The concentration of OH^- is reduced by a factor of 6 by the dilution; it becomes 7.44×10^{-4} mol L^{-1}. The pOH is the negative logarithm of this (3.128), and the pH is $14.00 - 3.128 = 10.87$.

8-12 The total volume of the solution is $264.9 + 127.3 = 392.2$ mL. The original chemical amount of KOH was

$$264.9 \text{ mL solution} \times \left(\frac{0.065 \text{ millimole KOH}}{1 \text{ mL solution}} \right) = 17.2185 \text{ mmol KOH}$$

By a similar calculation, 9.2929 mmol of HCl was originally present. The strong acid HCl and strong base KOH react instantly upon mixing. The process consumes essentially all of the HCl but leaves $17.2185 - 9.2929 = 7.9256$ mmol of KOH in excess. This KOH is completely dissociated to 7.9256 mmol of OH^- (and 7.9256 mmol of K^+). The amount of OH^- from the only other source, the autoionization of water, is many thousands of times less and can be ignored. Thus, the solution contains 7.9256 mmol of OH^- in 392.2 mL of water; the

solution is 0.02021 mol L^{-1} in OH^-. The pOH is $-\log(0.02021) = 1.6945$, and
$pH = 14.00 - pOH = \boxed{12.31}$.[6]

8-14 From the definition of pK_a, a smaller pK_a means a stronger acid. Hence, lactic acid is a weaker acid than pyruvic acid and $\boxed{\text{hypochlorous acid}}$ is weaker than either.

8-16 (a) The acid ionization of niacin proceeds by the reaction

$$C_5H_4NCOOH(aq) + H_2O(l) \rightleftharpoons C_5H_4NCOO^-(aq) + H_3O^+(aq)$$

(b) In aqueous solutions, the product of K_a of an acid and K_b of its conjugate base is K_w. Hence

$$K_b = \frac{K_w}{K_a} = \frac{1.0 \times 10^{-14}}{1.5 \times 10^{-5}} = \boxed{6.7 \times 10^{-10}}$$

(c) The K_a for the pyridinium ion appears in Table 8-2 (text page 332). It is 5.6×10^{-6}, smaller than the K_a of niacin. Niacin is a stronger acid than pyridinium ion, which means than the conjugate base of the pyridinium ion, which is pyridine, is a stronger base than the conjugate base of niacin; $\boxed{\text{niacinate}}$ ion is a weaker base.

A difficulty in the problem is figuring out the relationship of pyridine to pyridinium ion. The similarity of endings in "pyridinium" and "ammonium" helps.

8-18 (a) The nitrite ion from soluble KNO_2 hydrolyzes:

$$NO_2^-(aq) + H_2O(aq) \rightleftharpoons HNO_2(aq) + OH^-(aq)$$

(b) The conjugate acid of NO_2^- ion is HNO_2, which has K_a equal to 4.6×10^{-4}. The K_b of NO_2^- ion is therefore

$$K_b = \frac{K_w}{K_a} = \frac{1.0 \times 10^{-14}}{4.6 \times 10^{-4}} = \boxed{2.2 \times 10^{-11}}$$

8-20 (a) The pK_a of ethane is very large. Hence, ethane is an exceedingly weak acid. It is so weak that it is a non-acid in aqueous solution in any practical sense.

(b) The conjugate base of ethane has the formula $C_2H_5^-$. This formula is arrived at simply by subtracting H^+ from the formula of ethane.

---

[6]Extra significant figures were carried along in the intermediate stages in this calculation. The answer 12.31 has two significant figures. See text page A-24.

(c) The "ethide ion," the conjugate base of ethane, is an exceedingly strong base. It would extract H^+ from water to generate ethane:

$$C_2H_5^- + H_2O(aq) \longrightarrow C_2H_6(g) + OH^-(aq)$$

8-22 The equation given in the problem is the sum of the chemical equation for the third acid ionization reaction of phosphoric acid

$$HPO_4^{2-}(aq) + H_2O(l) \rightleftharpoons H_3O^+(aq) + PO_4^{3-}(aq)$$

and the reverse of the equation for the first acid ionization of carbonic acid (H_2CO_3)

$$H_2CO_3(aq) + H_2O(l) \rightleftharpoons H_3O^+(aq) + HCO_3^-(aq)$$

Hence, the desired equilibrium constant is

$$K = \frac{K_{a3,H_3PO_4}}{K_{a1,H_2CO_3}} = \frac{2.2 \times 10^{-13}}{4.3 \times 10^{-7}} = \boxed{5.1 \times 10^{-7}}$$

8-24 (a) According to Figure 8-6 (text page 337), cresol red starts its change from its basic to acidic color at a pH of 8.8, but thymolphthalein starts at the less acidic pH of 10.6. The color changes mark the pH at which the concentrations of the acid forms of the indicators start to become important. The basic form of thymolphthalein requires a lesser concentration of H_3O^+ to cause it to change to the acid form. It is accordingly a stronger base than the basic form of cresol red.

(b) In the solution $\boxed{8.8 < \text{pH} < 9.4}$ (see Figure 8-6, text page 337.)

8-26 Assume that the three acids in the problem are all monoprotic acids. If they are, and if the autoionization of water is negligible as a source of hydronium ion in the solutions, it can be shown that

$$[H_3O^+]^2 + K_a[H_3O^+] - c_aK_a = 0$$

where c_a is the concentration of the acid and K_a is its acidity constant.

(a) For pyruvic acid, insert $ca = 0.20$ M and $K_a = 10^{-2.49} = 0.00324$ into the preceding quadratic equation and solve for $[H_3O^+]$. The result is 0.0238M. This corresponds to a pH of $\boxed{1.62}$ and a fractional dissociation of the pyruvic acid equal to $0.0238/0.20 = \boxed{0.119}$.

(b) For lactic acid, insert $ca = 0.20$ M and $K_a = 10^{-3.85} = 0.000141$ into the equation and solve for $[H_3O^+]$. The result is 0.00532 M. This corresponds to a pH of $\boxed{2.27}$ and a fractional dissociation of the lactic acid of only $0.00532/0.20 = \boxed{0.027}$.

(c) For hypochlorous acid, insert $ca = 0.20$ M and $K_a = 10^{-7.53} = 2.85 \times 10^{-8}$ into the equation and solve for $[H_3O^+]$. The result is 7.68×10^{-5} M. This corresponds to a pH of $\boxed{4.11}$ and a fractional dissociation that is the least of all: $2.85 \times 10^{-8}/0.20 = \boxed{0.00038}$.

8-28 The molar mass of ascorbic acid is 176.126 g mol^{-1}. The concentration of ascorbic acid in the 100 mL of water is its chemical amount, 2.839×10^{-3} mol (computed by dividing 500×10^{-3} g by $\mathcal{M}$), divided by the volume of the solution (0.100 L). It is $\boxed{0.0284 \text{ mol L}^{-1}}$.

The acidity equilibrium is

$$HC_6H_7O_6(aq) + H_2O(l) \rightleftharpoons H_3O^+(aq) + C_6H_7O_6^-(aq)$$

for which the equilibrium expression is

$$\frac{[H_3O^+][C_6H_7O_6^-]}{[HC_6H_7O_6]} = K_a = 8.0 \times 10^{-5}$$

The concentration of ascorbate ion equals that of the hydronium ion, as long as the contribution of hydronium from the autoionization of water is negligible. Take this to be the case, and let $[H_3O^+]$ equal x. Then

$$\frac{x^2}{0.0284 - x} = 8.0 \times 10^{-5}$$

Solving this quadratic equation gives $x = 1.5 \times 10^{-3}$. The pH of the solution is $-\log(1.5 \times 10^{-3}) = \boxed{2.82}$.

8-30 (a) The K_a of benzoic acid is 6.5×10^{-5} (see text Table 8-2) and its analytical concentration is 0.027 mol L^{-1}. Using the pattern of Example 8-6 (text page 340) gives $[H_3O^+]$ as 1.3×10^{-3} mol L^{-1}, for a pH of $\boxed{2.88}$.

(b) A solution of acetic acid of pH 2.88 is to be prepared. The K_a of acetic acid is 1.8×10^{-5}. Because acetic acid is weaker than benzoic acid (by a factor of 3.6, the ratio of the K_a's of the two), more than 0.027 mol of it is required to lower the pH of a liter of pure water from 7.0 to 2.88. Let the required concentration of acetic acid be c_a. Then, at equilibrium

$$1.8 \times 10^{-5} = \frac{(1.3 \times 10^{-3})(1.3 \times 10^{-3})}{(c_a - 1.3 \times 10^{-3})}$$

Solving gives $c_a = \boxed{0.095 \text{ mol L}^{-1}}$. This is close to (but not equal to) 3.6 times the amount of benzoic acid that gives pH 2.88 in a liter of water.

8-32 Pentafluorobenzoic acid is strong among weak acids. If the equilibrium concentration of H_3O^+ is x, then

$$[C_6F_5COO^-] = x \quad \text{and} \quad [C_6F_5COOH] = 0.100 - x$$

Substitution into the equilibrium expression gives

$$\frac{[H_3O^+][C_6F_5COO^-]}{[C_6F_5COOH]} = K_a = 0.033 = \frac{x^2}{(0.100 - x)}$$

This equation can be solved by the quadratic formula or by approximation to give $x = 0.0433$. Note that x may *not* be neglected relative to 0.100 in the denominator of the fraction. The pH is the negative logarithm of 0.0433, or $\boxed{1.36}$. The $[H_3O^+]$ from the autoionization of water is negligible compared to $0.0433 \text{ mol L}^{-1}$.

8-34 From the pH of the solution (2.42) is obtained $[H_3O^+] = 3.80 \times 10^{-3} \text{ mol L}^{-1}$. This is also essentially the concentration of the 2-germaacetate ion at equilibrium; the concentration of the un-ionized 2-germaacetic acid is then $0.050 - (3.8 \times 10^{-3}) = 0.0462 \text{ mol L}^{-1}$. Substitution of these values into the equilibrium expression gives $K_a = \boxed{3.1 \times 10^{-4}}$. This acid is $\boxed{\text{stronger}}$ than acetic acid.

8-36 The original concentration of methylamine is $0.070/0.800 = 0.0875 \text{ mol L}^{-1}$. Some of this methylamine ionizes. The K_b-equilibrium is

$$CH_3NH_2(aq) + H_2O(l) \rightleftharpoons OH^-(aq) + CH_3NH_3^+(aq)$$

for which the equilibrium expression is

$$\frac{[OH^-][CH_3NH_3^+]}{[CH_3NH_2]} = K_b = 4.4 \times 10^{-4}$$

The concentration of methylammonium ion equals that of the hydroxide ion, as long as the contribution of hydroxide from the autoionization of water is negligible. Take this to be the case, and let $[OH^-]$ equal x. Then

$$\frac{x^2}{0.0875 - x} = 4.4 \times 10^{-4}$$

Neglecting x in the denominator of this equation and solving gives 6.20×10^{-3} for x. This is about 7% of 0.0875. Guessing somewhat smaller values of x and substitution in the original equation soon gives an x of 5.98×10^{-3}, the same answer obtained using the quadratic formula. The pOH of the solution is $-\log(5.98 \times 10^{-3}) = 2.223$, and the pH is $14.00 - 2.223 = \boxed{11.78}$.

8-38 The weak base reacts with water: $B(aq) + H_2O(l) \rightleftharpoons OH^-(aq) + BH^+(aq)$. If 0.0145 mol of B was originally present per liter, and 0.012% of it reacted to reach equilibrium, then at equilibrium:

$$[OH^-] = [BH^+] = (1.2 \times 10^{-4})(0.0145) = 1.74 \times 10^{-6} \text{ mol L}^{-1}$$

This assumes that the autoionization of water is an insignificant source of OH^- ion. The concentration of unreacted B is $0.9988 \times 0.0145 \approx 0.0145$ mol L^{-1}. Substitute these equilibrium concentrations into the proper equilibrium expression

$$K_b = \frac{[BH^+][OH^-]}{[B]} = \frac{(1.74 \times 10^{-6})^2}{0.0145} = \boxed{2.1 \times 10^{-10}}$$

Note the reasonable assumption that each molecule of the weak base accepts only one H^+; diacidic weak bases do however exist.

8-40 (a) An aqueous solution of $AlCl_3$ is acidic, thanks to the hydrolysis of the Al^{3+} ion.
(b) An aqueous solution of NH_4ClO_4 is acidic.
(c) An aqueous solution of AgF is basic.
(d) An aqueous solution of $SrCl_2$ is neutral.

8-42 The equilibrium

$$NH_4^+(aq) + F^-(aq) \rightleftharpoons NH_3(aq) + HF(aq)$$

is the K_a equilibrium for NH_4^+ minus the K_a equilibrium for HF. Therefore, its K equals the K_a for the first divided by the K_a for the second:

$$K = \frac{K_a(NH_4^+)}{K_a(HF)} = \frac{5.6 \times 10^{-10}}{6.6 \times 10^{-4}} = \boxed{8.5 \times 10^{-7}}$$

8-44 The solution contains 75.00×0.0460 mmol of the strong acid $HClO_4$ and is treated with 150.00×0.0230 mmol of the strong base KOH. The chemical amounts of the acid and base are equal so that the two exactly neutralize each other to form water containing K^+ and ClO_4^- ions. Neither of these ions acts detectably as an acid or base in water. Hence the pH of the solution equals $\boxed{7.0}$.

8-46 This is a buffer solution. The NH_4Cl dissociates completely in water, so the "original" concentration of NH_4^+ (after dissolution but before acid-base reactions) equals the chemical amount of NH_4^+ divided by the volume of the solution. The chemical amount of NH_4^+ is 3.62 g/53.49 g mol^{-1}, and the volume of the solution is 0.600 L. Hence the molarity of NH_4^+ ion is 0.113 mol L^{-1}. If there were no interaction involving the NH_4^+ ion, the NH_3, and the water, then the concentration of NH_3 would be $(400.0/600.0)(0.100) = 0.0667$ mol L^{-1}. The ammonium ion and the ammonia do interact in a equilibrium governed by the equation:

$$\frac{[NH_3][H_3O^+]}{[NH_4^+]} = K_a = 5.6 \times 10^{-10}$$

but the interaction does not change the concentrations of the two species significantly from the values just calculated. It therefore suffices simply to put these values in the expression and compute $[H_3O^+]$. It is 9.5×10^{-10} mol L^{-1}; the pH is $\boxed{9.02}$.

8-48 The reaction of the weak base "bis" in water can be represented:

$$\text{bis}(aq) + H_2O(l) \rightleftharpoons \text{bisH}^+(aq) + OH^-(aq)$$

for which the equilibrium expression is

$$\frac{[OH^-][\text{bisH}^+]}{[\text{bis}]} = K_b$$

The concentrations of the bis and its conjugate acid are essentially equal under the conditions described in the problem because the amount of the HCl that has been added is just enough to convert half of the bis to bisH$^+$ and leave half unreacted. It is true that both of these species then react with water, but the changes in amount caused by these interactions are negligible. Therefore, $K_b = [OH^-]$ and $pK_b = pOH$. The pOH is 8.8, and the pH is $14.0 - 8.8 = \boxed{5.2}$.

8-50 The pK_a for the first acid ionization of carbonic acid is 6.36. The pK_a for the second acid ionization of phosphoric acid is 7.21. From the standpoint of controlling the pH pf the artificial blood, the $\boxed{H_2PO_4^-/HPO_4^{2-} \text{ system}}$ would be better because the applicable pK_a is closer to the desired pH.

8-52 This is the titration of a strong acid with a strong base. Before any base is added, the concentration of H_3O^+ in the solution is 0.1439 M, because the HBr is completely ionized. The pH is 0.8419.

It is easy to verify that it requires 31.14 mL of 0.1219 M NaOH to titrate the 26.38 mL of 0.1439 M HBr to the equivalence point. At the equivalence point, the pH is 7.00. When the titration is 1.00 mL short of equivalence 30.14 mL of NaOH has been added. The unreacted H_3O^+ comprises 0.122 mmol (the difference between the chemical amount of H_3O^+ originally present and the chemical amount of NaOH added), and the volume of the solution is 26.38 mL + 30.14 mL = 56.52 mL. The concentration of the H_3O^+ is its chemical amount divided by this volume or 2.16×10^{-3} mol L^{-1}. The pH is $\boxed{2.666}$.

Going 1.00 mL past the equivalence point means adding a total of 32.14 mL of NaOH. The concentration of OH^- is the chemical amount of unreacted OH^- divided by the volume of the solution. This is 0.122 mmol divided by 58.52 mL or 2.085×10^{-3} M. The pOH is 2.681, and pH = 14.000 − pOH = $\boxed{11.319}$.

8-54 Before any titrating acid is added, the solution is 0.175 M in NH_3. Ammonia acts as a base:

$$NH_3(aq) + H_2O(l) \rightleftharpoons NH_4^+(aq) + OH^-(aq) \qquad \frac{[NH_4^+][OH^-]}{[NH_3]} = K_b = 1.8 \times 10^{-5}$$

The K_b given here comes from substituting 5.6×10^{-10}, the K_a of NH_4^+ ion, into $K_b = K_w/K_a$. If the equilibrium concentration of NH_4^+ is x, then the equilibrium concentration of NH_3 is $0.175 - x$ and that of OH^- is x. The other sources of OH^- are negligible. Substitution and solving for x (using the quadratic formula) gives $[OH^-] = 1.766 \times 10^{-3}$ mol L^{-1} and a pH of $\boxed{11.25}$.

The text (page 356) states that in the titration of a weak acid with a strong base, the pH at the half-equivalence point is very close to the pK_a of the weak acid: $pK_a = pH$. The analogous relationship $pK_b = pOH$ holds at half-equivalence in the titration of a weak base with a strong acid. Then, pH = 14.00 − pOH = $14.00 - 4.745 = \boxed{9.26}$.

It requires $(0.175/0.106)(140.0) = 231.13$ mL to titrate the solution to the equivalence point. At this point the total volume of the solution is 371.13 mL, and the mixture is a solution is dilute aqueous NH_4Cl with a nominal concentration of NH_4^+ ion equal to $0.175(140.00/371.13) = 0.0660$ mol L^{-1}. The concentration of NH_4^+ ion is not significantly lowered from its nominal value by its K_a-reaction with water because the K_a for NH_4 ion is small. Let x equal to the equilibrium concentration of H_3O^+, write the K_a expression, and substitute as usual:

$$\frac{[NH_3][H_3O^+]}{[NH_4^+]} = K_a = 5.6 \times 10^{-10} = \frac{x \cdot x}{0.0660 - x}$$

Solving gives $[H_3O^+] = 6.06 \times 10^{-6}$ mol L^{-1}. The pH is $\boxed{5.22}$.

When the titration is 1.00 mL past the equivalence point, the NH_4^+ ion is still present but is completely overshadowed as a source of H_3O^+ by the excess $HCl(aq)$. Only the HCl need be considered in figuring the pH. The first 231.13 mL of 0.106 mol L^{-1} HCl were neutralized, so the concentration of HCl in the solution is the amount of HCl contributed by the last 1.00 mL divided by the total volume of the solution. This is $(1.00/372.13)(0.106) = 2.85 \times 10^{-4}$ mol L^{-1}. This is also the concentration of H_3O^+ so the pH is $\boxed{3.55}$.

8-56 At the end-point of the titration, the chemist has a solution that is about 0.05 mol L^{-1} in $NH_4^+(aq)$ ion. This ion, a Brønsted-Lowry acid, acidifies the solution by the equilibrium

$$NH_4^+(aq) + H_2O(l) \rightleftharpoons NH_3(aq) + H_3O^+(aq) \qquad \frac{[NH_3][H_3O^+]}{[NH_4^+]} = K_a = 5.6 \times 10^{-10}$$

Let x equal the equilibrium concentration of the H_3O^+ and NH_3. Then

$$5.6 \times 10^{-10} = \frac{[NH_3][H_3O^+]}{[NH_4^+]} = K_a = 5.6 \times 10^{-10} = \frac{x^2}{0.05 - x}$$

The x in this equation is 5.3×10^{-6} so the pH is $\boxed{5.3}$. An appropriate indicator would be $\boxed{\text{methyl red}}$.

8-58 The chemical amount of HCl that has been added at the equivalence point is 6.7176 mmol. The aniline and HCl react in a 1 : 1 ratio. Hence the original amount of aniline was 6.7176 mmol. This amount of aniline was dissolved in 50.00 mL of solution. The original concentration of aniline was therefore $6.7176/50.00 = \boxed{0.1344 \text{ mol L}^{-1}}$.

The pOH at the half-equivalence point in the titration of the solution of aniline equals the pK_b of aniline. The pOH at half-equivalence is $14.00 - pH = 14.00 - 10.67 = 3.33$. Hence p$K_b$ equals 3.33 and $K_b = \boxed{4.7 \times 10^{-4}}$.

8-60 (a) The titration starts at low pH. It is therefore the titration of an acid by a base. Comparison of the shape of the curve to titration curves in the text establishes that the titration involves a weak acid and a strong base. The only weak acid-strong base combination listed is the titration of $\boxed{HNO_2 \text{ by NaOH}}$.

(b) Any indicator changing color within the range between pH 6 and 11. This includes bromothymol blue, cresol red, phenolphthalein, and thymolphthalein (see text Figure 8-6, text page 337).

(c) The pK_a of a weak acid equals the pH of the solution at the half-equivalence point in its titration. The half-equivalence point in this titration is at 20.0 mL at which point the pH equals 3.3 to 3.5 (by reading the graph). The K_a of the weak acid being titrated is therefore between 3 and 5×10^{-4}. The weak acid might be nitrous acid.

8-62 The successive equilibrium constants for the two-step ionization of phthalic acid (H_2Ph) differ by a factor of about 2500. Because the amounts of HPh^- and H_3O^+ produced by the first step are far larger than the amount consumed (in the case of HPh^-) or produced (in the case of H_3O^+) by the second step, the steps may be considered separately. Also, the autoionization of water is a negligible source of H_3O^+ in this solution. The first stage in the ionization is governed by the expression

$$\frac{[H_3O^+][HPh^-]}{[H_2Ph]} = K_{a1} = 1.26 \times 10^{-3} = \frac{x^2}{0.0100 - x}$$

where x is the equilibrium concentration of H_3O^+. Neglecting x relative to 0.0100 and solving the resulting equation gives $x = 2.85 \times 10^{-3}$. This number is *not* small compared to 0.0100 (it is 28.5% of it). Resorting to the quadratic formula then gives $x = 2.98 \times 10^{-3}$. The equilibrium concentrations of HPh^- and H_3O^+ are $\boxed{2.98 \times 10^{-3} \text{ mol L}^{-1}}$. The equilibrium concentration of H_2Ph is $0.0100 - 0.00298 = \boxed{0.0070 \text{ mol L}^{-1}}$.

The second stage of ionization is governed by the expression

$$\frac{[H_3O^+][Ph^{2-}]}{[HPh^-]} = K_{a2} = 3.10 \times 10^{-6} = \frac{y(2.98 \times 10^{-3})}{2.98 \times 10^{-3}}$$

where y is the equilibrium concentration of the Ph^{2-} ion. Solution of the equation is trivial; $[Ph^{2-}] = \boxed{3.10 \times 10^{-6} \text{ mol L}^{-1}}$.

8-64 This problem resembles problem 8-62 in that it treats a two-stage process in which the successive equilibrium constants differ so greatly that the stages can be treated separately. Now however, a weak base, the oxalate ion, is in solution. It reacts according to the equations

$$C_2O_4^{2-}(aq) + H_2O(l) \rightleftharpoons HC_2O_4^-(aq) + OH^-(aq) \qquad K_{b1} = K_w/K_{a2}$$
$$HC_2O_4^-(aq) + H_2O(l) \rightleftharpoons H_2C_2O_4(aq) + OH^-(aq) \qquad K_{b2} = K_w/K_{a1}$$

Let the final concentration of $HC_2O_4^-$ equal y. This is also the final concentration of OH^- because the first reaction is the only important source of OH^-. The concentration of $C_2O_4^{2-}$ at equilibrium is then $0.10 - y$. Therefore

$$\frac{y^2}{0.10 - y} = \frac{1.0 \times 10^{-14}}{6.4 \times 10^{-5}}$$

Solving for y gives 4.0×10^{-6}, and

$$[OH^-] = [HC_2O_4^-] = \boxed{4.0 \times 10^{-6}} \text{ mol L}^{-1}$$

The hydrogen oxalate ion ($HC_2O_4^-$) itself can serve as a weak base, as represented in the second chemical equation above. For the second reaction the equilibrium expression is

$$\frac{[OH^-][H_2C_2O_4]}{[HC_2O_4^-]} = K_{b2} = \frac{K_w}{K_{a1}} = \frac{1.0 \times 10^{-14}}{5.9 \times 10^{-2}}$$

Insertion of the known equilibrium concentrations of OH^- and $HC_2O_4^-$ gives $\boxed{[H_2C_2O_4] = 1.7 \times 10^{-13} \text{ mol L}^{-1}}$. The original concentration of the oxalate ion is 0.10 mol L^{-1}. Some of this oxalate reacts away, but not much. A tiny concentration of hydrogen oxalate ion (only 4.0×10^{-6} mol L^{-1}) remains and $\boxed{1.7 \times 10^{-13} \text{ mol L}^{-1}}$ of oxalic acid remains The final concentration of oxalate ion is still $\boxed{0.10 \text{ mol L}^{-1}}$. The very slight contribution of the second reaction to the concentration of OH^- is clear.

8-66 Dissolved carbonates exist in the rain drop as either carbonic acid, hydrogen carbonate ion or carbonate ion:

$$3.6 \times 10^{-5} = [H_2CO_3] + [HCO_3^-] + [CO_3^{2-}]$$

where all of the quantities on the right are to be determined. (Note the use of the convention of lumping dissolved CO_2 in with H_2CO_3, see marginal note on text page 359). The K_a-equilibria of carbonic acid provide additional relationships among these quantities:

$$\frac{[HCO_3^-][H_3O^+]}{[H_2CO_3]} = K_{a1} = 4.3 \times 10^{-7}$$

$$\frac{[CO_3^{2-}][H_3O^+]}{[HCO_3^-]} = K_{a2} = 4.8 \times 10^{-11}$$

The $[H_3O^+]$ is 1.0×10^{-4} mol L^{-1} in these equations. This means there are three equations in three unknowns. The easiest way to solve this system of equations is to observe that $[CO_3^{2-}]$ is likely to be quite small—almost all of the dissolved carbonate is tied up with hydrogen ion because hydrogen ion is abundant in this acidic solution. If $[CO_3^{2-}]$ is close to zero, then:

$$[H_2CO_3] + [HCO_3^-] \approx 3.6 \times 10^{-5} \quad \text{and} \quad \frac{[HCO_3^-](1.0 \times 10^{-4})}{[H_2CO_3]} \approx 4.3 \times 10^{-7}$$

Solving this pair of equations gives

$$[H_2CO_3] = \boxed{3.6 \times 10^{-5} \text{ mol L}^{-1}} \qquad [HCO_3^-] = \boxed{1.5 \times 10^{-7} \text{ mol L}^{-1}}$$

Now, substitution in the K_{a2} expression gives $[CO_3^{2-}] = 7.4 \times 10^{-14}$ mol L^{-1}. Almost all of the carbonate is present as carbonic acid, a small amount is present as hydrogen carbonate ion, and a vanishingly tiny fraction is present as CO_3^{2-} ion.

8-68 (a) In this reaction the Lewis acid HgI_2 accepts an electron pair from each of two I^- ions, which act as Lewis bases.

(b) The positively charged Co^{2+} ion, a Lewis acid, accepts electron pairs from the water molecules, which serve as Lewis bases.

(c) The Fe^{3+} ion acts as a Lewis acid and accepts electron pairs from the CN^- ions. The reaction can also be discussed in terms of the displacement of the Lewis base water from its combination with the Lewis acid Fe^{3+} by the stronger Lewis base CN^-.

8-70 (a) The slag-forming reaction is $CaO + SiO_2 \longrightarrow CaSiO_3$.

(b) The CaO is a Lewis base and the SiO_2 is a Lewis acid.

8-72 (a) This problem follows up the extension of the acid-base concepts suggested by problem 8-71. In the Brønsted system, the acid is the *donor* of a *positively* charged particle (the hydrogen ion); in this system (called the Lux-Flood acid-base system) the acid is the *acceptor* of a *negatively* charged particle (the oxide ion), and the base is the donor of the oxide ion.

(b) In the first reaction CaO donates O^{2-} to SiO_2, which accepts it. Therefore CaO is the base, and SiO_2 is the acid. In the second reaction the SiO_2 accepts an O^{2-} and is again the acid. The O^{2-} donor is the Ca_2SiO_4, which acts as the base. In the third reaction, the CaO donates O^{2-} to Ca_2SiO_4. The latter is an acid in this reaction, the reverse of its role in the second reaction.

8-74 The equation

$$2\,K(s) + 2\,H_2O(l) \longrightarrow 2\,KOH)(aq) + H_2(g)$$

represents the reaction better. If H_3O^+ ion were needed, the rapid reaction might exhaust it in the surroundings of the $K(s)$ and slow the reaction.

8-76 (a) The density of the nitric acid is 1.503 g cm^{-3}, which is 1.503 kg L^{-1}. The nitrate ion has a molality of 0.25 mol kg^{-1} in this solution. The molarity of the nitrate ion is then

$$\left(1.503 \,\frac{\text{kg}}{\text{L}}\right) \times \left(0.25 \frac{\text{mol NO}_3^-}{\text{kg}}\right) = 0.376 \text{ mol L}^{-1} = \boxed{0.38 \text{ mol L}^{-1}}$$

(b) In this case, $HNO_3(l)$ is the solvent, and H_2O is one of the products of the autoionization. The HNO_3, a pure solvent, is omitted from the autoionization expression, but the H_2O is included:

$$[\text{NO}_3^-][\text{NO}_2^+][\text{H}_2\text{O}] = K_{\text{HNO}_3}$$

If the concentration of nitrate ion is 0.38 mol L^{-1}, then the concentration of nitronium ion and of water are also 0.38 mol L^{-1}—the reaction is the only source of any of the three species. $K_{\text{HNO}_3} = (0.376)^3 = \boxed{5.3 \times 10^{-2}}$.

8-78 The pH of pure water depends on K_w according to:

$$\text{pH} = \text{p}K_w/2 = \frac{-\log K_w}{2}$$

Therefore, at 0°C, pure water has pH 7.472; at 25°C, 6.999, and at 60°C, 6.509. At the temperature rises, the pH diminishes—the degree of autoionization becomes greater with increasing temperature. The autoionization equilibrium is $\boxed{\text{endothermic}}$ because it shifts to favor its products with increasing temperature.

8-80 If "same color with methyl red" is taken to mean that the two solutions have the same pH, then, because acetic acid is a weak acid, the concentration of acetic acid must exceed the concentration of hydrochloric acid to bring the pH down to a particular value. The solution of acetic acid contains more moles of acid, so the solution of acetic acid can neutralize more moles of base.

8-82 The 0.100 M solutions of HCl and NH$_4$Cl are acidic, the 0.100 M solutions of Na$_3$PO$_4$ and NaCH$_3$COO are basic, and the 0.100M solution of KNO$_3$ is neutral. The compounds are, in order, a strong acid, the salt of a strong acid and a weak base, the salt of a weak acid and a strong base, the salt of a weak acid and a strong base, and the salt of a strong acid and a strong base.

8-84 (a) The most important equilibrium in the buffer solution is

$$HN_3(aq) + H_2O(l) \rightleftharpoons N_3^-(aq) + H_3O^+(aq) \qquad \frac{[\text{N}_3^-][\text{H}_3\text{O}^+]}{[\text{HN}_3]} = K_a$$

The K_a in this case is 1.9×10^{-5} according to text Table 8-2. The concentration of hydrazoic acid at equilibrium is essentially equal to 0.683 mol L^{-1}, its starting value, and the concentration of azide ion is essentially equal to its starting value, 0.593 mol L^{-1}. Putting these values into the equilibrium law, solving for [H$_3$O$^+$], and taking the negative logarithm gives a pH of $\boxed{4.66}$.

(b) $HN_3(aq) + OH^-(aq) \rightleftharpoons N_3^-(aq) + H_2O(l)$ $\boxed{\text{shifts right}}$.

(c) The pH of the buffer solution $\boxed{\text{increases}}$ slightly when NaOH(aq), a base, is added.

8-86 The pH shoots up suddenly upon the addition of 1 more drop of sodium hydroxide solution. This is the indication of an end-point; thus $4.71 + 0.01 = 4.72$ mL of 0.0410 M NaOH is just enough to bring the titration to the first equivalence point. At that point, the chemical amount of base that has been added equals the chemical amount of phosphoric acid that was present. The chemical amount of base is (0.0410 mol L^{-1} $\times$ 0.00472 L). This is therefore also the chemical amount of acid. The original concentration of the phosphoric acid is this quantity divided by 0.05000 L (50.00 mL) or $\boxed{3.87 \times 10^{-3} \text{ mol L}^{-1}}$.

8-88 The vinegar under analysis contains around 5 g of acetic acid per 100 g. Acetic acid is CH$_3$COOH ($\mathcal{M} = 60.05$ g mol^{-1}). Since the density of the vinegar is about 1.0 g cm^{-3}, Anne Dalton proposes to titrate 50.00 mL samples of about 0.83M acetic acid. She will need around 42 mL of 1.000 M base for each titration. If she detects the equivalence point to within ±0.02 mL, then her precision is $(\pm0.02/42) \times 100 \approx 0.05\%$.

Charlie Cannizzaro can expect pH values near 2.41, the calculated pH of a 0.83 M solution of acetic acid. An uncertainty of ±0.01 pH units corresponds in this range of pH to an uncertainty of about $\pm4\%$ in the concentration of acetic acid. This can be confirmed by inserting first 2.40 and then 2.42 in the expression

$$c_{CH_3COOH} = \frac{[H_3O^+][CH_3COO^-]}{K_a} + [H_3O^+] \approx \frac{10^{-2pH}}{1.8 \times 10^{-5}}$$

Anne Dalton's method is more precise. Its drawback is that it is slower and requires more skill. $\boxed{\text{Anne Dalton}}$ is hired.

8-90 The total chemical amount of HCl added is 18.393 mmol. The total chemical amount of NaOH added is 7.917 mmol. These numbers, obtained by multiplying the concentrations of the acid and base by their respective volumes in mL, are clearly not equal. The tablet supplies the extra base needed to neutralize

$18.393 - 7.917 = 10.476$ mmol of acid. The reaction is

$$CaCO_3(s) + 2\,H_3O^+(aq) \longrightarrow Ca^{2+}(aq) + CO_2(g) + 3\,H_2O(l)$$

Each mole of $CaCO_3$ takes up 2 mol of H_3O^+. Hence the tablet must contain $(10.476/2) = 5.238$ mmol of $CaCO_3$. The molar mass of $CaCO_3$ is 100.09 g mol^{-1} so the tablet contains 0.5243 g of $CaCO_3$. This is $\boxed{39.54\%}$ of the total.

8-92 The exhalation of the $CO_2(g)$ through the shell of the egg should lead to a $\boxed{\text{rise}}$ in the pH of the contents—a weak acid is being lost from the system.

8-94 The formula of the hydrogen tartrate anion is $HC_4H_4O_6^-$. The dissolution of potassium hydrogen tartrate gives this ion along with potassium ions. Hydrogen tartrate ion acts as a base in the reaction:

$$HC_4H_4O_6^-(aq) + H_3O^+(aq) \rightleftharpoons H_2C_4H_4O_6(aq) + H_2O(l)$$

to counteract the effect of an added acid and as an acid in the reaction:

$$HC_4H_4O_6^-(aq) + OH^-(aq) \rightleftharpoons C_4H_4O_6^{2-}(aq) + H_2O(l)$$

to counteract the effect of an added base.

8-96 A balanced equation for the reaction of aluminum hydroxide as a base is

$$Al(OH)_3(s) + 3\,H_3O^+(aq) \longrightarrow Al^{3+}(aq) + 3\,H_2O(l)$$

In this reaction, electron pairs of the oxygen atoms of the hydroxide groups on $Al(OH)_3$, which acts as a Lewis base, are shared to H^+ ions, which act as Lewis acids. When aluminum hydroxide reacts as an acid

$$Al(OH)_3(s) + OH^-(aq) \longrightarrow Al(OH)_4(aq)$$

it accepts an electron pair from a hydoxide ion at the Al center.

Answers to Even-Numbered Problems, Chapter 9

9-2 Comparison of the two formulas reveals the hydration reaction consumes 3/2 mol of water for every mole of plaster of paris. Thus:

$$25.0 \text{ kg CaSO}_4 \cdot \tfrac{1}{2}\text{H}_2\text{O} \times \left(\frac{1 \text{ mol CaSO}_4 \cdot \tfrac{1}{2}\text{H}_2\text{O}}{0.14515 \text{ kg CaSO}_4 \cdot \tfrac{1}{2}\text{H}_2\text{O}} \right) \times \left(\frac{3 \text{ mol H}_2\text{O}}{2 \text{ mol CaSO}_4 \cdot \tfrac{1}{2}\text{H}_2\text{O}} \right)$$

$$\times \left(\frac{0.01802 \text{ kg H}_2\text{O}}{1 \text{ mol H}_2\text{O}} \right) \times \left(\frac{1 \text{ L H}_2\text{O}}{1 \text{ kg H}_2\text{O}} \right) = \boxed{4.65 \text{ L H}_2\text{O}}$$

9-4 Even "bone-dry" residues can contain water of hydration, as indeed do dry bones. The 1.00 g sample amounts to 8.308×10^{-3} mol of $MgSO_4$. It picks up 0.898 g of water when the solution is evaporated in one temperature range but picks up only 0.150 g of water when the solution is evaporated in the second temperature range. These amounts are 0.0499 and 0.00833 mol. The chemical amount of water associated in the solid is 6.00 times the chemical amount of $MgSO_4$ in the first case and 1.00 times the chemical amount of $MgSO_4$ in the second case. The two solids contain $\boxed{6}$ and $\boxed{1}$ waters of crystallization per $MgSO_4$ respectively.

9-6 The 255 g of $AgNO_3$ is 1.50 mol of $AgNO_3$. The dissolution of this much silver nitrate in 100 g of water is just like the dissolution of 15.0 mol of silver nitrate in 1.00 kg of water. The graph (on text page 376) shows that 1.00 kg of water at 95°C easily accommodates 15.0 mol of $AgNO_3$, but that cooling the solution to $\boxed{\text{about 30°C}}$ would cause $AgNO_3(s)$ to tend to precipitate.

9-8 The molar mass of codeine equals 299.37 g mol^{-1}. If 1.00 g of codeine dissolves in 120 mL, then 8.33 g dissolves in 1.00 L of water. The molar mass of codeine equals 299.37 g mol^{-1} so that there is 0.0278 mol of codeine per 1.00 L of water. This is also $\boxed{0.0278 \text{ mol kg}^{-1}}$, taking the density of water as 1.00 kg L^{-1}. At the higher temperature, the solubility of the codeine *rises* to $\boxed{0.0557 \text{ mol kg}^{-1}}$; the dissolution reaction of codeine in water is $\boxed{\text{endothermic}}$ because it is shifted to the right at higher temperature.

9-10 The molar mass of $La_2(SO_4)_3$ is 566.0 g mol^{-1} so the 5.0 g mentioned in the problem equals 8.83×10^{-3} mol. Each mole of this salt gives two moles of positive ions and three moles of negative ions if completely dissociated. That is, ν for this salt equals 5. The volume of the solution is 40.0 L, so that the concentration of ions is:

$$c_{\text{ions}} = \frac{8.83 \times 10^{-3} \text{ mol } La_3(SO_4)_3}{40.0 \text{ L}} \times \left(\frac{5 \text{ mol ions}}{1 \text{ mol } La_2(SO_4)_3} \right)$$

$$= \boxed{1.1 \times 10^{-3} \text{ mol L}^{-1}}$$

The concentration of H_3O^+ and OH^- ions (as well as the concentrations of any products from the hydrolysis of SO_4^{2-} and La^{3+} ions) can be neglected.

9-12 (a) $Ag_2(SO_4)(s) \rightleftharpoons 2\,Ag^+(aq) + SO_4^{2-}(aq)$ $\quad K_{sp} = [Ag^+]^2[SO_4^{2-}]$
(b) $Br_2(l) \rightleftharpoons Br_2(aq)$ $\quad K = [Br_2]$
(c) $Cu(OH)_2(s) \rightleftharpoons Cu^{2+}(aq) + 2\,OH^-(aq)$ $\quad K_{sp} = [Cu^{2+}][OH^-]^2$

(d) $CaF_2(s) \rightleftharpoons Ca^{2+}(aq) + 2\,F^-(aq)$ $K_{sp} = [Ca^{2+}][F^-]^2$

9-14 Ionic substances mostly dissociate into ions as they dissolve. The equation and K_{sp} expression for such a reaction in the case of lead antimonate are

$$Pb_3(SbO_4)_2(s) \rightleftharpoons 3\,Pb^{2+}(aq) + 2\,SbO_4^{3-}(aq) \qquad K_{sp} = [Pb^{2+}]^3[SbO_4^{3-}]^2$$

9-16 The K_{sp} equilibrium is $BaSO_4(s) \rightleftharpoons Ba^{2+}(aq) + SO_4^-(aq)$. From the chemical equation, one mole of barium(II) ions and one mole of sulfate ions form in solution for every one mole of solid that dissolves. This assumes that neither of the two ions reacts significantly with water to form other species. Let x equal the concentration of $Ba^{2+}(aq)$ at equilibrium. The equilibrium concentration of $SO_4^{2-}(aq)$ is also x. Then:

$$K_{sp} = 1.1 \times 10^{-10} = [Ba^{2+}][SO_4^-] = x^2$$

Solving gives a concentration of $\boxed{1.0 \times 10^{-5} \text{ mol L}^{-1}}$ for both species.

9-18 The dissolution equilibrium and K_{sp} expression for TlSCN are:

$$TlSCN(s) \rightleftharpoons Tl^+(aq) + SCN^-(aq) \qquad K_{sp} = [Tl^+][SCN^-] = 1.82 \times 10^{-4}$$

If neither of the ions reacts significantly to form other species, then the concentrations of the two are equal at equilibrium:

$$[Tl^+] = [SCN^-] = \sqrt{K_{sp}} = \sqrt{1.82 \times 10^{-4}} = 1.35 \times 10^{-2} \text{ mol L}^{-1}$$

The molar mass of TlSCN is 262.46 g mol^{-1} so 3.54 g of it dissolves per liter, or $\boxed{0.354 \text{ g per 100 mL}}$.

9-20 (a) The K_{sp} equilibrium is $Hg_2Cl_2(s) \rightleftharpoons Hg_2^{2+}(aq) + 2\,Cl^-(aq)$. According to this chemical equation, one mole of mercury(I) (Hg_2^{2+}) ions forms in solution for every one mole of solid that dissolves. Let x equal the concentration of mercury(I) ion present at equilibrium. Then the equilibrium concentration of chloride ion is $2x$. The solubility-product expression is

$$K_{sp} = 2 \times 10^{-18} = [Hg_2^{2+}][Cl^-]^2 = x\,(2x)^2 = 4x^3$$

Solving for x gives a concentration of $\boxed{8 \times 10^{-7} \text{ mol L}^{-1}}$ for the mercury(I) ion and $\boxed{2 \times 10^{-6} \text{ mol L}^{-1}}$ for the chloride ion.

(b) A major assumption here is that the product ions do not interact with each other in solution. In fact however, the Hg_2^{2+} and Cl^- ions associate to a considerable extent in water. This is another way of saying that the dissolution equilibrium $Hg_2Cl_2(s) \rightleftharpoons Hg_2Cl_2(aq)$ is also important. It is so important that the ion concentrations computed on the basis of the K_{sp} bear little relationship to the solubility of this substance, which equals 0.025 mol L^{-1} at room temperature.

9-22 The molar mass of lead(II) iodate is 557.0 g mol^{-1}. Its solubility is

$$\text{solubility} = \frac{0.00896 \text{ g}}{0.400 \text{ L}} \times \frac{1 \text{ mol}}{557.0 \text{ g}} = 4.02 \times 10^{-5} \text{ mol L}^{-1}$$

Assume that neither the lead nor the iodate ion reacts further in the solution. Then the solubility of the salt equals the concentration of the Pb^{2+} ion and half the concentration of the IO_3^- ion at equilibrium. From the K_{sp} expression:

$$K_{sp} = [Pb^{2+}][IO_3^-]^2 = (4.02 \times 10^{-5})(2 \times 4.02 \times 10^{-5})^2 = \boxed{2.6 \times 10^{-13}}$$

9-24 The problem is similar to problem 9-22. The molar mass of the silver dichromate is 431.73 g mol^{-1}, so the solubility of the salt is 1.31×10^{-4} mol L^{-1}. The value of K_{sp} is obtained by substitution in the K_{sp} expression. It is $\boxed{9.0 \times 10^{-12}}$.

9-26 The 0.090 g of PbI_2 (molar mass 461.0 g mol^{-1}) equals 1.95×10^{-4} mol. The concentration of Pb^{2+} ion in the hot solution is therefore 1.95×10^{-4} mol L^{-1}, and the concentration of I^- ion is twice this, 3.90×10^{-4} mol L^{-1}. These molarities hardly change as the solution is cooled, but the K_{sp} of the dissolution-precipitation equilibrium does change. At 25°C, it becomes equal to 1.4×10^{-8} (see Table 9-1 on text page 380). Does the reaction quotient Q exceed K_{sp} at 25°C? It equals:

$$Q = [Pb^{2+}][I^-]^2 = (1.95 \times 10^{-4})(3.90 \times 10^{-4})^2 = 2.97 \times 10^{-11}$$

Clearly, the answer is $\boxed{\text{no}}$. No precipitate forms.

9-28 The finding that a precipitate forms at equilibrium in a system is not the same as observing a precipitate any time soon in an actual experiment: precipitation reactions can be slow. This fact accounts for the careful wording of the question. The calcium chloride and sodium fluoride solutions dilute each other as they are mixed. Immediately after mixing, but before reaction starts:

$$[Ca^{2+}] = \frac{2}{3}\left(1.0 \times 10^{-3} \text{ mol L}^{-1}\right) \quad \text{and} \quad [F^-] = \frac{1}{3}\left(6 \times 10^{-5} \text{ mol L}^{-1}\right)$$

Substituting these concentrations in the reaction quotient expression for the dissolution of CaF_2 gives a Q of about 2.7×10^{-13}. Because this is less than K_{sp}, $\boxed{\text{no precipitate}}$ of CaF_2 can form.

9-30 The strontium nitrate and sodium chromate solutions dilute each other upon mixing. The concentration of Sr^{2+} is 140/1000 of 0.0010 mol L^{-1} immediately after mixing, and the concentration of CrO_4^{2-} is 860/1000 of 0.0050 mol L^{-1}. Substituting these concentrations in the Q-expression for this dissolution gives a Q of about 6.0×10^{-7}. Because this is less than K_{sp}, $\boxed{\text{no precipitate}}$ forms.

9-32 The combined volume of the solutions is 8.10 L. After mixing but before any reaction occurs, the concentrations of Ag^+ and I^- ions are

$$[Ag^+] = \left(\frac{1.50}{8.10}\right) 0.080 = 0.0148 \text{ mol L}^{-1} ; \quad [I^-] = \left(\frac{6.60}{8.10}\right) 0.10 = 0.0815 \text{ mol L}^{-1}$$

The two ions react in a 1 : 1 ratio, so Ag^+ ion is the limiting reactant. At equilibrium, essentially all of the Ag^+ is removed from solution as $AgI(s)$. The remaining concentration of $I^-(aq)$ is $0.0815 - 0.0148 = 0.0667$ mol L^{-1}. Inserting this value into the K_{sp} expression gives

$$[Ag^+][I^-] = K_{sp} = 1.5 \times 10^{-16} = [Ag^+][0.0667]$$

Solving gives an equilibrium concentration of Ag^+ of $\boxed{2.25 \times 10^{-15} \text{ mol L}^{-1}}$. The fraction of the silver ions that remains in solution is $\boxed{1.52 \times 10^{-13}}$, which is less than one part in a trillion.

9-34 The reaction responsible for the precipitate is $Sr^{2+}(aq) + 2\,F^-(aq) \rightleftharpoons SrF_2(s)$. The chemical amount of F^- ion is 4.00 mmol, and the chemical amount of Sr^{2+} is 3.20 mmol. Imagine that all of the F^- ion reacts (it is the limiting reactant). If it does, then 1.20 mmol of Sr^{2+} remains, and the concentration of the remaining Sr^{2+} is 1.20 mmol/120 mL $= 0.010$ mol L^{-1}. Now, suppose that the $SrF_2(s)$ starts to redissolve. Some Sr^{2+} comes into solution, but the amount is negligible compared to the amount of unreacted excess Sr^{2+}. The dissolution of SrF_2 is however the *only* source of $F^-(aq)$. The dissolution proceeds until:

$$[Sr^{2+}][F^-]^2 = K_{sp} = 2.8 \times 10^{-9}$$

Inserting $[Sr^{2+}] = 0.010$ mol L^{-1} gives $\boxed{5.3 \times 10^{-4} \text{ mol L}^{-1}}$ as the equilibrium concentration of the fluoride ion.

9-36 The common-ion effect greatly reduces the solubility of the AgCl, which is only very slightly soluble in pure water in the first place. Let x represent this solubility. Then, $[\text{Cl}^-] = (0.150 + x)$ mol L^{-1} and $[\text{Ag}^+] = x$ mol L^{-1}. Substitute in the K_{sp} expression:

$$K_{sp} = [\text{Ag}^-][\text{Cl}^-] = x(0.150 + x) = 1.6 \times 10^{-10} \quad \text{and} \quad x = 1.1 \times 10^{-9} \text{ mol L}^{-1}$$

This means that 1.1×10^{-10} mol of AgCl dissolves in 100 mL of the 0.150 M NaCl. Taking the molar mass of AgCl as 143.3 g mol^{-1} gives a gram solubility of $\boxed{1.5 \times 10^{-8} \text{ g per 100 mL}}$.

9-38 (a) Let x equal the molar solubility of the silver arsenate. From the K_{sp} expression;

$$K_{sp} = 1.0 \times 10^{-22} = [\text{Ag}^+]^3[\text{AsO}_4^{3-}] = (3x)^3 x \quad \text{whence } x = \boxed{1.4 \times 10^{-6} \text{ mol L}^{-1}}$$

(b) Now, use $[\text{Ag}^+] = 0.100$ mol L^{-1} in the equilibrium expression. The molar solubility x equals the concentration of AsO_4^{3-}. From the K_{sp} expression:

$$1.0 \times 10^{-22} = [\text{Ag}^+]^3[\text{AsO}_4^{3-}] = (0.100)^3 x \quad \text{whence} \quad x = \boxed{1.0 \times 10^{-19} \text{ mol L}^{-1}}$$

9-40 (a) Assume that neither the Mg^{2+} ion nor the OH^- ion from the dissolution of the Mg(OH)_2 interacts further in the solution. Then, if x is the solubility of the Mg(OH)_2, the concentration of Mg^{2+} ion is x and the concentration of OH^- ion is $2x$ mol L^{-1}. At equilibrium,

$$K_{sp} = 1.2 \times 10^{-11} = x\,(2x)^2 \quad \text{from which} \quad x = \boxed{1.4 \times 10^{-4} \text{ mol L}^{-1}}$$

(b) If the pH is held at 9 by a buffer, then the $[\text{OH}^-]$ is being held at 10^{-5} mol L^{-1}. This concentration is *less* than the concentration of OH^- ion arising from the dissolution of the Mg(OH)_2 in pure water. Consequently, maintaining this pH should increase the solubility of the Mg(OH)_2. After the dissolution of Mg(OH)_2 comes to equilibrium, the concentration of OH^- ion remains at 10^{-5} mol L^{-1} because of the action of the buffer. Let x again represent the solubility of the salt. Then:

$$K_{sp} = [\text{Mg}^{2+}][\text{OH}^-]^2 = 1.2 \times 10^{-11} = x(10^{-5})^2$$

and the solubility is $\boxed{0.12 \text{ mol L}^{-1}}$.

9-42 If the solubility of the $AgOH(s)$ equals 0.001 mol L^{-1}, then the concentration of $Ag^+(aq)$ ion also equals 0.001 mol L^{-1}, assuming that no side-reactions remove $Ag^+(aq)$. Insertion of this value into the K_{sp} expression

$$K_{sp} = 1.5 \times 10^{-8} = [Ag^+][OH^-]$$

gives $[OH^-] = 1.5 \times 10^{-5}$. This corresponds to a pOH of 4.82 and a pH of $\boxed{9.18}$. It is assumed that the temperature is 25°C.

9-44 (a) If $[Cl^-]$ exceeds 3.2×10^{-10} mol L^{-1}, then AgCl precipitates. If it exceeds 4.9×10^{-2} mol L^{-1}, $PbCl_2(s)$ precipitates. These two concentrations are calculated by substitution into the two K_{sp} expressions. If $[Cl^-]$ is kept below $\boxed{4.9 \times 10^{-2} \text{ mol } L^{-1}}$, only the AgCl precipitates.

(b) If $[Cl^-]$ is held at 4.9×10^{-2} mol L^{-1} then $Ag^+(aq)$ in equilibrium with it and solid AgCl, but no solid $PbCl_2$ is present. The concentration of $Ag^+(aq)$ is only 3.3×10^{-9} mol L^{-1}. This is $\boxed{6.5 \times 10^{-9}}$ (a mere 6 or 7 billionths) of the 0.50 mol L^{-1} originally present.

9-46 The dissolution-precipitation equilibrium is best represented by the equation

$$CuS(s) + 2\,H_2O \rightleftharpoons Cu^{2+}(aq) + OH^-(aq) + HS^-(aq) \quad [Cu^{2+}][OH^-][HS^-] = K$$

The K for this reaction is tabulated in text Table 9-2 (text page 390) as 5×10^{-37}. The $[OH^-]$ is 1.0×10^{-11} mol L^{-1}, and the $[H_2S]$ is 0.1 mol L^{-1}, according to the problem. Substitution allows calculation of the equilibrium concentration of the Cu^{2+} ion. It is $\boxed{5 \times 10^{-25} \text{ mol } L^{-1}}$. This also equals the molar solubility of the $CuS(s)$ assuming that one mole of Cu^{2+} forms in solution for every mole of solid that dissolves.

9-48 The dissolution reaction is

$$MnS(s) + H_2O(l) \rightleftharpoons Mn^{2+}(aq) + HS^-(aq) + OH^-(aq)$$

for which

$$K = 3.0 \times 10^{-14} = [Mn^{2+}][HS^-][OH^-]$$

Lowering the concentration of H_3O^+ (raising the pH) increases the concentration of OH^- ion. By LeChatelier's principle, this lowers the equilibrium concentration of Mn^{2+} ion. Changing the pH also affects this equilibrium:

$$H_2S(aq) + H_2O(l) \rightleftharpoons HS^-(aq) + H_3O^+(aq) \quad \frac{[HS^-][H_3O^+]}{[H_2S]} = K_a = 9.1 \times 10^{-8}$$

Solve the K_a-expression (the second equilibrium expression above) for $[HS^-]$ and substitute the result in the first expression. Also, eliminate $[OH^-]$ in the first expression by using the K_w-expression, The $[Mn^{2+}]$ in this solution must equal 0.050 mol L^{-1}, and the $[H_2S]$ equals 0.10 mol L^{-1}; insert these two numerical values:

$$3.0 \times 10^{-14} = [Mn^{2+}]\left(\frac{9.1 \times 10^{-8}[H_2S]}{[H_3O^+]}\right)\left(\frac{1.0 \times 10^{-14}}{[H_3O^+]}\right)$$

$$3.0 \times 10^{-14} = 0.05\left(\frac{9.1 \times 10^{-8}(0.10)}{[H_3O^+]}\right)\left(\frac{1.0 \times 10^{-14}}{[H_3O^+]}\right)$$

Solving gives $[H_3O^+] = 1.2 \times 10^{-5}$ mol L^{-1} for a pH of $\boxed{4.9}$. At higher pH's K is exceeded, and solid MnS must eventually form from this solution.

A calculation using the K for $CdS(aq)$ and a $[H_3O^+]$ equal to 1.2×10^{-5} mol L^{-1} shows that at pH 4.9 the concentration of Cd^{2+} in equilibrium with $CdS(s)$ is only $\boxed{1.3 \times 10^{-15} \text{ mol } L^{-1}}$.

9-50 Write the equation for the breaking up of the complex ion, which is the reverse of the equation for its formation:

$$TlCl_4^-(aq) \rightleftharpoons Tl^{3+}(aq) + 4\,Cl^-(aq) \qquad \frac{[Tl^{3+}][Cl^-]^4}{[TlCl_4^-]} = K = \frac{1}{1 \times 10^{18}}$$

The small K means that the complex ion is heavily favored at equilibrium. Suppose that the Tl^{3+} ion from the $Tl(NO_3)_3$ reacts to completion with Cl^- ion from dissolved NaCl to form the complex. The Cl^- ion is in excess, so:

$$[TlCl_4^-] = 0.080 \text{ mol } L^{-1}$$

$$[Cl^-] = 0.50 - 4(0.080) = 0.18 \text{ mol } L^{-1}$$

Now, imagine that equilibrium is attained by the partial break-up of the complex. Because K is so small, the concentrations of the Cl^- and $TlCl_4^-$ ions change very little from the values just listed. Substitute them into the K-expression to obtain the equilibrium concentration of Tl^{3+}:

$$K = 1 \times 10^{-18} = \frac{[Tl^{3+}][Cl^-]^4}{[TlCl_4^-]} = \frac{[Tl^{3+}](0.18)^4}{0.080}$$

Solving gives $[Tl^{3+}] = \boxed{8 \times 10^{-17} \text{ mol } L^{-1}}$. The equilibrium concentration of $[TlCl_4^-]$ is $0.080 - 8 \times 10^{-17} = \boxed{0.080 \text{ mol } L^{-1}}$.

9-52 The solution of $FeCl_3$ will be $\boxed{\text{acidic}}$. After dissociation of the salt, the Cl^- ion is ineffective as a base, but the Fe^{3+} ion reacts to a significant extent with water to generate H_3O^+:

$$\boxed{Fe^{3+}(aq) + 2\,H_2O(l) \rightleftharpoons FeOH^{2+}(aq) + H_3O^+(aq)}$$

The reaction can also be represented:

$$Fe(H_2O)_6^{3+}(aq) + H_2O(l) \rightleftharpoons Fe(H_2O)_5OH^{2+}(aq) + H_3O^+(aq)$$

9-54 The reaction is

$$Fe(H_2O)_6^{2+}(aq) + H_2O(l) \rightleftharpoons Fe(H_2O)_5OH^+(aq) + H_3O^+(aq)$$

for which the K_a-expression is

$$K_a = 3 \times 10^{-6} = \frac{[H_3O^+][Fe(H_2O)_5OH^+]}{[Fe(H_2O)_6^{2+}]}$$

The equilibrium concentrations of $Fe(H_2O)_5OH^+$ and H_3O^+ are essentially equal. Represent them by x. Then

$$3 \times 10^{-6} = \frac{x \cdot x}{0.10 - x}$$

from which $x = [H_3O^+] = 5.5 \times 10^{-4}$ mol L^{-1}. The corresponding pH is $\boxed{3.3}$ which is less acidic that the pH 1.6 of 0.10 M $Fe(NO_3)_3$.

9-56 Treat the solution of $Ni(NO_3)_2$ as containing a weak acid that donates one hydrogen ion. The equation is

$$Ni(H_2O)_6^{2+}(aq) + H_2O(l) \rightleftharpoons Ni(H_2O)_5OH^+(aq) + H_3O^+(aq)$$

Assuming that the concentration of the $Ni(H_2O)_5OH^+$ ion equals that of the H_3O^+ ion, then both equal 1.0×10^{-5} mol L^{-1}. Substitute in the K_a expression:

$$K_a = \frac{(1 \times 10^{-5})^2}{(0.10 - 1 \times 10^{-5})} = \boxed{1 \times 10^{-9}}$$

9-58 The $BaSO_4(s)$ is more insoluble than the $CaSO_4(s)$. The two nitrates are both quite soluble, so the original concentrations of the Ba^{2+} and Ca^{2+} ions are 0.150 and 0.0800 mol L^{-1} respectively. Substitution of the Ca^{2+} concentration in the K_{sp} expression for $CaSO_4$ gives the concentration of SO_4^{2-} required before $CaSO_4$ can precipitate. It is $\boxed{3.0 \times 10^{-4} \text{ mol } L^{-1}}$. At this concentration of sulfate ion, the equilibrium concentration of Ba^{2+} is only 3.7×10^{-7} mol L^{-1}, according to the K_{sp} expression for $BaSO_4$. The separation is nearly total.

9-60 One way is dissolve the crystal, or a portion of it, in water and add $F^-(aq)$ ion (in the form of dissolved NaF). A precipitate of CaF_2 indicates that the crystal was $Ca(NO_3)_2$. A precipitate upon additional of CO_3^{2-} ion also indicates that the crystal contained Ca^{2+} ion. Another way is to dissolve the crystal and perform flame tests on the solution.

9-62 Add 6 M HCl to the solution. A precipitate means than either or both Hg_2^{2+} and Ag^+ ions are present. No precipitate means that only Cr^{3+} ion is present. If there is a precipitate, separate it, and add base (NaOH) to the supernatant liquid (filtrate). A precipitate at this point means that Cr^{+3} is present. Add concentrated aqueous ammonia to the original precipitate (the one formed by the addition of HCl). If it dissolves completely, Ag^+ ion is present and Hg_2^{2+} ion is absent. If it turns black, then Hg_2^{2+} ion is present. Separate the black precipitate and add HCl to neutralize the filtrate. A white precipitate means that Ag^+ ion is also definitely present.

9-64 The precipitate with hydrochloric acid means the presence of one or more ions from group 1. The lack of a precipitate with H_2S at pH 9 means that all ions from both groups 2 and 3 must be absent since H_2S treatment under basic conditions brings down the acid-insoluble sulfides as well as the base-insoluble sulfides. The precipitate with ammonium carbonate at pH 9.5 means the presence of one or more ions from group 4. No conclusions can be drawn about the presence or absence of ions from group 5.

9-66 The chemical equation is

$$\boxed{MgNH_4PO_4{\cdot}6H_2O(s) \rightleftharpoons Mg^{2+}(aq) + NH_4^+(aq) + PO_4^{3-}(aq) + 6\,H_2O(l)}$$

The six moles of water of crystallization are included on the right side of the equation to maintain a formal balance. They join the rest of the water in the solution, however, and are not included in the K_{sp} expression:

$$\boxed{[Mg^{2+}][NH_4^+][PO_4^{3-}] = K_{sp} = 2.3 \times 10^{-13}}$$

9-68 Assume that subsequent reactions of the product ions do not significantly affect the equilibrium concentrations of the ions formed in the two solubility equilibria. Then, the molar solubility of SrF_2 is $(K_{sp}/4)^{1/3}$ mol L^{-1}, and the molar solubility of $SrSO_4$ is $(K_{sp})^{1/2}$ mol L^{-1}. Substituting the K_{sp}'s, which are given, allows the molar solubilities to be calculated: for SrF_2 the result is $\boxed{8.9 \times 10^{-4} \text{ mol L}^{-1}}$; for $SrSO_4$ the result is $\boxed{5.3 \times 10^{-5} \text{ mol L}^{-1}}$. The gram solubilities are obtained by multiplying the molar solubilities by the proper molar masses. They are $\boxed{0.11 \text{ g L}^{-1}}$ and $\boxed{0.0097 \text{ g L}^{-1}}$ respectively. The point of the problem is that K_{sp} and the other measures of solubility are *not* directly proportional to each other.

9-70 The dissolution equilibrium is represented:

$$AgClO_4(s) \rightleftharpoons Ag^+(aq) + ClO_4^-(aq)$$

The large concentration of ClO_4^- present in 60 percent perchloric acid solution reduces the solubility of the $AgClO_4$ by shifting the above reaction to the left. This is an example of $\boxed{\text{the common-ion effect}}$.

9-72 Write the solubility-product expression

$$[Ca^{2+}][F^-]^2 = K_{sp} = 3.9 \times 10^{-11}$$

With $[Ca^{2+}] = 0.0020 \text{ mol L}^{-1}$, the largest $[F^-]$ that can be tolerated before CaF_2 tends to precipitate is $\boxed{14 \times 10^{-5} \text{ mol L}^{-1}}$. This exceeds the recommended $F^-(aq)$ concentration.

9-74 Solutions of Ag_2SO_4 and $BaCl_2$, both of which are dissociated into ions, are mixed. The concentrations of dissolved species after the mixing, but before any reactions occur are computed by taking the dissociation into ions and the 2-to-1 dilution of the solutions into account:

$$[Ba^{2+}] = 0.020 \quad [Ag^+] = 0.020 \quad [Cl^-] = 0.040 \quad [SO_4^{2-}] = 0.010 \text{ mol L}^{-1}$$

Two dissolution-precipitation equilibria take place simultaneously:

$$BaSO_4(s) \rightleftharpoons Ba^{2+}(aq) + SO_4^{2-}(aq) \quad \text{and} \quad AgCl(s) \rightleftharpoons Ag^+(aq) + Cl^-(aq)$$

Assume that the Ba^{2+} and SO_4^{2-} react to completion, The Ba^{2+} ion is in excess. Therefore, at completion the concentration of SO_4^{2-} ion is zero, and $\boxed{0.010 \text{ mol L}^{-1}}$ Ba^{2+} remains. A small portion of the precipitate now dissolves as the reaction comes to equilibrium. It is so small that the concentration of Ba^{2+} is hardly changed from 0.010 mol L^{-1}. Substituting this concentration into the K_{sp} expression for barium sulfate then gives $[SO_4^{2-}] = \boxed{1.1 \times 10^{-8} \text{ mol L}^{-1}}$, which is indeed very close to zero.

Likewise, Cl^- is in excess relative to Ag^+ in the precipitation of $AgCl(s)$. Assuming the reaction goes to completion, the Cl^- concentration is $\boxed{0.020 \text{ mol L}^{-1}}$. Again, coming to equilibrium hardly changes this. Substitution into the K_{sp} expression for AgCl then gives $[Ag^+] = \boxed{8.0 \times 10^{-9} \text{ mol L}^{-1}}$.

9-76 The equation $CaO(s) \rightleftharpoons Ca^{2+}(aq) + O^{2-}(aq)$ is exactly analogous to equations formerly written for the dissolution of metal sulfides such as $CuS(s)$ in water. Like such equations it is a poor representation of the chemical events taking place during dissolution since it fails to note the reaction of O^{2-} as a strong base. A better equation for the dissolution of $CaO(s)$ in water is

$$\boxed{CaO(s) + H_2O(l) \rightleftharpoons Ca^{2+}(aq) + 2\,OH^-(aq)} \qquad \boxed{[Ca^{2+}][OH^-]^2 = K_1}$$

The dissolution of $CaO(s)$ in acid is something yet again. It is better represented

$$\boxed{CaO(s) + 2\,H_3O^+(aq) \rightleftharpoons Ca^{2+}(aq) + 3\,H_2O(l)} \qquad [Ca^{2+}][H_3O^+]^2 = K_2$$

The equilibrium constant for the second reaction is related to K_1 (for the dissolution of $CaO(s)$ in plain water) by $\boxed{K_2 = K_1/K_w{}^2}$.

9-78 The equilibrium expression is

$$K = \frac{[\text{K·crypt}]}{[\text{crypt}][\text{K}^+]} = 3.1 \times 10^{11}$$

Assume that all of the K^+ reacts with the cryptand. If it does, then $[\text{K·crypt}] = 0.10$ mol L^{-1}. If a small concentration x of the K·crypt then dissociates as equilibrium is reached, the final concentration of K^+ is x, the final concentration of crypt is x, and the final concentration of Kcrypt is $0.10 - x$. Substitution in the equilibrium expression gives:

$$K = 3.1 \times 10^{11} = \frac{0.10 - x}{x \cdot x}$$

Solving for x gives $[\text{K}^+] = \boxed{5.7 \times 10^{-7} \text{ mol } L^{-1}}$.

9-80 According to text Table 9-3, the pH of a 0.1 mol L^{-1} solution of $Pb(NO_3)_2$ ion equals 3.6. The lowering of the pH from 7 is caused by hydrolysis of the Pb^{2+} ion. The NO_3^- ion is ineffective as a base and need not be considered. The hydrolysis reaction is:

$$Pb^{2+}(aq) + 2\,H_2O(l) \rightleftharpoons H_3O^+(aq) + Pb(OH)^+(aq) \qquad \frac{[H_3O^+][Pb(OH)^+]}{[Pb^{2+}]} = K_a$$

A pH of 3.6 corresponds to $[H_3O^+] = 2.51 \times 10^{-4}$ mol L^{-1}; this is the equilibrium concentration of the H_3O^+. The equilibrium concentration of $Pb(OH)^+(aq)$ is approximately equal to this same value, and the equilibrium concentration of

$Pb^{2+}(aq)$ is its original concentration minus this. Substitution into the K_a expression gives:

$$K_a = \frac{[H_3O^+][Pb(OH)^+]}{[Pb^{2+}]} = \frac{(2.51 \times 10^{-4})^2}{0.10 - 2.51 \times 10^{-4}} = \boxed{6.3 \times 10^{-7}}$$

Answers to Even Numbered Problems, Chapter 10

10-2 $\boxed{\text{Yes}}$, the heat from burning petroleum products can be traced back to the sun. The plants that furnished the organic matter that became oil grew originally by harvesting sunlight to drive the creation of high-energy compounds.

10-4 Flatirons had to be heavy to provide $\boxed{\text{enough heat capacity}}$ to maintain the required temperature for more than a few seconds. A larger m in the equation $q = mc_s\Delta T$ makes for a smaller ΔT for a given heat flow q.

10-6 The molar heat capacities are the specific heat capacities multiplied by the molar masses

$$c_P = \mathcal{M}c_s$$

as shown on text page 414. Analysis of the units confirms this relationship. The molar heat capacities (at constant pressure) for the gaseous halogens come out to be:

	c_s(J K^{-1}g^{-1})	$\mathcal{M}$(g mol^{-1})	c_P(J K^{-1}mol^{-1})
$F_2(g)$	0.824	38.0	31.3
$Cl_2(g)$	0.478	70.9	33.9
$Br_2(g)$	0.225	159.8	36.0
$I_2(g)$	0.145	253.8	36.8

The molar heat capacities increase steadily going down the table. The molar heat capacity of gaseous astatine should fall $\boxed{\text{between 37 and 38 J K}^{-1}\text{mol}^{-1}}$.

10-8 The rule of Dulong and Petit states that the molar heat capacities (c_P) of the (solid) metallic elements equal approximately 25 J K^{-1}mol^{-1}. Since $c_P = \mathcal{M}c_s$, divide the c_P of a metallic element by its molar mass to estimate its specific heat capacity. The estimates are $\boxed{0.491,\ 0.359,\text{ and } 0.232 \text{ J K}^{-1}\text{g}^{-1}}$ for V, Ga, and Ag, respectively:

10-10 Substitute the molar mass of uranium into the relationship between molar and specific heat capacity derived on text page 414:

$$c_s = \frac{c_P}{\mathcal{M}} = \frac{27.49 \text{ J K}^{-1}\text{mol}^{-1}}{238.03 \text{ g mol}^{-1}} = \boxed{0.1155 \text{ J K}^{-1}\text{g}^{-1}}$$

10-12 In the special case of heat exchange between equal masses of two substances at different temperatures:

$$\frac{c_{s1}}{c_{s2}} = -\frac{\Delta T_2}{\Delta T_1}$$

In this problem, this expression becomes

$$0.10 = -\frac{(T_f - 20.0)}{(T_f - 92.0)}$$

Note the adherence to the convention that a change in a quantity is the final value minus the initial. Solving gives $T_f = \boxed{26.5°\text{C}}$.

10-14 (a) The calorimeter constant is the heat absorbed by the calorimeter divided by the change in the temperature of the calorimeter:

$$1770\frac{\text{J}}{1.67 \text{ K}} = \boxed{1.06 \times 10^3 \text{ J K}^{-1}}$$

(b) The heat evolved in the dissolution of the NaOH equals the heat absorbed by the calorimeter:

$$q = 1.06 \times 10^3 \text{ J K}^{-1} \times 1.04°\text{C} \times \left(\frac{1 \text{ K}}{1°\text{C}}\right) = \boxed{1.10 \times 10^3 \text{ J}}$$

10-16 The system consisting of the pouch and its contents absorbs negative heat during the change because its surroundings become warmer. As stated on text page 419, $\Delta H = q_P$, the pressure is essentially constant in this change; ΔH is $\boxed{\text{negative}}$.

10-18 (a)

$$\frac{-683 \text{ kJ}}{1 \text{ mol Br}_2} \times \frac{1 \text{ mol Br}_2}{2(79.904) \text{ g Br}_2} = \boxed{-4.27 \text{ kJ g}^{-1}}$$

(b)

$$\frac{472 \text{ kJ}}{4 \text{ mol Fe}_3\text{O}_4} \times \frac{1 \text{ mol Fe}_3\text{O}_4}{231.54 \text{ g Fe}_3\text{O}_4} = \boxed{0.510 \text{ kJ g}^{-1}}$$

(c)

$$\frac{806 \text{ kJ}}{2 \text{ mol NaHSO}_4} \times \frac{1 \text{ mol NaHSO}_4}{120.07 \text{ g NaHSO}_4} = \boxed{3.36 \text{ kJ g}^{-1}}$$

10-20 The reaction absorbs negative heat at constant pressure, so its ΔH is -670 J.

$$\left(\frac{-670 \text{ J}}{1.00 \text{ g CuCl}_2}\right) \times \left(\frac{134.452 \text{ g CuCl}_2}{1 \text{ mol CuCl}_2}\right) = \boxed{-90.1 \times 10^3 \text{ J mol}^{-1}}$$

10-22 (a) Sublimation requires an input of heat; this change has a $\boxed{\text{positive}}$ ΔH.

(b) Vaporization has a $\boxed{\text{positive}}$ ΔH.

(c) The melting of any solid has a $\boxed{\text{positive}}$ ΔH.

10-24 (a) The condensation of the methane generates heat in the surroundings. Hence the q of the system consisting of the methane is negative:

$$\Delta H_{\text{cond}} = \frac{-555 \text{ J}}{1.00 \text{ g CH}_4} \times \frac{16.04 \text{ g CH}_4}{1 \text{ mol CH}_4} \times 2.00 \text{ mol CH}_4 = \boxed{-17800 \text{ J mol}^{-1}}$$

(b) The ΔH for the vaporization is just the negative of the ΔH for the condensation because the processes are each other's reverses: $\boxed{+17800 \text{ J mol}^{-1}}$.

10-26 In this process the liquid loses heat as it freezes, making ΔH negative:

$$\left(\frac{-28.8 \text{ kJ}}{1 \text{ mol NaCl}}\right) \times \left(\frac{1 \text{ mol NaCl}}{58.44 \text{ g NaCl}}\right) \times \left(56.2 \times 10^3 \text{ g NaCl}\right) = \boxed{-2.77 \times 10^4 \text{ kJ}}$$

10-28 The vaporization of 4 mol of $P(l)$ to form $P_4(g)$ can be imagined to proceed by the combustion of liquid phosphorus (the second reaction in the problem) followed by the "uncombustion" of the solid P_4O_{10} (the reverse of the first reaction in the problem). The ΔH_{vap} equals the sum of the ΔH's of the steps taken on this route:

$$\Delta H = -2977.0 - (-3035.7) = \boxed{+58.7 \text{ kJ}}$$

10-30 The reaction of interest can be constructed as three times the first reaction added to the second reaction. The way to establish this is to focus on the compounds that must appear on the left in the reaction of interest and trace back to learn the manipulations required to *put* them on the left from wherever they are in the reactions with known ΔH. The enthalpy changes combine in the same way as the equations:

$$\Delta H = \Delta H_2 + 3\Delta H_1 = 461.05 + 3(520.9) = \boxed{+2023.8 \text{ kJ}}$$

10-32 The conversion of monoclinic to rhombic sulfur can be imagined to proceed by the combustion of monoclinic sulfur to $SO_2(g)$ and then the "un-combustion" of $SO_2(g)$ to rhombic sulfur. The ΔH is the sum of the ΔH's for these steps:

$$\Delta H = -9.376 - (-9.293) = \boxed{-0.083 \text{ kJ g}^{-1}}$$

10-34   In every case the applicable equation is:

$$\Delta H^\circ = \sum_{\text{products}} \Delta H_f^\circ - \sum_{\text{reactants}} \Delta H_f^\circ$$

Enthalpies of formation are tabulated on a "per amount" basis. The ΔH_f°'s in Appendix D are given per mole, so each must be multiplied by the number of moles appearing in the balanced equation.

(a)

$$2 \text{ mol}\left(33.18\frac{\text{kJ}}{\text{mol}}\right) - 1 \text{ mol}\left(0\frac{\text{kJ}}{\text{mol}}\right) - 2 \text{ mol}\left(90.25\frac{\text{kJ}}{\text{mol}}\right) = \boxed{-114.14 \text{ kJ}}$$

(b) $\Delta H^\circ = 2(-110.52) - 1(-393.51) = \boxed{+172.47 \text{ kJ}}$

(c) $\Delta H^\circ = 3(-241.82) + 2(33.18) - 7/2(0) - 2(-46.11) = \boxed{-566.88 \text{ kJ}}$

(d) $\Delta H^\circ = 1(0) + 1(-110.52) - 1(-241.82) - 1(0) = \boxed{131.30 \text{ kJ}}$

10-36 (a) The reaction is $2\,Al(s) + Fe_2O_3(s) \longrightarrow 2\,Fe(s) + Al_2O_3(s)$. The standard enthalpy change is computed by combining ΔH_f°'s from Appendix D:

$$\Delta H^\circ = 2\underbrace{(0)}_{Fe(s)} + 1\underbrace{(-1675.7)}_{Al_2O_3(s)} - 2\underbrace{(0)}_{Al(s)} - 1\underbrace{(-824.2)}_{Fe_2O_3(s)} = \boxed{-851.5 \text{ kJ}}$$

(b) According to the result in part (a), 851.5 kJ of heat is given off at constant pressure when one mole of $Fe_2O_3(s)$ reacts. One mole of Fe_2O_3 is 159.69 g, so the amount of heat given off by the reaction of 3.21 g of $Fe_2O_3(s)$ is $\boxed{17.1 \text{ kJ}}$.

10-38 (a) The reaction is a dissolution:

$$NH_4NO_3(s) \longrightarrow NH_4^+(aq) + NO_3^-(aq)$$

Calculate the $\Delta H°$ using the values in Appendix D for the $\Delta H_f°$'s of the two ions and the solid salt:

$$\Delta H° = 1\underbrace{(-132.51)}_{NH_4^+(aq)} + 1\underbrace{(-205.0)}_{NO_3^-(aq)} - 1\underbrace{(-365.66)}_{NH_4NO_3(s)} = \boxed{+28.05 \text{ kJ}}$$

(b) The 15.0 g of NH_4NO_3 is 0.187 mol of NH_4NO_3. Dissolving this much ammonium nitrate to give the ions in their standard states (a concentration of 1 mol L^{-1}) would have a $\Delta H°$ of 5.26 kJ. In the problem, the final concentrations of the two ions are 2.5, not 1 mol L^{-1}, so, in principle, a somewhat different ΔH is expected. In the absence of ΔH data for the dilution from 2.5 to 1 M, simply conclude that the reaction absorbs *about* 5.3 kJ from its immediate surroundings, which are 100 g of solution. The heat capacity of these surroundings is approximately 418 J K^{-1}, so their temperature change as they give up the 5.3 kJ to the system is -12.6 K. The final temperature of the solution is then $\boxed{7.4°C}$.

(c) The dissolution of ammonium nitrate in water could be used in $\boxed{\text{cold packs}}$ for first aid.

10-40 The combustion reaction is represented

$$C_6H_{10}(l) + \frac{17}{2}O_2(g) \longrightarrow 6\,CO_2(g) + 5\,H_2O(l)$$

for which the standard enthalpy change is -3731.7 kJ. This $\Delta H°$ equals the standard enthalpy of formation of the products minus the standard enthalpy of formation of the reactants. According to the equation, 6 mol of CO_2 forms, so the $\Delta H_f°$ of the $CO_2(g)$ is 6 times its molar enthalpy of formation; similar statements apply to the 5 mol of $H_2O(l)$, the 17/2 mol of gaseous oxygen, and the 1 mol of liquid cyclohexene. The molar $\Delta H_f°$ of liquid cyclohexene is not known, but the other molar $\Delta H_f°$'s all appear in Appendix D. Combining those values, which are in kJ mol^{-1}, gives:

$$-3731.7 \text{ kJ} = 6\underbrace{(-393.51)}_{CO_2(g)} + 5\underbrace{(-285.83)}_{H_2O(l)} - 1\underbrace{(\Delta H_f°)}_{C_6H_{10}(l)} - 17/2\underbrace{(0.00)}_{O_2(g)}$$

Solving for the unknown gives the molar enthalpy of formation of liquid cyclohexene as $\boxed{-58.5 \text{ kJ mol}^{-1}}$.

10-42 The sublimation of 0.5610 mol of uranium hexafluoride can be represented

$$0.5610\ UF_6(s) \longrightarrow 0.5610\ UF_6(g)$$

The standard enthalpy change associated with the sublimation equals 0.5610 mol times the molar ΔH_f° of the gas minus 0.5610 mol times the molar ΔH_f° of the solid:

$$\Delta H^\circ = (0.5610\ \text{mol})\big(-2063.7 - (-2197.0)\big)\ \text{kJ mol}^{-1} = \boxed{+74.8\ \text{kJ}}$$

10-44 The atomization reaction is represented:

$$CF_3CHCl_2(g) \longrightarrow 2\,C(g) + 3\,F(g) + H(g) + 2\,Cl(g)$$

The reaction breaks 1 mol of C—C bonds, 3 mol of C—F bonds, 1 mol of C—H bonds, and 2 mol of C—Cl bonds. Breaking the bonds costs enthalpy as follows:

$$\Delta H_f^\circ = 1\underbrace{(348\ \text{kJ})}_{C-C} + 3\underbrace{(441\ \text{kJ})}_{C-F} + 1\underbrace{(436\ \text{kJ})}_{C-H} + 2\underbrace{(328\ \text{kJ})}_{C-Cl} = \boxed{2763\ \text{kJ}}$$

10-46 (a)

$$\Delta H^\circ \approx 3\underbrace{(436\ \text{kJ})}_{H-H} + 1\underbrace{(945\ \text{kJ})}_{N\equiv N} - 6\underbrace{(391\ \text{kJ})}_{N-H} = \boxed{-93\ \text{kJ}}$$

(b)

$$\Delta H^\circ \approx 2\underbrace{(436\ \text{kJ})}_{H-H} + 1\underbrace{(945\ \text{kJ})}_{N\equiv N} - 4\underbrace{(391\ \text{kJ})}_{N-H} - 1\underbrace{(161\ \text{kJ})}_{N-N} = \boxed{+92\ \text{kJ}}$$

(c)

$$\Delta H^\circ \approx 1\underbrace{(436\ \text{kJ})}_{H-H} + 1\underbrace{(945\ \text{kJ})}_{N\equiv N} - 2\underbrace{(391\ \text{kJ})}_{N-H} - 1\underbrace{(418\ \text{kJ})}_{N=N} = \boxed{+181\ \text{kJ}}$$

10-48 (a) The structures are

(b) The reaction to form one mole of chloroethane and one mole of hydrogen gas from one mole of methane and one mole of chloromethane can be imagined to proceed by the breaking of the seven C—H bonds and the one C—Cl bond in the methane and chloromethane followed by recombination of the atoms into the new species. The recombination requires formation of five C—H bonds, one C—Cl bond, one C—C bond, and one H—H bond. Combine the $\Delta H°$'s for all these changes to obtain $\Delta H°$ of the reaction: Note the use of positive signs for bond breaking and negative signs for bond making:

$$\Delta H° = 7 \underbrace{\Delta H°}_{C-H} -5 \underbrace{\Delta H°}_{C-H} + \underbrace{\Delta H°}_{C-Cl} - \underbrace{\Delta H°}_{C-Cl} - \underbrace{\Delta H°}_{C-C} - \underbrace{\Delta H°}_{H-H}$$

$$= 2 \underbrace{\Delta H°}_{C-H} -1 \underbrace{\Delta H°}_{C-C} -1 \underbrace{\Delta H°}_{H-H}$$

$$= 2 \underbrace{(413 \text{ kJ})}_{C-H} - \underbrace{(348 \text{ kJ})}_{C-C} -1 \underbrace{(436 \text{ kJ})}_{H-H} = \boxed{42 \text{ kJ}}$$

The reaction forming one mole of ethane from ethylene and hydrogen can be imagined to proceed by breaking all the bonds in the reactants and constructing one mole of ethane from the resulting gaseous atoms.

$$\Delta H° = 4 \underbrace{(413 \text{ kJ})}_{C-H} + \underbrace{(615 \text{ kJ})}_{C=C} + \underbrace{(436 \text{ kJ})}_{H-H} -6 \underbrace{(413 \text{ kJ})}_{C-H} - \underbrace{(348 \text{ kJ})}_{C-C} = \boxed{-123 \text{ kJ}}$$

10-50 (a) The left side of the equation shows 6 mol of Hg—Cl bonds and 8 mol of Al—Cl bonds; the right side shows the same number and kind of bonds (note that 2 mol of compound forms). The bond enthalpy of a single bond between a specific pair of elements is nearly constant from compound to compound, so $\Delta H°$ is expected to be close to zero.

(b) The proposed structure contains 4 Hg—Cl and 1 Hg—Hg bond. The $\Delta H°$ for the reaction given in the problem would differ substantially from zero if the proposed structure were correct because the number and kind of bonds would both change a lot. The fact that actual $\Delta H°$ is close to zero therefore argues against this structure.

10-52 The work done in a change of volume at constant pressure is $w = -P_{ext}\Delta V$. The change in the volume is $V_2 - V_1 = 800 - 150 = 650$ mL $= 0.650$ L, and P is 0.98 atm so $w = -0.64$ L atm. As explained on text page 436, 1 L atm is 101.325 J. The work is therefore $\boxed{-65 \text{ J}}$.

10-54 The work done on the system is

$$w = -P_{\text{ext}}\Delta V = -0.86 \text{ atm} (3.75 - 1.25 \text{ L}) = -2.15 \text{ L atm}$$

Convert to joules by multiplying by 101.325 J per L atm to obtain $\boxed{218 \text{ J}}$.

10-56 The volume of the system does not change. Since $\Delta V = 0$, $\boxed{w = 0}$, regardless of the pressure increase.

10-58 (a) The battery (not the toy) is the system of interest. Apply the first law to it. The battery performs work on the surroundings so it absorbs negative work: $w = -117.0$ J. It also absorbs negative heat (evolves heat to the surroundings) so $q = -3.0$ J. By the first law, $\Delta E = -120.0$ J.

(b) Recharging the battery restores it to the way it was before it was discharged. The final state is indistinguishable from the original state, so $\Delta E_{\text{overall}} = 0$. But:

$$\Delta E_{\text{overall}} = 0 = \Delta E_{\text{discharge}} + \Delta E_{\text{charge}} = (q + w)_{\text{discharge}} + (q + w)_{\text{charge}}$$

Substituting the known quantities (in units of joules) gives

$$0 = (-3.0 + -117.0)_{\text{discharge}} + (q + 210.0)_{\text{charge}} \quad \text{from which} \quad q_{\text{charge}} = \boxed{-90.0 \text{ J}}$$

The sign of q during the charge cycle is negative because the battery evolves heat to the surroundings as it is charged.

10-60 By definition, $\Delta H = \Delta E + \Delta(PV)$. Then:

$$\Delta H^\circ = \Delta E^\circ + P\Delta V$$

because the naughts imply a constant pressure. The desired quantity is $P\Delta V = \Delta H^\circ - \Delta E^\circ$. To evaluate the term on the left, assume that the gases from the explosion of the TNT are ideal gases and that the volumes of the solid TNT and the solid carbon (a product) are small. Then

$$P\Delta V \approx \Delta n_{\text{g}} RT = (15 - 0) \text{ mol}(8.315 \text{ J K}^{-1}\text{mol}^{-1})(298.15 \text{ K}) = \boxed{37.2 \times 10^3 \text{ J}}$$

10-62 (a) The combustion of benzoic acid is represented

$$\boxed{C_6H_5COOH(s) + 15/2\, O_2(g) \longrightarrow 7\, CO_2(g) + 3\, H_2O(l)}$$

(b) The ΔE° of a reaction is the amount of heat absorbed by the reaction at constant volume. In general, the pressure changes when a reaction occurs at constant volume. In this case, the pressure inside the constant-volume calorimeter certainly increases. It is known (text page 438) that $\Delta E = q_V$. The heat absorbed by the calorimeter is

$$q_V = C\Delta T = (9382 \text{ J K}^{-1}) \times (2.15 \text{ K}) = 2.017 \times 10^4 \text{ J}$$

The heat absorbed by the reaction within the calorimeter is the negative of this. Therefore, for the reaction of benzoic acid:

$$\left(\frac{-2.017 \times 10^4 \text{ J}}{0.800 \text{ g}}\right) \times \left(\frac{122.12 \text{ g}}{1 \text{ mol}}\right) = -3.079 \times 10^6 \text{ J mol}^{-1}$$

$$\Delta E^\circ = -3079 \text{ kJ mol}^{-1} \times 1.000 \text{ mol} = \boxed{-3079 \text{ kJ}}$$

(c) Write $\Delta H^\circ = \Delta E^\circ + \Delta n_g RT$ where Δn_g is the change in the number of moles of gas on the two sides of the equation. It equals $-1/2$ mol in the combustion of one mole of solid benzoic acid. Substituting gives

$$\begin{aligned}
\Delta H^\circ &= \Delta E^\circ = \Delta n_g RT \\
&= -3079 \times 10^3 \text{ J} + (-\frac{1}{2}\text{ mol})(8.315 \text{ J K}^{-1}\text{ mol}^{-1})(298.15 \text{ K}) \\
&= \boxed{-3080 \text{ kJ}}
\end{aligned}$$

(d) The necessary relationship is

$$\Delta H^\circ = \sum_{\text{products}} \Delta H_f^\circ - \sum_{\text{reactants}} \Delta H_f^\circ$$

where the quantity on the left is -3080 kJ. and all but one of the terms on the right are known:

$$-3080 \text{ kJ} = 7\underbrace{(-393.51)}_{CO_{2(g)}} + 3\underbrace{(-285.83)}_{H_2O_{(l)}} - \underbrace{(\Delta H_f^\circ)}_{C_6H_5COOH_{(s)}}$$

$$\Delta H_f^\circ(C_6H_5COOH(s)) = \boxed{-532 \text{ kJ mol}^{-1}}$$

10-64 If the molar mass of indium were 76 g mol^{-1} then the molar heat capacity of indium would be

$$c_P = \mathcal{M}c_s = 76 \text{ g mol}^{-1}(0.233 \text{ J K}^{-1}\text{g}^{-1}) = 17.7 \text{ J K}^{-1}\text{mol}^{-1}$$

This is about one-third less than the rule of Dulong and Petit requires. The actual molar mass of In is 114.8 g mol^{-1}, which gives a molar heat capacity much more nearly in accord with the rule of Dulong and Petit.

10-66 The difference between the melting and boiling points of water is 100 K. The heat required to raise the temperature of 1.00 g of water across this range is

$$q = mc_s\Delta T = (1.00\text{ g})(4.18\text{ J K}^{-1}\text{g}^{-1})(100\text{ K}) = 418\text{ J}$$

This assumes that the specific heat capacity of water remains constant over the range (in fact it changes somewhat). According to the statement of Lavoisier and Laplace, the heat required to melt 1.00 g of ice is 3/4 of this or $\boxed{314\text{ J}}$.

10-68 The energy released by the combustion of gasoline in fueling the car over a kilometer is

$$1\text{ km} \times \left(\frac{1\text{ L gasoline}}{8.0\text{ km}}\right) \times \left(\frac{0.68\text{ kg}}{1\text{ L}}\right) \times \left(\frac{48 \times 10^3\text{ kJ}}{1\text{ kg}}\right) = 4080\text{ kJ}$$

The energy released in the oxidation of food is about (100 / 0.3) kJ or 330 kJ. Driving releases $\boxed{\text{about 12 times more energy}}$.

10-70 Let the system consist of the ice and the water and assume that no heat leaks in or out through the walls of the glass. Then

$$q_{\text{ice}} + q_{\text{water}} = 0$$

The ice at 0.0°C melts to form water at 0.0°C. The water from the melted ice has $\Delta T = 0$ (absorbs no heat) and can be omitted in the heat balance. The heat absorbed by the melting of the ice is the specific enthalpy of fusion multiplied by the mass of the ice. The heat absorbed by the water is the specific heat capacity of the water times its temperature change times its mass:

$$\Delta H_{\text{fus}} m_{\text{ice}} + (c_s m \Delta T)_{\text{water}} = 0$$

$$(333\text{ J g}^{-1})m_{\text{ice}} + (4.18\text{ J K}^{-1}\text{g}^{-1})(150\text{ g})(-25\text{ K}) = 0$$

Solving gives $m_{\text{ice}} = \boxed{47\text{ g}}$.

10-72 The standard enthalpy of water vapor at 400°C, $\boxed{H^\circ_{400}}$ is greater. The q for converting "cool" water vapor at 300 to "hot" water vapor at 400°C is certainly positive at a constant pressure of 1 atm, and q equals ΔH between the two states.

10-74 The problem discloses essential ΔH° data without referring to them by name. The ΔH° of the reaction is the difference between the ΔH°_f's of the products and the reactants. It is

$$2 \underbrace{(96.23)}_{\text{HgBr}(g)} - 1 \underbrace{(-206.77)}_{\text{Hg}_2\text{Br}_2(s)} = \boxed{+399.2\text{ kJ}}$$

10-76 Specifying the physical states of the reactants and product is essential:

$$N_2(g) + 4\,H_2(g) + Pt(s) + 3\,F_2(g) \longrightarrow (NH_4)_2PtF_6(s)$$

10-78 (a) The point of drawing all the Lewis structures is to identify the type and number of bonds in the compounds in parts (b) through (d). The O_2 has an $O{=}O$ double bond; CO_2 has two $C{=}O$ double bonds; $H{-}O{-}H$ has two $O{-}H$ single bonds; CH_4 has four $C{-}H$ single bonds, C_2H_5OH has five $C{-}H$ single bonds, one $C{-}C$ single bond, one $C{-}O$ single bond, and one $O{-}H$ single bond; C_8H_{18} has 18 $C{-}H$ single bonds and seven $C{-}C$ single bonds.

(b) For the burning of one mole of gaseous methane

$$\Delta H^\circ = -\underbrace{4(463)}_{O-H} - \underbrace{2(728)}_{C=O} + \underbrace{4(413)}_{C-H} + \underbrace{2(498)}_{O=O} \approx \boxed{-660 \text{ kJ}}$$

(c) For the burning of one mole of gaseous octane

$$\Delta H^\circ = -\underbrace{18(463)}_{O-H} - \underbrace{16(728)}_{C=O} + \underbrace{18(413)}_{C-H} + \underbrace{7(348)}_{C-C} + \underbrace{25/2(498)}_{O=O} \approx \boxed{-3887 \text{ kJ}}$$

(d) For the burning of one mole of gaseous ethanol

$$\Delta H^\circ = -\underbrace{6(463)}_{O-H} - \underbrace{4(728)}_{C=O} + \underbrace{5(413)}_{C-H} + \underbrace{1(348)}_{C-C} + \underbrace{1(351)}_{C-O} + \underbrace{1(463)}_{O-H} + \underbrace{3(498)}_{O=O}$$

$$\approx \boxed{-969 \text{ kJ}}$$

10-80 Using the assumption that no energy is lost to the surroundings of the water

$$-mg\Delta h = c_s m \Delta T$$

The m's cancel out. Before inserting the given values into this equation, convert c_s to units of $J\ K^{-1}kg^{-1}$ to assure that all other units also cancel out correctly. Then:

$$-(9.81 \text{ m s}^{-2})(-100 \text{ m}) = (4180 \text{ J K}^{-1}\text{kg}^{-1})\Delta T$$

From this expression $\Delta T = \boxed{+0.235 \text{ K}}$. Note that $1\text{ J} = 1\text{ kg m}^2\text{s}^{-2}$.

10-82 The problem requires the careful application of the definitions, the use of the ideal gas law, and the use of the first law of thermodynamics.

(a) Substituting in $PV = nRT$ gives $P = \boxed{1.000 \text{ atm}}$.

(b) The q is given explicitly in the problem. It is $\boxed{+3.000 \text{ kJ}}$. That is, the system absorbs 3.000 kJ of heat.

(c) The strong rigid container neither expands nor contracts during the heating. There is therefore no pressure-volume work. Because there is no other mechanical linkage to the surroundings, $\boxed{w = 0}$.

(d) The temperature change of the gas is given by

$$\Delta T = \frac{q}{nc_P} = \frac{3000 \text{ J}}{(1.000 \text{ mol})(12.47 \text{ J K}^{-1}\text{mol}^{-1})} = 240.6 \text{ K}$$

so the final T is $\boxed{513.7 \text{ K}}$.

(e) Use the ideal gas equation: $P = \boxed{1.881 \text{ atm}}$.

(f) The first law states that $\Delta E = q + w$. Because both q and w are known it is easy to obtain ΔE: $\Delta E = 3000 + 0 = \boxed{3000 \text{ J}}$.

(g) The enthalpy change $\Delta H = \Delta E + \Delta(PV)$. The pressure and volume before the heating and also after the heating are known. Therefore it is possible to compute $\Delta(PV)$ by direct substitution. It is 19.73 L atm, which is 2000 J. Hence, $\Delta H = 3000 + 2000 = 5000$ J or $\boxed{5.000 \text{ kJ}}$.

(h) The law of conservation of energy applies to energy, not enthalpy. The change in enthalpy is greater than the heat input because the gas is under higher pressure after the heating. Recall that $q_P = \Delta H$ requires constant pressure.

Answers to Even-Numbered Problems, Chapter 11

11-2 (a) This problem combats the misconception that exothermic processes must always be spontaneous. The freezing of ice above 0°C is exothermic but non-spontaneous. The reaction:

$$CaMgSi_2O_6(s) + 2\,CO_2(g) \longrightarrow CaMg(CO_3)_2(s) + 2\,SiO_2(s)$$

which is the reverse of the reaction in Example 11-7 (text page 469) is exothermic but nonspontaneous above 195°.

(b) At room temperature, the melting of all elemental metals except mercury is endothermic but not spontaneous. The reaction

$$CO_2(g) \longrightarrow C(s) + O_2(g)$$

(like other "reverse combustion" processes) is also endothermic and not spontaneous.

11-4 As the temperature is increased the principle of Berthelot and Thomsen becomes ⎡less nearly correct⎤ as the $T\Delta S$ term grows to dominance.

11-6 (a) Label the three equal volumes L (left), C (center), and R (right). Each of the four molecules could be in any one of the three and thus has three equally probable ways of going into the overall volume. The number of microstates is therefore 3^4 or ⎡81⎤.

(b) There are 15 distinguishable states of the system. The 15 are given in the following table in which the first three rows of each column specify the number of particles in each of the three volumes L, C, and R, and the fourth row gives the number of *ways* that such a state can be constructed:

Left (L)	4	0	0	2	2	0	2	1	1	3	3	1	1	0	0
Center (C)	0	4	0	2	0	2	1	2	1	1	0	3	0	1	3
Right (R)	0	0	4	0	2	2	1	1	2	0	1	0	3	3	1
No. Microstates	1	1	1	6	6	6	12	12	12	4	4	4	4	4	4

Notice that the sum of the numbers in the bottom row is 81. As the table shows, there is only one way in which all four molecules can be in volume L, but 81 ways in which the four can be distributed. The requested probability is ⎡1/81⎤. The probabilities of each of the other 14 possible states are computed similarly. For example, the 12 ways in which two particles can be in L, one in C, and one in R can be explicitly listed. If the particles are designated a, b, c, and d, the 12 ways are:

	1	2	3	5	6	7	8	9	10	11	12	
Left	ab	ab	ac	ac	ad	ad	bc	bc	bd	bd	cd	cd
Center	c	d	b	d	b	c	a	d	a	c	a	b
Right	d	c	d	b	c	b	d	a	c	a	b	a

These 12 distributions are indistinguishable because the particles are indistinguishable from each other. The probability of this distribution is therefore 12/81.

11-8 (a) Computer constructed: $\Delta S < 0$.   (b) Gas leaks out: $\Delta S > 0$.   (c) Solid sublimes: $\Delta S > 0$.

11-10 (a) Cooling a system without encountering a chemical or phase transformation always lowers its entropy; $\boxed{\Delta S < 0}$.

(b) A solid is converted chemically into several moles of gas; $\boxed{\Delta S > 0}$.

(c) The chemical amount of gas in a system increases; $\boxed{\Delta S > 0}$.

(d) A gas condenses to a liquid; $\boxed{\Delta S < 0}$.

11-12 (a) The $\Delta S°$ of the gas-phase reaction between acetic acid and ammonia to give methylamine, carbon dioxide and hydrogen equals the standard molar entropies (the $S°$'s) of the products each multiplied by its chemical amount in the chemical equation minus the $S°$'s of the reactants each multiplied by its chemical amount in the equation:

$$\Delta S° = \underset{H_2(g)}{1\,(130.57)} + \underset{CO_2(g)}{1\,(213.63)} + \underset{CH_3NH_2(g)}{1\,(243.30)}$$
$$- \underset{CH_3COOH(g)}{1\,(282.4)} - \underset{NH_3(g)}{1\,(192.34)} = \boxed{+112.8 \text{ J K}^{-1}}$$

where the numbers in parenthesis are standard molar entropies (with units of J K^{-1}mol^{-1}) and the other numbers are the numbers of moles from the balanced equation.

(b) The reactants are different substances altogether. They however comprise a mole of a solid and a mole of a liquid, and should have less entropy than the two moles of gaseous reactants in the equation in part (a). The products are identically the same and therefore have the same entropy. Hence the increase in entropy is $\boxed{\text{larger}}$ than in part (a).

11-14 The standard entropy changes ($\Delta S°$'s) of these reactions equal 1 mol times the tabulated molar $S°$ of the gaseous halogen and 1 mol times the tabulated molar $S°$ of gaseous hydrogen subtracted from 2 mol times the molar $S°$ of the hydrogen halide. The answers are

F	Cl	Br	I
14.10 J K^{-1}	20.07 J K^{-1}	21.26 J K^{-1}	21.81 J K^{-1}

The entropy change becomes more positive going down the table. However, $\Delta S°$ for the reaction forming $HF(g)$ is less than extrapolation from the values for the other reactions would predict.

11-16 The reaction involves ions, but the procedure for computing $\Delta S°$ is the same as the one used in cases involving solids, liquids and gases.

$$\Delta S° = \underbrace{1\,(239.3)}_{I_3^-(aq)} - \underbrace{1\,(111.3)}_{I^-(aq)} - \underbrace{1\,(137.2)}_{I_2(aq)} = \boxed{-9.2 \text{ J K}^{-1}}$$

11-18 A system containing the "chemically-mixed" boron halides has greater entropy than a system consisting exclusively of BCl_3 and BF_3. It has the same number of gas-phase molecules, but more distinguishable kinds of molecules, hence more microstates and higher entropy.

11-20 The claims of this looney advertisement $\boxed{\text{do not violate}}$ the first law of thermodynamics. There is plenty of energy in the air on even the coldest days. The claims do however violate the second law.

11-22 The fundamental criterion for the spontaneity in the decomposition of quartz or any other process is the sign of ΔS_{univ} for the process. The large positive entropy change quoted for the decomposition of quartz applies to the reactants and products only. The ΔS of the surroundings in the proposed process apparently is negative and large enough to prevent the process from being spontaneous.

11-24 The melting point of tetraphenylgermane is 505.65 K (converted from 232.5°C). The molar enthalpy of fusion is the specific enthalpy of fusion multiplied by the molar mass. It is

$$\Delta H_{fus}(381.03 \text{ g mol}^{-1}) \times (106.7 \text{ J g}^{-1}) = 4.066 \times 10^4 \text{ J mol}^{-1}$$

Combining these numbers gives the molar entropy of fusion:

$$\Delta S_{fus} = \frac{\Delta H_{fus}}{T_{fus}} = \frac{4.066 \times 10^4 \text{ J mol}^{-1}}{505.65 \text{ K}} = \boxed{+80.40 \text{ J K}^{-1}\text{mol}^{-1}}$$

Note that Trouton's rule holds roughly.

11-26 (a) At constant temperature and pressure, $\Delta G = \Delta H - T\Delta S$. The numerical values in the problem are for 1.00 mol of tin because the chemical equation has coefficients of 1. A temperature of -30°C equals 243.15 K. Then

$$\Delta G°_{243.15} = -2.1 \times 10^3 \text{ J} - (243.15 \text{ K})(-7.4 \text{ J K}^{-1}) = \boxed{-300 \text{ J}}$$

(b) If the amount of tin is increased from 1.00 to 2.50 mol, then ΔG is multiplied by the factor $2.50/1.00$, from -300 to $\boxed{-750 \text{ J}}$, because of the larger size of the system.

(c) The ΔG of the process at $-30°C$ is always negative, regardless of the amount of tin. Therefore, $\boxed{\text{yes}}$, tin in any amount tends to convert spontaneously from white tin to gray tin at this low temperature.

(d) At equilibrium at constant temperature and pressure, $\Delta G = 0$. To compute the temperature that makes this happen, write

$$T = \frac{\Delta G - \Delta H°}{-\Delta S°} = \frac{0 - \Delta H°}{-\Delta S°} = \frac{\Delta H°}{\Delta S°} = \frac{-2.1 \times 10^3 \text{ J}}{-7.4 \text{ J K}^{-1}} = \boxed{284 \text{ K}} = 11°C$$

This assumes that $\Delta H°$ and $\Delta S°$ are independent of temperature.

11-28 Trouton's rule states that the molar entropy of vaporization of most substances is close to 88 J K^{-1}mol^{-1}. The liquid form and gaseous forms of a substance are in equilibrium at 1 atm at the normal boiling point, so

$$T_b = \frac{\Delta H_{vap}}{\Delta S_{vap}} = \frac{16.15 \times 10^3 \text{ J mol}^{-1}}{88 \text{ J K}^{-1}\text{mol}^{-1}} = \boxed{184 \text{ K}}$$

11-30 The $\Delta G°$ of the reaction at 25°C is calculated from the standard Gibbs functions of formation of the reactants and products given in Appendix D. It is assumed that the $CaCO_3$ is aragonite in both cases. The two results are identical, confirming that it is immaterial whether the notation " $H_3O^+(aq)$" or "$H^+(aq)$" is used in a properly balanced equation.

(a)

$$\Delta G° = \underbrace{1\,(-553.58)}_{Ca^{2+}(aq)} + \underbrace{1\,(-394.36)}_{CO_2(g)} + \underbrace{3\,(-237.18)}_{H_2O(l)} - \underbrace{1\,(-1128.84)}_{CaCO_3(s)} - \underbrace{2\,(-237.18)}_{H_3O^+(aq)}$$

$$= \boxed{-56.28 \text{ kJ}}$$

(b)

$$\Delta G° = \underbrace{1\,(-553.58)}_{Ca^{2+}(aq)} + \underbrace{1\,(-394.36)}_{CO_2(g)} + \underbrace{1\,(-237.18)}_{H_2O(l)} - \underbrace{1\,(-1128.84)}_{CaCO_3(s)} - \underbrace{2\,(0.00)}_{H^+(aq)}$$

$$= \boxed{-56.28 \text{ kJ}}$$

11-32 The necessary data are

	ΔH_f°	S°	ΔG_f°
$CoCl_2 \cdot 6H_2O$	-2115.4 kJ mol^{-1}	343 J K^{-1}mol^{-1}	-1725.2 kJ mol^{-1}
$H_2O(g)$	-241.82	188.72	-228.59
$CoCl_2(s)$	-312.5	109.16	-269.8

The form of the computation is similar for all three quantities. The molar value from the table is multiplied by the coefficient of the substance in the balanced equation. These numbers are then added, after changing the signs of those on the reactant side of the equation. The three answers are

$$\Delta H^\circ = \boxed{-352.0 \text{ kJ}} \quad \Delta S^\circ = \boxed{-898 \text{ J K}^{-1}} \quad \Delta G^\circ = \boxed{-83.9 \text{ kJ}}$$

It can be confirmed that indeed $\Delta G^\circ = \Delta H^\circ - T\Delta S^\circ$ when T is 298.15 K.

11-34 At a high enough temperature, the $T\Delta S$ term swamps out the ΔH term and the sign of ΔG is the same as the sign of $-T\Delta S$. Hence, ΔG is positive in case (a) and (c) and negative in case (b) and (d).

11-36 Compute ΔH° and ΔS° for each reaction from the data in Appendix D. If ΔH° and ΔS° have the same sign, then it is possible to determine a temperature that makes ΔG° equal to 0. This temperature is accurate insofar as $\Delta H^\circ_{298.15}$ and $\Delta S^\circ_{298.15}$ are good approximations of ΔH° and ΔS° at other T's. If ΔH° and ΔS° are both positive, then the reaction is spontaneous at temperatures above the temperature of equilibrium; if both are negative, then the reaction is spontaneous at temperatures below the temperature of equilibrium. If ΔH° and ΔS° have opposite signs, then the reaction is either spontaneous at all temperatures (ΔH° negative and ΔS° positive) or non-spontaneous at all temperatures (ΔH° positive and ΔS° negative). A common error is to forget to convert entropy changes (usually in joules per kelvin) and enthalpy changes (usually in kilojoules) to matching units before starting the arithmetic.

(a) The reaction given for the preparation of phosgene has

$$\Delta H^\circ_{298.15} = -108.3 \text{ kJ} \qquad \text{and} \Delta S^\circ_{209.15} = -136.99 \text{ J K}^{-1}$$

The quotient $\Delta H^\circ / \Delta S^\circ$ equals the temperature at which $\Delta G^\circ = 0$; it is 790 K. $\boxed{\text{Below 790 K}}$ the reaction is spontaneous. Because of the frailty of the assumption that ΔH° and ΔS° are temperature-independent, using more that two significant figures is pointless.

(b) For the decomposition of $KClO_3$ as given in the problem, $\Delta H^\circ = -78.04$ kJ, and $\Delta S^\circ = 494.07$ J K^{-1}. The reaction is spontaneous $\boxed{\text{at all temperatures}}$.

(c) Using the ΔH_f° and S° values tabulated for graphite and for wüstite, $\Delta H^\circ = 155.75$ kJ, and $\Delta S^\circ = 161.61$ J K^{-1} for the reduction of iron(II) oxide to iron using carbon and generating carbon monoxide. The reaction is spontaneous $\boxed{\text{above about 960 K}}$.

11-38 (a) The $\Delta G_{298.15}^\circ$ of the reaction is calculated from the standard Gibbs functions of formation of the reactants and products. It is

$$\Delta G_{298.15}^\circ = \underbrace{2\,(0.0)}_{W(s)} + \underbrace{3\,(-394.36)}_{CO_2(g)} - \underbrace{3\,(0.0)}_{C(s)} - \underbrace{2\,(-764.08)}_{WO_3(s)} = \boxed{+345.08 \text{ kJ}}$$

In this computation, the units of the numbers in parentheses are kJ mol^{-1} and the units of the integers are mol. The process is *not* spontaneous at 298.15 K.

(b) The fact that a gas is generated from solids means that ΔS° of this reaction is positive. It follows that a high enough temperature will make the ΔG° at that temperature negative no matter what the value of ΔH°.

11-40 The reaction to give liquid formic acid is exothermic while the reaction to give gaseous formic acid is endothermic because the energy released as the independent molecules of HCOOH in the gaseous phase join in non-bonded association in the liquid state is fairly large compared to the energy change in breaking and making the chemical bonds to go from CO and H_2 to HCOOH. It is large enough to topple ΔH° from positive to negative. No doubt liquid HCOOH has an extensive network of hydrogen bonds, the strongest type of non-bonded attraction.

In the reaction, the same two moles of gases react to give either one mole of gaseous formic acid or one mole of liquid formic acid In both gases, the product has a lower entropy than the reactants. This "disfavors" the reactions by making T
So negative. The standard Gibbs function is positive because the $T\Delta S^\circ$ contribution to $\Delta G^\circ = \Delta H^\circ - T\Delta S^\circ$ dominates for these reactions at this temperature.

11-42 An acceptable equation is $C_2H_6(g) \rightleftharpoons C_2H_2(g) + 2\,H_2(g)$ for which:

$$\Delta G^\circ = \underbrace{2\,(0.0)}_{H_2(g)} + \underbrace{1\,(209.20)}_{C_2H_2(g)} - \underbrace{1\,(-32.89)}_{C_2H_6(g)} = \boxed{242.09 \text{ kJ}}$$

where the values in parentheses are molar ΔG_f°'s from Appendix D. This ΔG° is related to the K of the reaction at 298.15 K by $\Delta G^\circ = -RT \ln K$. The ΔG°

applies per mole of reaction as written. Hence:

$$\ln K = \frac{\Delta G^\circ}{-RT} = \frac{242.09 \times 10^3 \text{ J mol}^{-1}}{-(8.315 \text{ J K}^{-1} \text{ mol}^{-1})(298.15 \text{ K})} = -97.66$$

The equilibrium constant is the antiln of -97.66, which equals $\boxed{3.8 \times 10^{-43}}$. Notice that if the dehydrogenation reaction is represented with a doubled equation (coefficients 2, 2, and 4), then ΔG° comes out twice as big, and K is squared.

11-44 Compute the ΔG°'s of the reactions from the molar ΔG_f°'s of their reactants and products. These are tabulated in Appendix D. Then apply the relationship $\ln K = \Delta G^\circ / -RT$.

(a) $\Delta G^\circ = 2(-46.0) - 2(0.0) - 1(0.0) - 1(0.0) = -92.0$ kJ $\quad K = \boxed{1.3 \times 10^{16}}$.

(b) $\Delta G^\circ = (-228.59) + (-604.05) - (-898.56) = 65.92$ kJ $\quad K = \boxed{2.8 \times 10^{-12}}$.

(c) $\Delta G^\circ = 1(-301.9) - 4(-26.50) - 1(-147.06) = -48.84$ kJ $\quad K = \boxed{3.6 \times 10^{8}}$.

11-46 Calculate ΔG° at 298.15 K for this acid-ionization reaction and from it obtain K_a:

$$\Delta G^\circ = \underbrace{1\,(-1130.28)}_{\text{H}_2\text{PO}_4^-(aq)} + \underbrace{1\,(-237.18)}_{\text{H}_3\text{O}^+(aq)} - \underbrace{1\,(-1142.54)}_{\text{H}_3\text{PO}_4(aq)} - \underbrace{1\,(-237.18)}_{\text{H}_2\text{O}(l)} = +12.26 \text{ kJ}$$

$$\ln K_a = \frac{\Delta G^\circ}{-RT} = \frac{12.26 \times 10^3 \text{ kJ mol}^{-1}}{-(8.315 \text{ J K}^{-1} \text{ mol}^{-1})(298.15 \text{ K})} = -4.94$$

Solving gives K_a equal to $\boxed{0.0071}$. This is close to the value of 0.00752 listed in text Table 8-2.

11-48 (a) The equation and the equilibrium expression for the dissolution of the rhombic sulfur are

$$\text{S}(s) \rightleftharpoons \text{S}(aq) \qquad K = [\text{S}(aq)] = 1.9 \times 10^{-8} \quad \text{(at 25°C)}$$

The ΔG° of this process is calculated as follows:

$$\Delta G^\circ = -RT \ln K = -(8.315 \text{ J K}^{-1} \text{ mol}^{-1})(298.15 \text{ K})(-17.78)$$

$$= \boxed{+44.1 \times 10^3 \text{ J mol}^{-1}}$$

That is, if 1 mol of S(s) dissolves to give 1 molar aqueous sulfur, then the ΔG° of the reaction is +44.1 kJ. The standard state of aqueous sulfur is unattainable in practice, but the concept is still useful.

(b) Sulfur dissolves to a slight extent because some $S(aq)$ must be present to make $K > 0$.

11-50 (a) The problem asks for the ΔG°'s of two reactions at 873 K. For each:

$$\Delta G^\circ_{873} = -RT \ln K = -(8.315 \text{ J K}^{-1} \text{ mol}^{-1})(873 \text{ K}) \ln K$$

For the reaction giving 1 mol of $MgFe_2Cl_8(g)$, ΔG°_{873} equals $\boxed{-2.02 \text{ kJ}}$. The other reaction has ΔG°_{873} equal to $\boxed{+26.8 \text{ kJ}}$

(b) Compare the reciprocals of the K's given in the problem. At 873 K, $\boxed{MgFe_2Cl_8(g)}$ is thermodynamically more stable than $CaFe_2Cl_8(g)$ with respect to decomposition to $FeCl_3(g)$ and $MCl_2(s)$ (where M stands for the alkali earth element) because the decomposition of $MgFe_2Cl_8(g)$ has the smaller K.

11-52 (a) The equilibrium constant for the dissolution of chlorine in water increases with increasing temperature. This means that ΔH of the reaction is positive. The process is $\boxed{\text{endothermic}}$.

(b) The van't Hoff equation relates the K's of a reaction at different temperatures to the ΔH° of that reaction:

$$\ln \frac{K_2}{K_1} = \frac{\Delta H^\circ}{R} \left(\frac{1}{T_1} - \frac{1}{T_2} \right)$$

Neither K_2 nor K_1 is known, but their ratio equals 2. The temperatures are $T_2 = 303.15$ K and $T_1 = 283.15$ K. Substitution of these values and $R = 8.315$ J K^{-1} mol^{-1} gives

$$\ln 2 = \frac{\Delta H^\circ}{R} \left(\frac{1}{283.15} - \frac{1}{303.15} \right)$$

Solving reveals that ΔH° equals $+2.47 \times 10^4$ J mol^{-1}, that is, the ΔH° of the reaction as written in the problem is $\boxed{+25 \text{ kJ}}$.

11-54 The computation of ΔH° and ΔS° for the production of dimethyl ether from methanol follows the standard pattern:

$$\Delta H^\circ = \underbrace{1(-241.82)}_{H_2O(g)} + \underbrace{1(-213.63)}_{CH_3OCH_3(g)} - \underbrace{2(-200.66)}_{CH_3OH(g)} = \boxed{-24.55 \text{ kJ}}$$

$$\Delta S^\circ = \underbrace{1(188.72)}_{H_2O(g)} + \underbrace{1(266.27)}_{CH_3OCH_3(g)} - \underbrace{2(239.70)}_{CH_3OH(g)} = \boxed{-24.41 \text{ J K}^{-1}}$$

The reaction is exothermic. Therefore equilibrium production of dimethyl ether by this means is favored by $\boxed{\text{low temperatures}}$.

11-56 If an accurate ΔG°_{373} for the reaction can be obtained, then computation of K is straightforward. Do not use the ΔG°_f data in Appendix D because the result is $\Delta G^\circ_{298.15}$, which differs a lot from ΔG°_{373}. To get ΔG°_{373} calculate $\Delta H^\circ_{298.15}$ and $\Delta S^\circ_{298.15}$ for the change:

$$\Delta H^\circ = \underbrace{2\,(0.0)}_{\text{Cu}(s)} + \underbrace{1\,(-285.83)}_{\text{H}_2\text{O}(l)} - \underbrace{1\,(0.0)}_{\text{H}_2(g)} - \underbrace{1\,(-168.6)}_{\text{Cu}_2\text{O}(s)} = -117.23 \text{ kJ}$$

$$\Delta S^\circ = \underbrace{2\,(33.15)}_{\text{Cu}(s)} + \underbrace{1\,(69.91)}_{\text{H}_2\text{O}(l)} - \underbrace{1\,(130.57)}_{\text{H}_2(g)} - \underbrace{1\,(93.14)}_{\text{Cu}_2\text{O}(s)} = -87.50 \text{ J K}^{-1}$$

Then, assume that these 298.15 K values hold at 373 K

$$\Delta G^\circ_{373} = \Delta H^\circ - T\Delta S^\circ = -117.23 \times 10^3 \text{ J} - 373 \text{ K}(-87.50 \text{ J K}^{-1}) = -84.59 \text{ kJ}$$

$$\ln K_{373} = \frac{\Delta G^\circ_{373}}{-RT} = \frac{-84.59 \times 10^3 \text{ J}}{-(8.315 \text{ J K}^{-1} \text{ mol}^{-1})(373 \text{ K})} = 27.27$$

Taking the antiln of both sides gives $K_{373} = \boxed{7 \times 10^{11}}$.

11-58 The two K's are both for the reaction

$$\text{ClCH}_2\text{COOH}(aq) + \text{H}_2\text{O}(l) \rightleftharpoons \text{ClCH}_2\text{COO}^-(aq) + \text{H}_3\text{O}^+(aq)$$

but at two different temperatures. They are $K_1 = 1.528 \times 10^{-3}$ at $T_1 = 273.15$ K, and $K_2 = 1.230 \times 10^{-3}$ at $T_2 = 313.15$ K. The van't Hoff equation becomes

$$\ln\left(\frac{K_2}{K_1}\right) = \ln\left(\frac{1.230 \times 10^{-3}}{1.528 \times 10^{-3}}\right) = \frac{\Delta H^\circ}{R}\left(\frac{1}{273.15} - \frac{1}{313.15}\right)$$

Solving gives $\Delta H^\circ = -3.86 \times 10^3$ J mol^{-1}. The ΔH° of the acid dissociation reaction as written above is thus $\boxed{-3.86 \text{ kJ}}$.

11-60 The reaction is exothermic because K goes down with higher T. Insert the two temperatures (on the Kelvin scale) and the two K's into the van't Hoff equation:

$$\ln\left(\frac{93.1}{2780}\right) = \frac{\Delta H^\circ}{R}\left(\frac{1}{295.15} - \frac{1}{315.15}\right)$$

Solution of this equation gives $\Delta H^\circ = -1.313 \times 10^5$ J mol^{-1}, which means $\Delta H^\circ = \boxed{-131 \text{ kJ}}$ for the reaction as written (2 mol of reactant giving 1 mol of product).

To estimate ΔS° substitute $\Delta H^\circ = -1.313 \times 10^5$ J mol^{-1} into the expression:

$$\ln K = -\frac{\Delta H^\circ}{RT} + \frac{\Delta S^\circ}{R}$$

using either of the two K, T pairs. A common error is using units with kilojoules for ΔH° but joules for R. The result is $\Delta S^\circ = \boxed{-379 \text{ J K}^{-1}}$ for the reaction as written.

11-62 (a) The ΔH° of the autoionization of water

$$2\,\text{H}_2\text{O} \rightleftharpoons \text{H}_3\text{O}^+(aq) + \text{OH}^-(aq)$$

can be found from the data in the problem by solving for ΔH° in

$$\ln\left(\frac{9.614 \times 10^{-14}}{1.139 \times 10^{-15}}\right) = \frac{\Delta H^\circ}{8.3145 \text{ J K}^{-1} \text{ mol}^{-1}}\left(\frac{1}{273.15} - \frac{1}{333.15}\right)$$

The result is $+55.9435$, which rounds to $\boxed{55.9 \text{ kJ mol}^{-1}}$. This enthalpy change is per mole of the reaction, which is for 2 mol of H_2O reacting. Using the standard enthalpies of formation tabulated in Appendix D gives the slightly different and more precise answer of $+55.84$ kJ mol^{-1}.

(b)

$$\Delta S^\circ = R \ln K + \frac{\Delta H^\circ}{T}$$

$$= (8.3145 \text{ J K}^{-1} \text{ mol}^{-1}) \ln(1.139 \times 10^{-15}) + \frac{55.84 \times 10^3 \text{ J mol}^{-1}}{273.15 \text{ K}}$$

$$= \boxed{-81.66 \text{ J K}^{-1} \text{ mol}^{-1}}$$

The answer obtained by combining standard entropies from Appendix D equals -80.66 J K^{-1} mol^{-1}, which is quite close.

(c) The pH of water equals 7.000 when K_w equals exactly 1.00×10^{-14}. Substituting this and the standard enthalpy and entropy changes in

$$\ln K = -\frac{\Delta H^\circ}{RT} + \frac{\Delta S^\circ}{R}$$

gives

$$\ln(1.00 \times 10^{-14}) = -\frac{55.84 \times 10^3 \text{ J mol}^{-1}}{8.3145 \text{ J K}^{-1} \text{ mol}^{-1}}\left(\frac{1}{T}\right) + \frac{-80.66 \text{ J K}^{-1}\text{mol}^{-1}}{8.3145 \text{ J K}^{-1} \text{ mol}^{-1}}$$

Solving for T gives the desired temperature as $\boxed{298.02 \text{ K}}$, which is very close to 298.15 K, the standard-state temperature.

11-64 (a) The problem is solved in one step by substitution in the Clausius-Clapeyron equation (text page 475):

$$\ln\left(\frac{P_2}{P_1}\right) = \frac{\Delta H_{vap}}{R}\left(\frac{1}{T_1} - \frac{1}{T_2}\right)$$

The problem gives P_1 and T_1 (0.1316 atm, 343.25 K) and P_2 and T_2 (0.5263 atm, 373.95 K). Solution for ΔH_{vap} gives $\boxed{48.18 \text{ kJ mol}^{-1}}$.

(b) Set P_2 equal to 1.000 atm (at the normal boiling point) and set T_2 equal to T_b:

$$\ln\left(\frac{1.000}{0.1316}\right) = \frac{48.18 \times 10^3 \text{ J mol}^{-1}}{8.315 \text{ J K}^{-1} \text{ mol}^{-1}}\left(\frac{1}{343.25 \text{ K}} - \frac{1}{T_b}\right)$$

Solving for T_b gives $T_b = \boxed{390.1 \text{ K}}$, which is equivalent to 117.0°C. Substitution of the other P, T pair (0.5263 atm and 373.95 K) for P_2 and T_2 in the equation gives the same answer.

11-66 The vapor pressure of the water equals 1.00 atm at a temperature of 100°C (373.15 K). Let the boiling temperature at the reduced pressure of 0.50 atm equal T_2. Then, using the Clausius-Clapeyron equation:

$$\ln\left(\frac{0.50}{1.00}\right) = \frac{40.66 \times 10^3 \text{ J mol}^{-1}}{8.315 \text{ J K}^{-1} \text{ mol}^{-1}}\left(\frac{1}{373.15 \text{ K}} - \frac{1}{T_2}\right)$$

Solving for T_2 gives a temperature of 354.4 K, which is equivalent to $\boxed{81.3°C}$.

11-68 The equation for the formation of aqueous glucose from its elements

$$6\,C(s) + 3\,O_2(g) + 6\,H_2(g) \longrightarrow C_6H_{12}O_6(aq)$$

is the sum of the equations:

$$6\,C(s) + 6\,O_2(g) \longrightarrow 6\,CO_2(g) \qquad\qquad \Delta H° = -2361.06 \text{ kJ}$$

$$6\,H_2(g) + 3\,O_2(g) \longrightarrow 6\,H_2O(l) \qquad\qquad \Delta H° = -1714.98 \text{ kJ}$$

$$6\,CO_2(g) + 6\,H_2O(l) \longrightarrow C_6H_{12}O_6(aq) + 6\,O_2(g) \qquad \Delta H° = +2810 \text{ kJ}$$

The $\Delta H°$'s of the first and second equations were 6 mol times $\Delta H_f°$ of $CO_2(g)$ and 6 mol times $\Delta H_f°$ of $H_2O(l)$ respectively. The $\Delta H°$ of the third equation is computed in problem 11-67. It is 2810 kJ. Adding these three $\Delta H°$'s gives $\boxed{-1266 \text{ kJ}}$.

11-70 (a) The probability that the first keystroke generates a number less than or equal to 50 is 1/2. The probability of 10 successive numbers in the 0-50 range is the product of the probabilities of the single events: $(0.5)^{10} = \boxed{0.000976563}$.

(b) The probability is $(0.5)^{10^9}$ or $\boxed{10^{-3.01 \times 10^8}}$. This equals one chance in 1 followed by 301 million zeros.

(c) The probability is $(.99)^{10}$ or $\boxed{0.90438}$.

(d) The probability is $(0.99)^{10^9}$ or $\boxed{10^{-4.36 \times 10^6}}$. The point here is that although getting a 100 is a 100-to-1 long-shot on any given keystroke, it is essentially certain to happen by the time a billion chances are taken.

11-72 The chance that the first of the 500 atoms is in the left half of the optical trap is 1/2. Assume that the gas of sodium atoms behaves ideally so that the chance that a second sodium atom is in the left half of the optical trap is not influenced by the presence of the first. Then the chance that second atom is in the left half is also 1/2. The chance that both atoms are in the left half is $1/2 \times 1/2 = 1/4$. By extension, p, the chance that all 500 sodium atoms are in the left half, is

$$p = \left(\frac{1}{2}\right)^{500}$$

To compute p take the logarithm of both sides of this equation

$$\log p = 500 \log \frac{1}{2} = 500 \times (-0.30103) = -150.51$$

and then the antilog of both sides

$$p = 10^{-150.51} = 10^{0.49} \times 10^{-151} = \boxed{3 \times 10^{-151}}$$

Even with this very small sample of gas, the chance of "spontaneous congregation" on one side of the vessel is essentially zero.

11-74 (a) The chemical amount of gas increases (from 9 to 10 mol) so ΔS of the reaction should be $\boxed{\text{positive}}$.

(b) The chemical amount of gas decreases (from 9 to 4 mol) so ΔS of the reaction should be $\boxed{\text{negative}}$.

11-76 The container is insulated so no heat can pass into it from the surroundings or out of it to the surroundings: $q = 0$. Further, the container is rigid so that it can neither expand nor contract during the process. There is no other mechanical linkage or electrical linkage to the surroundings. Hence $w = 0$. By the first law then, $\Delta E = 0$. The total energy of the contents of the container $\boxed{\text{remains constant}}$ during the process.

At equilibrium the hot water will have cooled and that some or all of the ice will have melted spontaneously. The change occurs in isolation from the surroundings so $\Delta S_{\text{surr}} = 0$. The second law states that for this spontaneous process $\Delta S_{\text{univ}} > 0$. Hence $\boxed{\Delta S_{\text{system}} > 0}$.

11-78 (a) Higher temperature makes the conversion of rhombic to monoclinic sulfur a spontaneous process. According to the discussion on text page 468, both ΔH° and ΔS° must then be $\boxed{\text{positive}}$: if the two had unlike signs, then there would be no "cross-over temperature;" if the two were both negative, then the conversion would be favored by lower temperature.

(b) Equilibrium between rhombic and monoclinic sulfur at constant temperature and pressure means $\Delta G = 0$. Therefore, $0 = \Delta H - T\Delta S$. Rearranging and substituting the values from the problem gives

$$\Delta S_{\text{transition}} = \frac{\Delta H_{\text{transition}}}{T_{\text{transition}}} = \frac{400 \text{ J}}{368.5 \text{ K}} = \boxed{1.09 \text{ J K}^{-1}}$$

11-80 The vaporization of water is $H_2O(l) \rightleftharpoons H_2O(g)$ and

$$\Delta H^\circ = -241.82 - (-285.83) = 44.01 \text{ kJ} \qquad \Delta S^\circ = 188.72 - 69.91 = 118.81 \text{ J K}^{-1}$$

At equilibrium between the two phases, ΔG° equals zero. Therefore

$$\Delta G^\circ = \Delta H^\circ - T\Delta S^\circ \quad \text{implies } T = \frac{\Delta H^\circ}{\Delta S^\circ}$$

For the water vaporization

$$T = \frac{44.01 \times 10^3 \text{ J}}{118.81 \text{ J K}^{-1}} = \boxed{370.4 \text{ K}}$$

This differs from 373.15 K because the calculation uses ΔH° and ΔS° at 298.15 K, and both of these quantities change somewhat with the temperature.[7]

11-82 At equilibrium between solid and gaseous iodine:

$$\Delta G = 0 = \Delta H - T\Delta S \quad \text{so that} \quad T = \frac{\Delta H}{\Delta S}$$

---

[7]It is worth noting that the experimental normal boiling point of water is 99.975°C and not exactly 100° C under the current definition of the Celsius scale.

The ΔH and ΔS at the melting temperature are well approximated by $\Delta H°$ and $\Delta S°$ at 298.15 K. These are computed for the reaction $I_2(s) \rightleftharpoons I_2(g)$ from data in Appendix D in the usual way. Then:

$$T \approx \frac{\Delta H°}{\Delta S°} = \frac{62.44 \times 10^3 \text{ J}}{144.44 \text{ J K}^{-1}} = \boxed{432.3 \text{ K}}$$

Thus, solid and vapor are in equilibrium at 159.1°C) when the pressure of $I_2(g)$ is 1 atm. Since the melting point of $I_2(s)$ is 113.5° at $P = 1$ atm, the solid-gas equilibrium at 1 atm $\boxed{\text{cannot be achieved}}$; the solid melts first.

11-84 The problem provides data on the enthalpies of formation and Gibbs functions of formation of the two isomers of $Pt(NH_3)_2I_2$. The formation reactions are

$$Pt(s) + N_2(g) + 3\,H_2(g) + I_2(s) \rightleftharpoons cis\text{-}Pt(NH_3)_2I_2(s)$$

$$Pt(s) + N_2(g) + 3\,H_2(g) + I_2(s) \rightleftharpoons trans\text{-}Pt(NH_3)_2I_2(s)$$

Use the relationship $\Delta G° = \Delta H° - T\Delta S°$ to compute $\Delta S°$'s of the two reactions. For the *cis* isomer

$$\Delta S° = \frac{\Delta H° - \Delta G°}{T} = \frac{(-286.56 \text{ kJ}) - (-130.25 \text{ kJ})}{298.15 \text{ K}} = -0.52427 \text{ kJ K}^{-1}$$

For the *trans* isomer, a similar calculation gives $\Delta S° = -0.52135$ kJ K^{-1}. These $\Delta S°$'s are the standard entropies of the products minus the standard entropies of the reactants. Let x equal the standard molar entropy of the *cis*-compound. For the first reaction:

$$-524.27 \text{ J K}^{-1} = \underbrace{1\,(x)}_{cis} - \underbrace{1\,(41.63)}_{Pt(s)} - \underbrace{1\,(191.50)}_{N_2(g)} - \underbrace{3\,(130.57)}_{I_2(s)}$$

where the numbers in parentheses are molar $S°$ values (Appendix D) and the other numbers are the appropriate numbers of moles. Solving gives $S°$ of the *cis* compound: $\boxed{216.71 \text{ J K}^{-1}\text{mol}^{-1}}$. For the *trans*, a similar calculation gives $\boxed{219.63 \text{ J K}^{-1}\text{mol}^{-1}}$.

11-86 The $\Delta G°$ for this reaction is simply the negative of $\Delta G_f°$ of 1.00 mol of $NiO(s)$. Combining data from Appendix D in the usual way establishes that $\Delta G° = +211.7$ kJ per mole of the reaction as written. Substitute this value into $\ln K = -\Delta G°/RT$ to obtain $\ln K$ at 298.15 K:

$$\ln K = \frac{211.7 \times 10^3 \text{ J mol}^{-1}}{-(8.315 \text{ J K}^{-1} \text{ mol}^{-1})(298.15 \text{ K})} = -85.39$$

Taking the antiln gives K equal to 8×10^{-38}. For this heterogeneous equilibrium, $K = \sqrt{P_{O_2}}$, so that the partial pressure of oxygen in the system at equilibrium is K^2 or $\boxed{7 \times 10^{-75} \text{ atm}}$.

11-88 (a) In the van't Hoff equation for the equilibrium decomposition of barium nitride

$$\ln\left(\frac{K_2}{K_1}\right) = \frac{\Delta H^\circ}{R}\left(\frac{1}{T_1} - \frac{1}{T_2}\right)$$

the two K's are $K_1 = 4.5 \times 10^{-19}$ and $K_2 = 6.2 \times 10^{-12}$ and $T_1 = 1000$ K and $T_2 = 1200$ K. Substitution gives ΔH° equal to 8.2×10^5 J mol^{-1}. Hence, the ΔH° of the reaction written with coefficients of 1, 3 and 1 is $\boxed{820 \text{ kJ}}$.

(b) Another substitution of K's and T's in the van't Hoff equation gives a ΔH° equal to 16.4×10^5 J mol^{-1}. It follows that the ΔH° of the reaction written with coefficients of 2, 6 and 2 is $\boxed{1640 \text{ kJ}}$. It makes sense that a reaction involving twice as much material should have a ΔH° that is twice as large, and this is the point of the problem.

11-90 (a) The sublimation of $NbI_5(s)$ is described by

$$NbI_5(s) \rightleftharpoons NbI_5(g) \qquad\qquad K = P_{NbI_5}$$

where K depends on the temperature according to the equation (page 474 of the text):

$$\ln K = -\frac{\Delta H^\circ_{subl}}{RT} + \frac{\Delta S^\circ_{subl}}{R}$$

Substituting the equilibrium expression into this equation give

$$\ln P_{NbI_5} = -\frac{\Delta H^\circ_{subl}}{RT} + \frac{\Delta S^\circ_{subl}}{R}$$

The natural logarithm is 2.3026 times the base-10 logarithm

$$2.3026 \log P_{NbI_5} = -\frac{\Delta H^\circ_{subl}}{RT} + \frac{\Delta S^\circ_{subl}}{R}$$

The problem gives an expression for $\log P_{NbI_5}$ in terms of T. Insert it on the left

$$2.3026\left(\frac{-6762}{T} + 8.566\right) = -\frac{\Delta H^\circ_{subl}}{RT} + \frac{\Delta S^\circ_{subl}}{R}$$

Term-by-term comparison of the two sides of this equation shows that

$$\Delta H^\circ_{subl} = 2.3026(6762)R \quad \text{and} \quad \Delta S^\circ_{subl} = 2.3026(8.566)R$$

$$\Delta H^\circ_{\text{subl}} = \boxed{129.5 \times 10^3 \text{ J mol}^{-1}} \quad \text{and} \quad \Delta S^\circ_{\text{subl}} = \boxed{164.0 \text{ J K}^{-1}\text{mol}^{-1}}$$

Note that "−6762" and "−4653" in the problem actually are in kelvins. The "8.566" (and "5.43") are dimensionless.

(b) A similar analysis for the vaporization of the $NbI_5(l)$ gives

$$\Delta H^\circ_{\text{vap}} = \boxed{89.08 \times 10^3 \text{ J mol}^{-1}} \quad \text{and} \quad \Delta S^\circ_{\text{vap}} = \boxed{104 \text{ J K}^{-1}\text{mol}^{-1}}$$

(c) The normal boiling point of a liquid is defined as the temperature at which its vapor pressure equals 1 atm. Therefore:

$$\log 1 = 0 = \frac{-4653}{T_b} + 5.43 \quad \text{and} \quad T_b = \boxed{857 \text{ K}}$$

(d) According to the hint, the vapor pressures of the liquid and solid NbI_5 are equal at the melting point. Simply equate the expressions that give the temperature dependence of these quantities and subscript the temperature as the melting point:

$$\frac{-6762}{T_{\text{melt}}} + 8.566 = \frac{-4653}{T_{\text{melt}}} + 5.43$$

Solving gives $\boxed{673 \text{ K}}$ as the melting point.

11-92 In general, human activities increase the entropy of the objects and structures at or near the surface of the earth. Regions of low entropy (such as reclaimed pure metals, organized buildings and roads and every other useful human artifact) are created at the expense of the increase in the entropy of the larger surroundings.

11-94 (a) Identify the first reaction, which is the reduction of NO_2 with ammonia, as reaction 1. Obtain ΔG°_f's from Appendix D and use them:

$$\Delta G^\circ(1) = \underbrace{7\,(0.0)}_{N_2(g)} + \underbrace{12\,(-228.59)}_{H_2O(g)} - \underbrace{6\,(51.29)}_{NO_2(g)} - \underbrace{8\,(-16.48)}_{NH_3(g)} = \boxed{-2918.98 \text{ kJ}}$$

$$\ln K = \frac{\Delta G^\circ}{-RT} = \frac{-2918.98 \times 10^3 \text{ J}}{-(8.315 \text{ J K}^{-1} \text{ mol}^{-1})(298.15 \text{ K})} = 1177 \qquad \boxed{K_1 = 1.3 \times 10^{510}}$$

(b) The balanced equations for the two other reactions are

$$6\,NO(g) + 4\,NH_3(g) \longrightarrow 5\,N_2(g) + 6\,H_2O(g)$$

$$3\,N_2O(g) + 2\,NH_3(g) \longrightarrow 4\,N_2(g) + 3\,H_2O(g)$$

Call these reaction 2 and reaction 3 respectively. The changes in the standard Gibbs function accompanying them at 298.15 K are:

$$\Delta G°(2) = \underbrace{5\,(0.0)}_{N_2(g)} + \underbrace{6\,(-228.59)}_{H_2O(g)} - \underbrace{6\,(86.55)}_{NO(g)} - \underbrace{4\,(-16.48)}_{NH_3(g)} = \boxed{-1824.92 \text{ kJ}}$$

$$\Delta G°(3) = \underbrace{4\,(0.0)}_{N_2(g)} + \underbrace{3\,(-228.59)}_{H_2O(g)} - \underbrace{3\,(104.18)}_{N_2O(g)} - \underbrace{2\,(-16.48)}_{NH_3(g)} = \boxed{-965.35 \text{ kJ}}$$

The equilibrium constants are:

$$\ln K_2 = \frac{\Delta G°}{-RT} = \frac{-1824.92 \times 10^3 \text{ J}}{-(8.315 \text{ J K}^{-1} \text{ mol}^{-1})(298.15 \text{ K})} = 736 \qquad \boxed{K_2 = 9 \times 10^{310}}$$

$$\ln K_3 = \frac{\Delta G°}{-RT} = \frac{-965.35 \times 10^3 \text{ J}}{-(8.315 \text{ J K}^{-1} \text{ mol}^{-1})(298.15 \text{ K})} = 389 \qquad \boxed{K_3 = 8 \times 10^{160}}$$

The three reactions all have overwhelmingly large equilibrium constants at room temperature.

(c) The three reactions have Δn_g's of 5 mol, 1 mol and 2 mol respectively. Therefore, all are favored by low pressure. All three are spontaneous at 298.15 K and are exothermic: raising the temperature lowers their K's. The reactions are however far too slow to proceed significantly toward equilibrium at room temperature. The rapid removal of NO_x depends raising the temperature enough to have a fast reaction, yet still have large equilibrium conversion to N_2.

Answers to Even-Numbered Problems, Chapter 12

12-2 The reduction of U_3O_8 with calcium is represented

$$\boxed{U_3O_8(s) + 8\,Ca(s) \longrightarrow 3\,U(s) + 8\,CaO(s)}$$

The standard enthalpy change of this reaction at 298.15 K computed from standard enthalpies of formation at 298.15 K:

$$\Delta H° = 8 \underbrace{(-635)}_{CaO(s)} + 3 \underbrace{(0.00)}_{U(s)} - 1 \underbrace{(-3575)}_{U_3O_8(s)} - 8 \underbrace{(0.00)}_{Ca(s)} = -1505 \text{ kJ}$$

and the standard entropy change is computed from standard entropies at 298.15 K:

$$\Delta S^\circ = 8 \underbrace{(39.8)}_{\text{CaO}(s)} + 3 \underbrace{(50.2)}_{\text{U}(s)} - 1 \underbrace{(282.6)}_{\text{U}_3\text{O}_8(s)} - 8 \underbrace{(41.4)}_{\text{Ca}(s)} = -144.8 \text{ J K}^{-1}$$

Then

$$\Delta G^\circ_{298.15} = \Delta H^\circ - T\Delta S^\circ = -1505 - (298.15 \text{ K})(-144.8 \times 10^{-3} \text{ kJ K}^{-1})$$
$$= -1462 \text{ kJ}$$

The reaction produces 3 mol of uranium, so the standard enthalpy, and standard Gibbs function changes per mol of U are one-third as large:

$$\Delta H^\circ = \boxed{-502 \text{ kJ mol}^{-1}} \quad \text{and} \quad \Delta G^\circ = \boxed{-487 \text{ kJ mol}^{-1}}$$

12-4 The decomposition reaction is

$$\text{CuO}(s) \longrightarrow \text{Cu}(s) + 1/2\,\text{O}_2(g)$$

Calculate the ΔH° and ΔS° of this reaction from the data in Appendix D. The tabulated values are on a molar basis, so the entry for $\text{O}_2(g)$ must be multiplied by 1/2. Omitting this step is a common error.

$$\Delta H^\circ = 157.3 \text{ kJ} \qquad \Delta S^\circ = 93.03 \text{ J K}^{-1} \qquad T = \frac{\Delta H^\circ}{\Delta S^\circ} = \boxed{1690 \text{ K}}$$

12-6 According to its formula, chalcopyrite is 34.94% S by mass and 34.62% Cu. These facts are enough to solve the problem.

(a) The mass of copper producible from 1.0 metric tons of chalcopyrite is

$$0.3462 \times (1.0 \times 10^3 \text{ kg}) = \boxed{3.5 \times 10^2 \text{ kg}}$$

(b) Similarly, the mass of sulfur in the metric ton of chalcopyrite is 349.4 kg. If all of the sulfur is converted to SO_2 the yield is 699.5 kg of SO_2, which is 1.090×10^4 mol of SO_2. This chemical amount of SO_2 occupies $\boxed{2.4 \times 10^5 \text{ L}}$ at STP. Note that writing chemical equations is unnecessary.

12-8 All three reactions involve the reduction of one mole of $\text{Fe}_2\text{O}_3(s)$ to $\text{Fe}(s)$. The calculations require combining the appropriate Gibbs functions of formation from Appendix D to obtain $\Delta G^\circ_{298.15}$'s for the reactions as written. The results are $\boxed{150.7 \text{ kJ}}$ for the reduction by $\text{C}(s)$ with $\text{CO}_2(g)$ as a product, $\boxed{330.7 \text{ kJ}}$

for the reduction by $C(s)$ with $CO(g)$ as a product, and $\boxed{-29.43 \text{ kJ}}$ for the reduction by $CO(g)$ with $CO_2(g)$ as a product. The best of the three reducing agents under standard conditions is $\boxed{CO(g)}$ because the $\Delta G^\circ_{298.15}$ of its reaction is the most negative.

12-10 Both reactions are spontaneous in the direction given in the problem:

$$\Delta G^\circ = \underbrace{(0.00)}_{Mg(s)} + 2\underbrace{(-261.90)}_{Na^+(aq)} - 2\underbrace{(0.00)}_{Na(s)} - 1\underbrace{(-454.8)}_{Mg^{2+}(aq)} = -69.0 \text{ kJ}$$

$$\Delta G^\circ = 2\underbrace{(0.00)}_{Na(s)} + \underbrace{(-569.45)}_{MgO(s)} - \underbrace{(0.00)}_{Mg(s)} - 1\underbrace{(-375.48)}_{Na_2O(s)} = -193.97 \text{ kJ}$$

12-12 The skill developed here is useful in Chapter 13. Some students may choose to work the problem by simply following the procedure for balancing redox equations.

(a) oxidation: $12\,OH^-(aq) + 4\,PH_3(g) \longrightarrow P_4(s) + 12\,H_2O(l) + 12\,e^-$

reduction: $12\,e^- + 16\,H_2O(l) + 4\,CrO_4^{2-} \longrightarrow 4\,Cr(OH)_4^-(aq) + 16\,OH^-(aq)$

(b) oxidation: $2\,OH^-(aq) + Fe(s) \longrightarrow Fe(OH)_2(s) + 2\,e^-$

reduction: $2\,e^- + 2\,H_2O(l) + NiO_2(s) \longrightarrow Ni(OH)_2(s) + 2\,OH^-(aq)$

(c) oxidation: $2\,OH^-(aq) + 2\,NH_2OH(aq) \longrightarrow N_2(g) + 4\,H_2O(g) + 2\,e^-$

reduction: $e^- + NO_2(g) \longrightarrow NO_2^-(aq)$

12-14 In acidic solution, H^+ and H_2O may be added either as reactants or as products to achieve balance.

(a) $8\,MnO_4^-(aq) + 5\,H_2S(aq) + 14\,H^+(aq) \longrightarrow 8\,Mn^{2+}(aq) + 5\,SO_4^{2-}(aq) + 12\,H_2O(l)$

(b) $4\,Zn(s) + NO_3^-(aq) + 10\,H^+(aq) \longrightarrow 4\,Zn^{2+}(aq) + NH_4^+(aq) + 3\,H_2O(l)$

(c) $5\,H_2O_2(aq) + 2\,MnO_4^-(aq) + 6\,H^+(aq) \longrightarrow 5\,O_2(g) + 2\,Mn^{2+}(aq) + 8\,H_2O(l)$

(d) $2\,Sn(s) + 2\,NO_3^-(aq) + 10\,H^+(aq) \longrightarrow 2\,Sn^{4+}(aq) + N_2O(g) + 5\,H_2O(l)$

(e) $3\,UO_2^{2+}(aq) + Te(s) + 4\,H^+(aq) \longrightarrow 3\,U^{4+}(aq) + TeO_4^{2-}(g) + 2\,H_2O(l)$

(f) $2\,PbSO_4(s) + 2\,H_2O(l) \longrightarrow Pb(s) + PbO_2(s) + 2\,SO_4^{2-}(aq) + 4\,H^+(aq)$

12-16 In basic solution, OH^- and H_2O take part either as reactants or as products.

(a) $3\,OCl^-(aq) + I^-(aq) \longrightarrow IO_3^-(aq) + 3\,Cl^-(aq)$

(b) $2\,SO_3^{2-}(aq) + 2\,Be(s) \longrightarrow S_2O_3^{2-}(aq) + Be_2O_3^{2-}(aq)$

(c) $P_4(s) + 2\,H_2O(l) + 4\,OH^-(aq) \longrightarrow 2\,HPO_3^{2-}(aq) + 2\,PH_3(g)$

(d) $3H_2BO_3^-(aq) + 8Al(s) + 8OH^-(aq) + 7H_2O(l) \longrightarrow 3BH_4^-(aq) + 8H_2AlO_3^-(aq)$

(e) $2\,OH^-(aq) + 3\,O_2(g) + 2\,Sb(s) \longrightarrow 3\,H_2O_2(aq) + 2\,SbO_2^-(aq)$

(f) $2\,Sn(OH)_6^{2-}(aq) + Si(s) \longrightarrow 2\,HSnO_2^-(aq) + SiO_3^{2-}(aq) + 5\,H_2O(l)$

12-18 (a) One point of the problem is that the same reactants can give different products depending on the conditions:

$$Cu(s) + 2\,HNO_3(aq) + 2\,H^+(aq) \longrightarrow Cu^{2+}(aq) + 2\,NO_2(g) + 2\,H_2O(l)$$

$$3\,Cu(s) + 2\,HNO_3(aq) + 6\,H^+(aq) \longrightarrow 3\,Cu^{2+}(aq) + 2\,NO(g) + 4\,H_2O(l)$$

(b) $Cu(s) + HNO_3(aq) + 2\,H^+(aq) \longrightarrow Cu^{2+}(aq) + HNO_2(aq) + H_2O(l).$

12-20 The products of the reaction of bornite with aqueous copper(II) chloride are copper(I) chloride and sulfur (by comparison to the reactions given on text page 493). The balanced half-equations and overall equation are:

oxidation: $\quad Cu_5FeS_4(s) + 5\,Cl^-(aq) \longrightarrow 5\,CuCl(s) + Fe^{2+}(aq) + 4\,S(s) + 7\,e^-$

reduction: $\quad e^- + Cu^{2+}(aq) + Cl^-(aq) \longrightarrow CuCl(s)$

overall: $\quad Cu_5FeS_4(s) + 7\,CuCl_2(aq) \longrightarrow 12\,CuCl(s)FeCl_2(aq) + 4\,S(s)$

12-22 (a) $Ba(s) + O_2(g) \longrightarrow BaO_2(s)$

(b) $2\,Ce^{4+}(aq) + 2\,I^-(aq) \longrightarrow 2\,Ce^{3+}(aq) + I_2(aq)$

(c) $3\,Mn(OH)_3(s) + Cr(s) \longrightarrow 3\,Mn(OH)_2(s) + Cr(OH)_3(s)$

(d) $Hg_2Cl_2(s) + Co(s) \longrightarrow 2\,Hg(l) + CoCl_2(s)$

12-24 The equation for the disproportionation of $S_2O_3^{2-}$ ion in acid is $S_2O_3^{2-}(aq) + H^+(aq) \longrightarrow S(s) + HSO_3^-(aq).$

12-26 The quantity of charge in coulombs is the product of the steady current in amperes and the time in seconds: $Q = It.$ Applying this equation to the circuit involving flashlight bulb gives $Q = \boxed{10.8\ C}.$

12-28

$$t = \frac{Q}{I} = \frac{4.87 \times 10^5 \text{ C}}{41.2 \text{ A}} = \boxed{1.18 \times 10^4 \text{ s}}$$

12-30 The two half-reactions and the overall reaction are:

$$\text{anode:} \quad \boxed{\text{Ni}(s) + 2\,\text{OH}^-(aq) \longrightarrow \text{Ni(OH)}_2(s) + 2\,e^-}$$

$$\text{cathode:} \quad \boxed{6\,e^- + \text{BrO}_3^- + 3\,\text{H}_2\text{O} \longrightarrow \text{Br}^- + 6\,\text{OH}^-}$$

$$\text{overall:} \quad \boxed{3\,\text{Ni} + \text{BrO}_3^- + 3\,\text{H}_2\text{O} \longrightarrow \text{Br}^- + 3\,\text{Ni(OH)}_2}$$

12-32 The cell has two electrodes made of an non-reactive conductor (platinum, for example) dipping into an aqueous solution of NiI_2 and connected to the source of outside voltage.

12-34

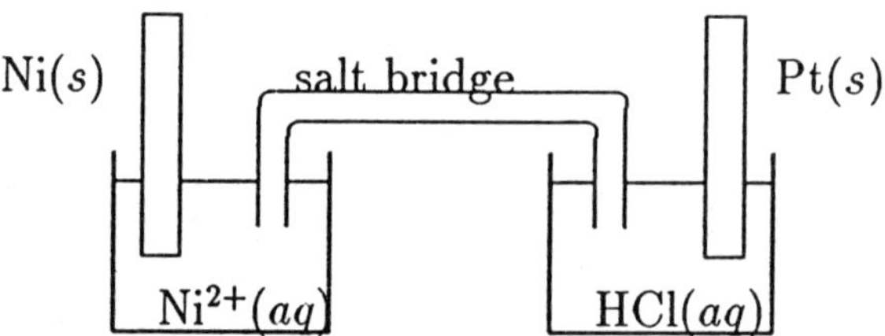

A stream of gaseous hydrogen plays over the platinum electrode. When the two electrodes are then connected by a wire, electrons flow from left to right through the wire. Negative ions migrate to the left in the salt bridge, and positive ions migrate to the right. The net reaction is:

$$\boxed{\text{Ni}(s) + 2\,\text{HCl}(aq) \longrightarrow \text{H}_2(g) + \text{NiCl}_2(aq)}$$

12-36 (a) Oxidation occurs at the lead electrode and reduction at the copper electrode. Electrons flow from left to right through the wire connecting the electrodes.

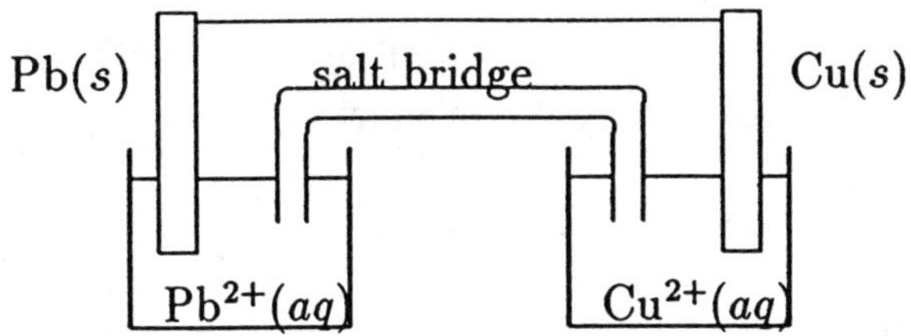

(b) Oxidation occurs at the silver electron and reduction occurs at the graphite electron. Electrons flow from left to right though the wire.

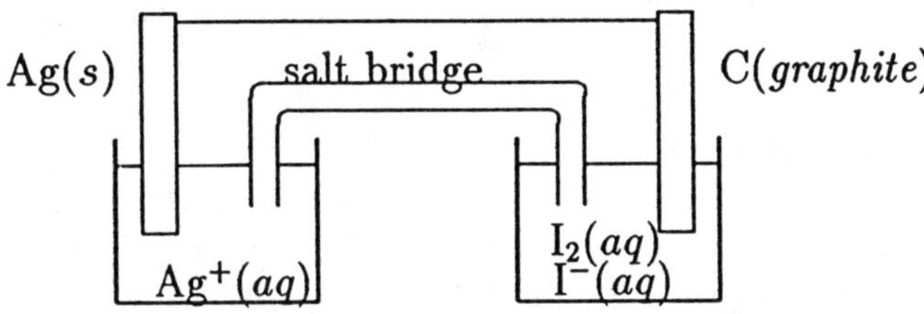

12-38 A current of 60 μA delivers 60×10^{-6} C to the screen each second. Dividing by the Faraday constant establishes that this is 6.21×10^{-10} mol of electrons, which is $\boxed{3.7 \times 10^{14} \ e^{-\text{'s}}}$.

12-40

(a) $\quad 2.00 \text{ mol Zn}^{2+} \times \left(\dfrac{2 \text{ mol } e^-}{\text{mol Zn}^{2+}} \right) \times \left(\dfrac{96485 \text{ C}}{\text{mol } e^-} \right) = \boxed{3.86 \times 10^5 \text{ C}}$

(b) $\quad 1.00 \text{ mol H}_2\text{O}_2 \times \left(\dfrac{2 \text{ mol } e^-}{\text{mol H}_2\text{O}_2} \right) \times \left(\dfrac{96485 \text{ C}}{\text{mol } e^-} \right) = \boxed{1.93 \times 10^5 \text{ C}}$

(c) $\quad 2.50 \text{ mol MnO}_4^- \times \left(\dfrac{3 \text{ mol } e^-}{\text{mol MnO}_4^-} \right) \times \left(\dfrac{96485 \text{ C}}{\text{mol } e^-} \right) = \boxed{7.24 \times 10^5 \text{ C}}$

(d) $\quad 1.70 \text{ mol H}_2\text{S} \times \left(\dfrac{8 \text{ mol } e^-}{2 \text{ mol H}_2\text{S}} \right) \times \left(\dfrac{96485 \text{ C}}{\text{mol } e^-} \right) = \boxed{6.56 \times 10^5 \text{ C}}$

12-42 The quantity of electricity is 0.9600 mol of electrons, obtained by dividing 92,630 C by the Faraday constant. Oxidizing one mole of Ni(s) to Ni²⁺(aq) requires passage of two moles of electrons, so the maximum amount of Ni(s) dissolved is $\boxed{0.4800 \text{ mol}}$. Note that an oxidation reaction *not* involving Ni might be responsible for the generation of the voltage by this cell. Hence the minimum amount of Ni(s) dissolved is none. The mass of Ni(s) corresponding to 0.4800 mol of Ni(s) is $\boxed{28.17 \text{ g}}$.

12-44 (a)

$$\boxed{Zn(s) + Cd^{2+}(aq) \longrightarrow Zn^{2+}(aq) + Cd(s)}$$

(b)

$$Q = It = (1.45 \text{ A})(2.60 \times 3600 \text{ s}) = \boxed{1.36 \times 10^4 \text{ C}}$$

$$= 1.36 \times 10^4 \text{ C} \times \left(\frac{1 \text{ mol } e^-}{96485 \text{ C}}\right) = \boxed{0.141 \text{ mol } e^-}$$

(c) Every 1 mol of electrons that passes through the cell oxidizes 1/2 mol of zinc. Hence 0.0703 mol of zinc or $\boxed{4.60 \text{ g}}$ of Zn is oxidized. This is the mass lost by the zinc electrode.

(d) Every 1 mol of electrons that passes through the cell reduces 1/2 mol of cadmium(II) ions. This is 0.0702 mol of Cd^{2+}, or $\boxed{7.91 \text{ g}}$ of Cd^{2+}. This mass of cadmium metal is gained by the cadmium electrode.

12-46 Write the balanced half-equations:

$$2H_2O \longrightarrow O_2 + 4\,H^+ + 4\,e^- \quad \text{and} \quad 4\,Cl^- \longrightarrow 2\,Cl_2 + 4\,e^-$$

The passage of 4 mol of e^- generates 1 mol of $O_2(g)$; the same 4 mol of e^- generates 2 mol of $Cl_2(g)$. Hence the volume of chlorine will be twice the volume of oxygen, or $\boxed{5.30 \text{ L}}$.

12-48 Multiplying the relative mass of hydrogen that is liberated by 1 gives close to the known atomic mass of H. Similarly, multiplying the relative masses of oxygen, chlorine, and tin that are liberated by 2 for oxygen, by 1 for chlorine and by 2 for tin gives numbers that are close to the respective relative atomic masses of those three elements. Faraday's second law (text page 509) states that a given amount of charge liberates masses of different substances in proportion to their molar masses divided by the number of moles of electrons required to produce or consume one mole of the substance. This divisor equals the change in oxidation state in the reaction. The absolute values of the oxidation states of the four elements are therefore 1 (H), 2 (O), 1 (Cl) and 2 (Sn). Hydrogen is liberated at the cathode and must come from the reduction of +1 hydrogen (H^+); other positive oxidation states of hydrogen are nearly unknown. The oxygen is liberated at the *anode*, so that the oxidation state of the oxygen is -2. Similarly, the oxidation state of chlorine is -1 and that of tin is $+2$.

12-50 At the cathode the potassium ion is reduced: $e^- + K^+ \longrightarrow K$; at the anode, hydroxide ion is oxidized to oxygen: $2\,OH^- \longrightarrow 1/2\,O_2 + H_2O + 2\,e^-$.

12-52 A steady current of 75,000 A for 7.0 days (6.048×10^5 s) means that 4.536×10^{10} C passes through the circuit. Dividing by the Faraday constant gives the chemical amount of electricity passing through the circuit. It is 4.70×10^5 mol e^-. Every 2 mol of electrons theoretically accounts for the deposition of 1 mol of Mg. The theoretical yield of Mg is therefore 2.35×10^5 mol, which is $\boxed{5.7 \times 10^6 \text{ g}}$.

12-54 If the temperature rise is sufficient to melt the vanadium, as seems likely, the equation is:

$$\boxed{V_2O_5(s) + 5\,Ca(l) \longrightarrow 2\,V(l) + 5\,CaO(s)}$$

Thus, 5 mol of Ca is required for every 2 mol of V. The 20.0 kg of V is 0.393 kmol of V so 0.982 kmol of Ca is theoretically required. This is $\boxed{39.3 \text{ kg}}$ of Ca.

12-56 A current of 1.5 A for 22 minutes means that 1.98×10^3 C, which is 0.0205 mol of electrons, passes the cell. Each mole of electron deposits one mole of silver on the spoon; since the molar mass of silver is 107.87 g mol^{-1}, this is 2.214 g of silver. The volume of this silver is 0.211 cm^3. This volume of silver covers the 16 cm^2 of spoon surface to an average thickness of $\boxed{0.013 \text{ cm}}$.

12-58 The equilibrium $CO_2(g) + C(s) \rightleftharpoons 2\,CO(g)$ is crucial in the reduction of iron ore in a blast furnace. At room temperature it lies far to the left, but at 1875° (2148 K) it is strongly shifted to the right. Compute the $\Delta H°$ and $\Delta S°$ from the data in Appendix D and then use the relationship

$$-RT \ln K = \Delta H° - T\Delta S°$$

to solve for K at the two temperatures. The preliminary results are $\Delta H° = 172.47$ kJ and $\Delta S° = 175.75$ J K^{-1}; the answers are $K_{298.15} = \boxed{9.2 \times 10^{-22}}$ and $K_{2148} = \boxed{9.7 \times 10^4}$. The second K is only an approximation because $\Delta H°$ and $\Delta S°$ do change between 298 and 2148 K. Note that the answer to this problem is given on text page 495.

12-60 Iron ore is Fe_2O_3 admixed with silica, alumina and other material; copper ore is usually a sulfide. Iron ore is reduced with $CO(g)$ in a blast furnace by reaction of coke (carbon) with a limited amount of air. Fluxes like CaO are added to make a fluid slag from the silica and alumina impurities. Blast furnace slag has often simply been dumped, but also finds use in making portland cement. The roasting of copper ores gives off sulfur dioxide, which must not be vented into the atmosphere and is often processed into sulfuric acid. It reduces the copper only partially (to the copper(I) sulfide) and oxidizes impurities. Roasting of copper ores takes place at relatively low temperatures. Removal of impurities goes on at higher temperature (about 1100°C) but still below the 1535°C at which iron is reduced in a blast furnace. The flux in this step of copper recovery is $CaCO_3$, and the slag contains impurity iron. The copper is still combined chemically in Cu_2S. This compound is broken down by heating in the air with the generation of more $SO_2(g)$.

12-62

$$4\,OCl^-(aq) + 2\,OH^-(aq) + S_2O_3^{2-} \longrightarrow 2\,SO_4^{2-}(aq) + 4\,Cl^-(aq) + H_2O(aq)$$

12-64 The black silver sulfide is reduced $2\,e^- + Ag_2S(s) \longrightarrow 2\,Ag(s) + S^{2-}(aq)$, and the elemental aluminum is oxidized $Al(s) \longrightarrow Al^{3+}(aq) + 3\,e^-$. The overall reaction is

$$3\,Ag_2S(s) + 2\,Al(s) \longrightarrow 2\,Al^{3+}(aq) + 6\,Ag(s) + 3\,S^{2-}(aq)$$

The actual process involves many other reactions occurring simultaneously and ensuing.

12-66 The reaction is the reduction of NO to N_2O and the oxidation of NH_2OH to N_2 *or* the reduction of NO to N_2 and the oxidation of NH_2OH to N_2O. The ambiguity makes no difference in the balancing of the equation, but would make a difference in the construction of an electrochemical cell (Chapter 13) to run the reaction.

oxidation: $2\,OH^-(aq) + 2\,NH_2OH(aq) \longrightarrow N_2(g) + 4\,H_2O(g) + 2\,e^-$

reduction: $e^- + H_2O(g) + 2\,NO(g) \longrightarrow N_2O(g) + 2\,OH^-(aq)$

oxidation: $4\,OH^-(aq) + 2\,NH_2OH(aq) \longrightarrow N_2O(g) + 5\,H_2O(g) + 4\,e^-$

reduction: $4\,e^- + 2\,H_2O(g) + 2\,NO(g) \longrightarrow N_2(g) + 4\,OH^-(aq)$

12-68 The equation for the bacteria's action is

$$2\,SeO_4^{2-}(aq) + 2\,H_2O(l) \longrightarrow 2\,Se(s) + 3\,O_2(g) + 4\,OH^-(aq)$$

It is tempting to focus narrowly on the formula of the selenate ion and reason that 2 mol of $O_2(g)$ must form in the reduction of 1 mol of SeO_4^{2-}. This is wrong because it does not take into account the charge on the selenate ion. The correct ratio is 3/2. This ratio is used in the following:

$$m_{O_2} = 10^{12}\,L \times \left(\frac{0.100\ g\ SeO_4^{2-}}{1\ L}\right) \times \left(\frac{1\ mol\ SeO_4^{2-}}{143\ g\ SeO_4^{2-}}\right)$$

$$\times \left(\frac{3\ mol\ O_2}{2\ mol\ SeO_4^{2-}}\right) \times \left(\frac{32.0\ g\ O_2}{1\ mol\ O_2}\right) = \boxed{3.4 \times 10^{10}\ g\ O_2}$$

12-70 The problem can be solved by a series of unit conversions. The maximum allowable concentration of Sn^{2+} is 10 ppm. Each 100 mL sample of rinse solution has a mass of 100 g because it is a dilute solution and consists mostly of water (density $= 1.00\ g\ mL^{-1}$). Then:

$$100\ mL \times \left(\frac{1.00\ g}{1.00\ mL}\right) \times \left(\frac{10 \times 10^{-6}\ g\ Sn^{2+}}{1\ g\ sample}\right) \times \left(\frac{1\ mol\ Sn^{2+}}{118.69\ g\ Sn^{2+}}\right)$$

$$\times \left(\frac{2\ mol\ e^-}{1\ mol\ Sn^{2+}}\right) \times \left(\frac{96485\ C}{1\ mol\ e^-}\right) \times \left(\frac{1\ s}{25.0 \times 10^{-3}\ C}\right) = \boxed{6.50\ s}$$

12-72 The ΔG° of the reaction $Mg(s) + H_2O(l) \longrightarrow H_2(g) + MgO(s)$ is -341 kJ.

The negative sign shows that water and magnesium $\boxed{\text{do tend to react}}$ in their standard states at room temperature. The tendency is even greater under the non-standard conditions in which the partial pressure of the product hydrogen is less than 1 atm, which is the case if water is simply poured over magnesium. The reaction is however quite slow because ordinary bulk samples of $Mg(s)$ are passivated by a coating of $MgO(s)$.

12-74

$$55.5\ kg\ crude\ Cu \times \left(\frac{98.3\ kg\ Cu}{100\ kg\ crude\ Cu}\right) \times \left(\frac{1\ mol\ Cu}{0.06355\ kg\ Cu}\right) \times \left(\frac{2\ mol\ e^-}{1\ mol\ Cu}\right)$$

$$\times \left(\frac{96485\ C}{1\ mol\ e^-}\right) \times \left(\frac{1\ s}{2.00 \times 10^3\ C}\right) = \boxed{8.3 \times 10^4\ s}$$

This is about 1 day.

12-76 The problem states that 260 mg of the metallic element (call it M) is deposited during 12.1 minutes. Express this as a ratio and then compute the molar mass of M by applying unit-factors as follows:

$$\frac{260 \text{ mg M}}{12.1 \text{ min}} \times \left(\frac{1 \text{ g M}}{1000 \text{ mg M}}\right) \times \left(\frac{1 \text{ min}}{60 \text{ s}}\right) \times \left(\frac{1 \text{ s}}{2.0 \text{ C}}\right)$$
$$\times \left(\frac{96485 \text{ C}}{1 \text{ mol } e^-}\right) \times \left(\frac{n \text{ mol } e^-}{\text{mol M}}\right) = \frac{17.3(n) \text{ g M}}{\text{mol M}}$$

The n here equals the number of moles of electrons required to reduce one mole of M from the fused salt to the metal. It is an integer. The precision of the observed current is only two significant figures. This restricts the precision of the answer to two significant digits, that is, to $17.3 \pm 0.9\,(n)$ g mol^{-1}. With this in mind, a table is constructed to identify possibilities for M by their molar masses. The table runs up to $n = 8$, which corresponds to a 8-electron reduction:

n	Range of Molar Mass	Possible Elements
1	16.3—18.2 g mol^{-1}	none
2	33.7—35.5	Cl
3	51.0—52.8	Cr
4	68.3—70.1	Ga
5	85.6—87.4	Rb, Sr
6	102.9—104.7	Rh
7	120.2—122.0	Sb
8	137.5—139.3	none

In the list of possibilities, Cl is not a metal, Ga does not have a +4 oxidation state, Rb and Sr do not have a +5 oxidation state, and Sb does not have a +7 oxidation state. The answers are therefore $\boxed{\text{Cr}}$, which does have the +3 oxidation state, and $\boxed{\text{Rh}}$, which does have the +6 oxidation state (in RhF$_6$).

12-78 The equation

$$[\text{Co(H}_2\text{O})_6]^{2+} + \text{HSO}_5^- \longrightarrow [\text{Co(H}_2\text{O})_5]^{2+} + 2\text{H}^+ + \text{SO}_4^-$$

is unbalanced with respect to the electrical charge represented on the two sides and also unbalanced with respect to the element oxygen (11 O's are shown on the left, and 10 O's on the right). A balanced version is:

$$[\text{Co(H}_2\text{O})_6]^{2+} + \text{HSO}_5^- \longrightarrow [\text{Co(H}_2\text{O})_5]^{2+} + \text{SO}_4^- + \text{H}_2\text{O}$$

In this reaction, the hydrogen peroxosulfate ion (see text page 875) oxidizes the hexaaquacobalt(II) ion to pentaaquahydroxocobalt(III) ion. The other product is the peroxosulfite ion. This problem may be easier for beginners than for those who automatically "correct" the formulas of the S-containing ions to those of the hydrogen sulfate ion and the sulfite ion.

Answers to Even-Numbered Problems, Chapter 13

13-2 The electrical work that the two batteries expend in the surroundings is the negative of the work that the surroundings perform on the two batteries. A formula for the latter is given on text page 527. Thus,

$$w_{\text{frombattery}} = -w_{\text{elec}} = -(-It\Delta\mathcal{E})$$

$$= (1.1 \text{ A})(10 \text{ h})(2.9 \text{ V}) = 31.9 \text{ V A h} = \boxed{0.0319 \text{ kW-h}}$$

The energy costs $2.00, which is means it costs $62 per kW-h. This is 620 times more expensive than the energy consumed burning the porch light in problem 13-1.

13-4 The point of the problem is to establish how very long it would take to extract the maximum possible work from an electrochemical cell. Solve the equation on text page 527 for t, and insert w_{elec}, I, and $\Delta\mathcal{E}$:

$$t = \frac{w_{\text{elec}}}{-I\Delta\mathcal{E}} = \frac{-1.00 \text{ J}}{-(1.60 \times 10^{-18} \text{ A})(2.00 \text{ V})} = \boxed{3.13 \times 10^{17} \text{ s}}$$

This is about 10 billion years.

13-6 (a) The convention is to write the oxidation (anode) half-cell on the left and the reduction (cathode) half-cell on the right. Therefore the redox reaction in this cell is

$$\text{Mn}(s) + \text{Cr}^{2+}(aq) \longrightarrow \text{Mn}^{2+}(aq) + \text{Cr}(s)$$

This process involves transfer of 2 mol of electron per mole of reaction as written ($n = 2$). The standard cell voltage is computed as follows:

$$\Delta\mathcal{E}° = \mathcal{E}°(\text{Cr}^{2+}|\text{Cr}) - \mathcal{E}°(\text{Mn}^{2+}|\text{Mn}) = -0.905 - (-1.029) = \boxed{+0.124 \text{ V}}$$

where the standard reduction potentials comes from Appendix E. The standard cell voltage is a direct measure of the standard Gibbs function of the reaction:

$$\Delta G° = -n\mathcal{F}\Delta\mathcal{E}° = -2 \text{ mol}\left(\frac{96,485 \text{ C}}{\text{mol}}\right)(0.124 \text{ V}) = -23.93 \times 10^3 \text{ C V}$$

$$= \boxed{-23.93 \text{ kJ}}$$

(b) The procedure is similar that used in the previous part:

$$\text{Mn}(s) + 2\,\text{Mn}^{3+}(aq) \longrightarrow 3\,\text{Mn}^{2+}(aq) \quad n = 2$$

$$\Delta\mathcal{E}^\circ = \mathcal{E}^\circ(\text{Mn}^{3+}|\text{Mn}^{2+}) - \mathcal{E}^\circ(\text{Mn}^{2+}|\text{Mn}) = 1.51 - (-1.029) = \boxed{+2.54 \text{ V}}$$

$$\Delta G^\circ = -n\mathcal{F}\Delta\mathcal{E}^\circ = -2\,\text{mol}(96,485\,\text{C mol}^{-1})(2.54\,\text{V}) = -490 \times 10^3 \text{ C V}$$
$$= \boxed{-490 \text{ kJ}}$$

(c) The procedure is similar to that used in the previous parts:

$$2\,\text{Ag}(s) + \text{I}_2(s) \longrightarrow 2\,\text{Ag}^+(aq) + 2\,\text{I}^-(aq) \quad n = 2$$

$$\Delta\mathcal{E}^\circ = \mathcal{E}^\circ(\text{Ag}^+|\text{Ag}) - \mathcal{E}^\circ(\text{I}_2|\text{I}^-) = 0.7996 - (0.535) = \boxed{+0.265 \text{ V}}$$

$$\Delta G^\circ = -n\mathcal{F}\Delta\mathcal{E}^\circ = -2\,\text{mol}\left(\frac{96,485\,\text{C}}{1\,\text{mol}}\right)(2.646\,\text{V}) = -51.1 \times 10^3 \text{ C V}$$
$$= \boxed{-51.1 \text{ kJ}}$$

13-8 Use the equations on text page 528:

$$\Delta G = -n\mathcal{F}\Delta\mathcal{E} = w_{\text{elec, max}}$$

The $\Delta\mathcal{E}$ is 0.48 V, and $\mathcal{F}$ is 96,485 C mol^{-1}. If one mole of zinc were oxidized, then n would equal 2 mol, and $w_{elec,max}$ would be -92.6 kJ. But 1.000 g of zinc is only 0.0153 mol. Therefore insert $n = 0.0306$ mol into the formula. This shows that the maximum work done *on* the cell is -1.4 kJ. The maximum work done by the cell is the negative of this, or $\boxed{+1.4 \text{ kJ}}$ per gram of zinc consumed.

13-10 (a) The reduction of Fe^{3+} to Fe^{2+} has $\mathcal{E}^\circ = 0.770$ V, and the reduction of Cd^{2+} to Cd has $\mathcal{E}^\circ = -0.4026$ V. In this galvanic cell with all species in their standard states, the half-reaction Fe$^{3+} + e^- \longrightarrow$ Fe^{2+} takes place at the cathode, and the half-reaction Cd $\longrightarrow$ Cd$^{2+} + 2\,e^-$ takes place at the anode. Only in this way is $\Delta\mathcal{E}^\circ$ of the cell positive.

(b) $\quad \Delta\mathcal{E}^\circ = \mathcal{E}^\circ\,(\text{cathode}) - \mathcal{E}^\circ\,(\text{anode}) = 0.770 - (-0.4026) = \boxed{1.173 \text{ V}}$.

13-12 (a) The increase in the concentration of Cl^- ion means that $Cl_2(g)$ is reduced as the cell operates spontaneously. The half-reaction at the cathode, where reduction occurs, is accordingly $\boxed{Cl_2 + 2e^- \longrightarrow 2\,Cl^-}$. At the anode, the half-reaction is $\boxed{Ga \longrightarrow Ga^{3+} + 3\,e^-}$.

(b) The overall voltage equals $\Delta\mathcal{E}^\circ = 1.918$ V. It equals the standard reduction potential of the reduction half-reaction minus the standard reduction potential of the oxidation half-reaction The standard reduction potential for the couple $Cl_2|Cl^-$ is 1.3583 V (Appendix E). Hence:

$$\Delta\mathcal{E}^\circ = 1.918 = \mathcal{E}^\circ\,(\text{cathode}) - \mathcal{E}^\circ\,(\text{anode}) = 1.3583 - \mathcal{E}^\circ\,(Ga^{3+}|Ga)$$

Solving gives $\boxed{-0.560 \text{ V}}$ as the standard reduction potential for the couple $Ga^{3+}(aq)|Ga(s)$.

13-14 The $\Delta\mathcal{E}^\circ$ of the disproportionation reaction is

$$\Delta\mathcal{E}^\circ = \mathcal{E}^\circ\,(\text{cathode}) - \mathcal{E}^\circ\,(\text{anode}) = (2.87) - (-0.37) = 3.24 \text{ V}$$

The positive voltage means that $\boxed{\text{yes}}$, the reaction is spontaneous. Note how a negative standard reduction potential of the anode half-reaction favors overall spontaneity.

13-16 To remove Mn^{3+} in solutions of Mn^{2+}, $\boxed{\text{add some Mn}}$. The reaction

$$Mn(s) + Mn^{3+}(aq) \longrightarrow 2\,Mn^{2+}(aq)$$

occurs spontaneously under standard-state conditions:

$$\Delta\mathcal{E}^\circ = 1.51 - (-1.029) = 2.54 \text{ V}$$

Even when the $Mn^{3+}(aq)$ is very dilute (a non-standard concentration), the reaction remains spontaneous as written.

13-18 An acidic solution of sodium perchlorate would be an $\boxed{\text{oxidizing agent}}$.

13-20 At pH 0, the ClO^- is essentially completely converted to $HClO$. The standard reduction potentials for ClO_3^-, $HClO$ and Cl_2 in acidic aqueous media are 1.47, 1.63, and 1.3583 V respectively, according to Appendix E. This means that at pH 0 a solution of $\boxed{\text{NaClO}}$ is the strongest bleach, despite the fact that the change in oxidation number of the Cl is only -2.

At pH 0 the reduction potential of $O_3(g)$ to $O_2(g)$ is 2.07 V (given on text page 536). Therefore $\boxed{\text{ozone is stronger}}$ as a bleach than any of the chlorine-containing species.

13-22 The species that are produced by reductions having standard reduction potentials algebraically *less* than that of $Pt(OH)_2(s)$ will be oxidized; the others will not be oxidized.

(a) The reduction of $Ni(OH_2)(s)$ to $Ni(s)$ has $\mathcal{E}°$ equal to -0.66 V, so $Ni(s)$ $\boxed{\text{is oxidized}}$ by $Pt(OH)_2(s)$.

(b) The reduction of $O_2(g)$ to $OH^-(aq)$ has $\mathcal{E}°$ equal to $+0.401$ V, so $OH^-(aq)$ $\boxed{\text{is not oxidized}}$ by $Pt(OH)_2(s)$.

(c) The reduction of $PbO(s)$ to $Pb(s)$ has $\mathcal{E}°$ equal to -0.576 V, so $Pb(s)$ $\boxed{\text{is oxidized}}$ by $Pt(OH)_2(s)$.

(d) The reduction of $F_2(g)$ to $F^-(aq)$ has $\mathcal{E}°$ equal to $+2.87$ V, so $F^-(aq)$ $\boxed{\text{is not oxidized}}$ by $Pt(OH)_2(s)$.

Note that the proper comparison is to the reduction potential for the half-reaction occurring in basic solution. It is incorrect for example in part (a) to make a comparison to 0.23 V, which is the standard reduction potential for $Ni^{2+}(aq)|Ni(s)$, not $Ni(OH)_2(s)|Ni(s)$.

13-24 (a) Sulfur $(S(s))$ $\boxed{\text{does tend}}$ spontaneously to disproportionate at pH 14 to give $HS^-(aq)$ and $S_2O_3^{2-}(aq)$. The $\Delta\mathcal{E}°$ of the reaction is $-0.51-(-0.74) = 0.23$ V.

(b) According to the reduction potentials given in the problem $S(s)$ is oxidized to $S_2O_3^{2-}(aq)$ more readily than $HS^-(aq)$ is oxidized to $S(s)$ at pH 14; the $S(s)$ is a $\boxed{\text{stronger}}$ reducing agent under these conditions.

13-26 Calculate the standard potential difference of the cell using data from Appendix E. At the anode is the $Ag^+(aq)|Ag(s)$ couple; at the cathode is the $Cl_2(aq)|Cl^-(aq)$ couple. For the overall cell reaction

$$2\,Ag(s) + Cl_2(g) \longrightarrow 2Ag^+(aq) + 2\,Cl^-(aq) \qquad \Delta\mathcal{E}° = (1.3583 - 0.7996)\ \text{V}$$

This standard voltage must be corrected for the non-standard concentrations in this cell. At 25°C:

$$\Delta\mathcal{E} = \Delta\mathcal{E}° - \frac{0.0592\ \text{V}}{n}\log Q = 0.5587\ \text{V} - \frac{0.0592\ \text{V}}{2}\log\left(\frac{[Ag^+]^2[Cl^-]^2}{P_{Cl_2}}\right)$$

$$= 0.5587\ \text{V} - \frac{0.0592\ \text{V}}{2}\log\left(\frac{(0.25)^2(0.016)^2}{1.00}\right) = \boxed{0.701\ \text{V}}$$

13-28 The cells in the three parts of this problem are the same as in the three parts of problem 13-6. Take the standard cell voltages from problem 13-6, construct the reaction quotients according to the reactions written in that problem, and set up the following equations using the Nernst relation:

(a)

$$\Delta \mathcal{E} = 0.124 \text{ V} - \frac{0.0592 \text{ V}}{2} \log \left(\frac{[\text{Mn}^{2+}]}{[\text{Cr}^{2+}]} \right)$$

$$= 0.124 \text{ V} - \frac{0.0592 \text{ V}}{2} \log \left(\frac{0.002}{0.075} \right) = \boxed{0.171 \text{ V}}$$

(b)

$$\Delta \mathcal{E} = 2.54 \text{ V} - \frac{0.0592 \text{ V}}{2} \log \left(\frac{[\text{Mn}^{2+}]^2_{\text{cathode}} [\text{Mn}^{2+}]_{\text{anode}}}{[\text{Mn}^{3+}]^2} \right)$$

$$= 2.54 \text{ V} - \frac{0.0592 \text{ V}}{2} \log \left(\frac{(0.95)^2)(0.010)}{(0.050)^2} \right) = \boxed{2.52 \text{ V}}$$

(c)

$$\Delta \mathcal{E} = 0.265 \text{ V} - \frac{0.0592 \text{ V}}{2} \log \left(\frac{[\text{I}^-]^2}{[\text{Ag}^+]^2} \right)$$

$$= 0.265 \text{ V} - \frac{0.0592 \text{ V}}{2} \log \left(\frac{(0.075)^2}{(0.10)^2} \right) = \boxed{0.272 \text{ V}}$$

13-30 Use the Nernst equation and the standard reduction potential from Appendix E. For the half-reaction:

$$\text{I}_2(s) + 2\,e^- \longrightarrow 2\,\text{I}^-(aq) \qquad \mathcal{E}^\circ = 0.535 \text{ V}$$

$$\mathcal{E} = \mathcal{E}^\circ - \frac{0.0592 \text{ V}}{n_{\text{hc}}} \log[\text{I}^-]^2$$

$$\mathcal{E} = 0.535 \text{ V} - \frac{0.0592 \text{ V}}{2} \log[1.5 \times 10^{-6}]^2 = \boxed{0.880 \text{ V}}$$

13-32 The reduction potential computed in problem 13-30 is algebraically larger than the potential for the reduction of Fe^{3+} to Fe^{2+} (0.770 V), so the $\boxed{I_2(s)|I^-(aq)}$ half-cell acts as the cathode half-cell. If all of the reactants were in their standard states, this half-cell would be the anode. The very low concentration of $I^-(aq)$ in this cell causes the reversal.

13-34 The redox reaction in the cell is:

$$Cu(s) + Br_2(l) \longrightarrow Cu^{2+}(aq) + 2\,Br^-(aq)$$

in which $Cu(s)$ is oxidized and $Br_2(l)$ is reduced. Compute the standard potential difference associated with this reaction using data from Appendix E:

$$\Delta\mathcal{E}° = \mathcal{E}°\,(\text{cathode}) - \mathcal{E}°\,(\text{anode}) = 1.065 - 0.3402 = 0.725 \text{ V}$$

Use the Nernst equation at 25°C to obtain the unknown concentration:

$$\Delta\mathcal{E} = \Delta\mathcal{E}° - \frac{0.0592 \text{ V}}{n} \log Q$$

$$0.963 \text{ V} = 0.725 \text{ V} - \frac{0.0592 \text{ V}}{2} \log\left([Cu^{2+}][Br^-]^2\right)$$

Inserting $[Cu^{2+}] = 1.00 \text{ mol L}^{-1}$ and solving for $[Br^-]$ gives $[Br^-] = 10^{-4.02} = \boxed{9.5 \times 10^{-5} \text{ mol L}^{-1}}$.

13-36 (a) The equations for the reduction (at the cathode) and oxidation (at the anode) are respectively:

$$Cr_2O_7^{2-}(aq) + 14\,H^+(aq) + 6\,e^- \longrightarrow 2\,Cr^{3+}(aq) + 7\,H_2O(l)$$

$$6\,I^-(aq) \longrightarrow 3\,I_2(s) + 6\,e^-$$

Appendix E provides the standard reduction potentials. Combine them to obtain the standard voltage of the cell:

$$\Delta\mathcal{E}° = \mathcal{E}°\,\text{cathode} - \mathcal{E}°\,(\text{anode}) = 1.33 - 0.535 = 0.795 \approx \boxed{0.80 \text{ V}}$$

(b) Write the Nernst equation at 25°C for the cell and insert the known voltages and concentrations:

$$\Delta\mathcal{E} = \Delta\mathcal{E}° - \frac{0.0592 \text{ V}}{n} \log Q = \Delta\mathcal{E}° - \frac{0.0592 \text{ V}}{6} \log\left(\frac{[Cr^{3+}]^2}{[Cr_2O_7^{2-}][I^-]^6[H^+]^{14}}\right)$$

$$0.870 \text{ V} = 0.795 \text{ V} - \frac{0.0592 \text{ V}}{6} \log\left(\frac{[Cr^{3+}]^2}{(1.5)\,(0.40)^6\,(1.0)^{14}}\right)$$

Solving gives $[Cr^{3+}] = \boxed{1.2 \times 10^{-5} \text{ mol L}^{-1}}$.

13-38 (a)

$$\text{Cathode}: \quad \text{MnO}_4^-(aq) + 8\,\text{H}^+(aq) + 5\,e^- \longrightarrow \text{Mn}^{2+}(aq) + 4\,\text{H}_2\text{O}(l)$$

$$\text{Anode}: \quad \text{Fe}^{2+}(aq) \longrightarrow \text{Fe}^{3+}(aq) + e^-$$

$$\text{MnO}_4^-(aq) + 8\,\text{H}^+(aq) + 5\,\text{Fe}^{2+}(aq) \longrightarrow \text{Mn}^{2+}(aq) + 4\,\text{H}_2\text{O}(l) + 5\,\text{Fe}^{3+}(aq)$$

(b) The standard potential difference of the cell (using the $\mathcal{E}°$'s from Appendix E) is $1.491 - 0.770 = 0.721$ V. Write the Nernst equation at 25°C

$$\Delta\mathcal{E} = \Delta\mathcal{E}° - \frac{0.0592}{5}\log\frac{[\text{Mn}^{2+}]\,[\text{Fe}^{3+}]^5}{[\text{H}^+]^8[\text{MnO}_4^-][\text{Fe}^{2+}]^5}$$

and substitute this $\Delta\mathcal{E}°$ and the given initial concentrations:

$$\Delta\mathcal{E} = 0.721 - \frac{0.0592}{5}\log\big((0.01)^{-8}\big) = 0.721 - \frac{0.0592}{5}(16) = \boxed{0.532 \text{ V}}$$

(c) The overall reaction has $n = 5$ and $\Delta\mathcal{E}° = 0.721$ V. Substitute these values as shown in Example 13-8 on text page 545:

$$\log_{10} K = \frac{n}{0.0592 \text{ V}}\Delta\mathcal{E}° = \frac{5}{0.0592 \text{ V}}0.721 \text{ V} = 60.9 \qquad K = 10^{60.9} = \boxed{8 \times 10^{60}}$$

13-40 The standard potential difference of the reaction is the standard reduction potential for the conversion of mercury(II) to mercury(I) minus the standard reduction potential for the conversion of gold(III) to gold(0): $\Delta\mathcal{E}° = 0.905 - 1.42 = -0.515$ V. The chemical amount of electrons transferred in the reaction written in the problem is 6, so $n = 6$. At 25°C:

$$\log K = \frac{n}{0.0592 \text{ V}}\Delta\mathcal{E}° = \frac{6}{0.0592 \text{ V}}(-0.515 \text{ V}) = -52.2$$

The equilibrium constant for the reaction is $\boxed{K = 6 \times 10^{-53}}$.

The second question really concerns the reverse of the reaction written in the problem. The equilibrium constant of the reverse reaction is the reciprocal of the K just computed:

$$K_{\text{reverse}} = \frac{1}{6 \times 10^{-53}} = \frac{[\text{Hg}^{2+}]^6}{[\text{Au}^{3+}]^2[\text{Hg}_2^{2+}]^3}$$

After the Hg_2^{2+} and Au^{3+} are mixed, but before the reaction starts, their concentrations are both 0.50 mol L^{-1}. The equilibrium constant is very large, so the reaction proceeds until the Hg_2^{2+} ion, which is the limiting reactant, is essentially all consumed. Suppose that *all* of the Hg_2^{2+} reacts. The concentration of Hg^{2+} formed then equals 1.00 mol L^{-1}, and the concentration of excess Au^{3+} is 0.166 mol L^{-1}, by the stoichiometry of the reaction. Let the concentration of Hg_2^{2+} actually left at equilibrium equal x. Then $[Au^{3+}] = 0.166 + 2/3x$ mol L^{-1}, and $[Hg^{2+}] = 1.00 - x/2$ mol L^{-1} at equilibrium. Substitute these expressions into the equilibrium expression:

$$\frac{1}{6 \times 10^{-53}} = \frac{(1.00 - x/2)^6}{(0.166 + 2/3x)^2 (x)^3}$$

Because x is certainly very small compared to 0.166, this equation becomes

$$\frac{1}{6 \times 10^{-53}} \approx \frac{(1.00)^6}{(0.166)^2 (x)^3}$$

Solving gives $x = \boxed{1 \times 10^{-17} \text{ mol L}^{-1}}$. This is the equilibrium concentration of Hg_2^{2+} ion. The equilibrium concentration of Au^{3+} ion is $\boxed{0.166 \text{ mol L}^{-1}}$, and the equilibrium concentration of Hg^{2+} ion is $\boxed{1.00 \text{ mol L}^{-1}}$.

13-42 (a) The reaction in the problem equals the second of the following half-reactions subtracted from the first

$$Hg^{2+}(aq) + e^- \longrightarrow \frac{1}{2}Hg_2^{2+}(aq) \qquad \frac{1}{2}Hg_2^{2+}(aq) + e^- \longrightarrow Hg(l)$$

The $\Delta\mathcal{E}^\circ$ for the reaction is accordingly the $\mathcal{E}^\circ$ for the second half-reaction subtracted from the $\mathcal{E}^\circ$ for the first. Taking values from Appendix E, $\Delta\mathcal{E}^\circ = 0.905 - 0.7961 = \boxed{0.109 \text{ V}}$. Although the half-equations in Appendix E are written with coefficients that are twice the coefficients in the above, the standard potential difference is the same; it does not depend on the amount of substance that reacts. Inserting this number into the relationship (at 25°C) between the equilibrium constant and the standard potential difference gives:

$$\log_{10} K = \frac{n}{0.0592}\Delta\mathcal{E}^\circ = \frac{1}{0.0592 \text{ V}}(0.109 \text{ V}) = 1.84 \qquad \boxed{K = 69}$$

(b) In a solution of mercury(I) chloride (Hg_2Cl_2), there is at equilibrium a small concentration of mercury(II) (the Hg^{2+} ion) formed according to the reverse of the equation in the problem. The addition of ammonia to a solution of Hg_2Cl_2 precipitates this mercury(II) as $HgNH_2Cl$. Removal of mercury(II) from solution shifts the equilibrium to the left (LeChatelier's principle). The shift produces elemental mercury as well.

13-44 The standard potential difference for the reaction is

$$\Delta\mathcal{E}^\circ = \mathcal{E}^\circ(Ag^+|Ag) - \mathcal{E}^\circ(H^+|H_2) = 0.7996 - (0) = 0.7996 \text{ V}$$

Write the Nernst equation at 25°C

$$\Delta\mathcal{E} = \Delta\mathcal{E}^\circ - \frac{0.0592}{n} \log \frac{[H^+]}{[Ag^+]P_{H_2}^{1/2}}$$

and substitute all of the known values

$$1.030 = 0.7996 - \frac{0.0592}{1} \log \frac{[H^+]}{(1.00)(1.00)}$$

From this equation, $[H^+]$ is 1.28×10^{-4} mol L^{-1}. The pH in the buffer solution in the cell is $\boxed{3.89}$. Now, compute the K_a of the benzoic acid by substitution in the acid ionization equilibrium expression

$$K_a = \frac{[H_3O^+][C_6H_5COO^-]}{[C_6H_5COOH]} = \frac{(0.000128)(0.050)}{(0.10)} = \boxed{6.4 \times 10^{-5}}$$

13-46 (a) Elemental lead is oxidized, and hydrogen ion is reduced. The standard potential difference is

$$\Delta\mathcal{E}^\circ = \mathcal{E}^\circ(H^+|H_2) - \mathcal{E}^\circ(Pb^{2+}|Pb) = 0 - (-0.1263) = \boxed{0.1263 \text{ V}}$$

(b) Write the Nernst equation at 25°C for this cell and substitute all of the known values

$$\Delta\mathcal{E} = \Delta\mathcal{E}^\circ - \frac{0.0592}{n} \log \frac{[Pb^{2+}]P_{H_2}}{[H^+]^2}$$

$$0.22 = 0.1263 - \frac{0.0592}{2} \log \frac{[Pb^{2+}]1.0}{[1.00]^2}$$

Solving gives the $[Pb^{2+}]$ as 6.8×10^{-4} mol L^{-1}.

(c) Presumably the lead concentration in part (b) is in equilibrium with the 0.15 M Cl$^-$ ion. If so, the K_{sp} expression applies:

$$K_{sp} = [Pb^{2+}][Cl^-]^2 = (6.8 \times 10^{-4})(0.15)^2 = \boxed{1.5 \times 10^{-5}}$$

13-48 The overall reaction is

$$Zn(s) + HgO(s) + H_2O(l) \longrightarrow Zn(OH)_2(s) + Hg(l)$$

for which

$$\Delta G^\circ = 1(-553.5) + 1(0) - 1(0) - 1(-58.56) - 1(-237.18) = -257.76 \text{ kJ}$$

Compute the $\Delta\mathcal{E}^\circ$ by substitution, recalling that a joule is a volt-coulomb:

$$\Delta\mathcal{E}^\circ = -\frac{\Delta G^\circ}{n\mathcal{F}} = -\frac{-257.76 \times 10^3 \text{ J}}{(2 \text{ mol})(96485 \text{ C mol}^{-1})} = \boxed{1.336 \text{ V}}$$

13-50 The reaction in the battery is

$$Zn(s) + HgO(s) + H_2O(l) \longrightarrow Zn(OH)_2(s) + Hg(l)$$

The equation implies that the reduction of 1 mol of HgO in the battery occurs by passage of 2 mol of electrons through an outside circuit. The 0.50 g of HgO (molar mass 216.59 g mol^{-1}) is 2.3×10^{-3} mol of HgO. Hence, 4.6×10^{-3} mol of electrons pass the circuit. Multiplying by the Faraday constant converts this to $\boxed{4.5 \times 10^2 \text{ C}}$.

13-52 The maximum possible work absorbed by the cell (at constant T and P) equals the charge Q from problem 13-50 times the potential difference through which each portion of that charge is moved. The potential difference of a zinc-mercuric oxide cell fortunately hardly changes in the course of its discharge. Therefore the maximum work *out* of the cell is:

$$-w_{\text{elec,max}} = Q\Delta\mathcal{E} = (4.455 \times 10^2 \text{ C})(1.34 \text{ V}) = \boxed{600 \text{ J}}$$

For this cell ΔG is nearly equal to ΔG° because the reactants and products stay in standard states throughout the reaction. The maximum work out of the cell can therefore also be computed using

$$-w_{\text{elec,max}} = -\Delta G \approx \Delta G^\circ$$

The standard Gibbs function of the cell reaction was obtained in the course of solving problem 13-48. It is -257.76 kJ per mol of HgO reacting. For 0.50 g of HgO(s), ΔG° equals -0.595 kJ. Thus, the maximum electric work *from* the cell is, again, $+600$ J.

13-54 Partial discharge of the lead-acid battery in the vain effort to start the car reduces the concentration of sulfuric acid within the battery. This raises the freezing point of the solution. After the battery re-cools (it generates some internal heat in the starting effort), it can now freeze despite the somewhat warmer ambient temperature.

13-56 The $\Delta G°$ for the oxidation of 1.00 mol of $CO(g)$ to $CO_2(g)$ is -257.21 kJ, regardless of how the reaction is performed. If the reaction is performed with 100% efficiency at standard conditions, then $\Delta G° = -w$. The cell absorbs -257.21 kJ of work, that is, $\boxed{257.21 \text{ kJ}}$ of work is obtained.

13-58 The standard cell voltage for the rusting of iron by the overall equation

$$Fe(s) + 1/2\,O_2(g) + 2\,H^+(aq) \longrightarrow Fe^{2+}(aq) + H_2O(l)$$

is $\Delta\mathcal{E}° = \mathcal{E}°_{\text{cathode}} - \mathcal{E}°_{\text{anode}} = 1.229 - (-0.409) = \boxed{1.638 \text{ V}}$. There is a considerable driving force for the rusting of iron at a pressure of $O_2(g)$ of 1 atm and a pH of 0. Making the water even more acidic increases the driving force.

13-60 The successful use of titanium as a sacrificial anode proves that it is more easily oxidized than iron. Assuming that $Ti(s)$ corrodes to $Ti^{3+}(aq)$, the standard reduction potential for Ti^{3+} to $Ti(s)$ is therefore $\boxed{\text{more negative than } -0.409 \text{ V}}$, which is the standard potential for the reduction of $Fe^{2+}(aq)$ to $Fe(s)$.

13-62 First, compute the volume of zinc that will be coated onto the steel. This volume is the thickness of the coating times the width of the steel times the length times two (for the two sides) and comes out to 0.0500 m^3. Next comes the mass of the zinc. Convert the volume to cm^3 and multiply by 7.133 g cm^{-3}, the density of zinc, to get a mass of 3.566×10^5 g of Zn. This is 5.454×10^3 mol of Zn. Each mole of Zn requires 2 mol of electrons to plate it out, and a mole of electrons is 96,485 coulombs. The total amount of electricity used is therefore 1.053×10^9 C. The energy used is the amount of electricity multiplied by the voltage, which is 3.5 V in this case. The energy is thus 3.68×10^9 V C which is 3.68×10^9 J. This energy must be divided by 0.9 because the galvanizing is only 90% efficient. This raises the energy consumption to 4.09×10^9 J, equivalent to 1.14×10^3 kW-hr, which costs $\boxed{\$114}$. It is not necessary to know the current unless it is desired to calculate how long the galvanizing will take.

13-64 (a) The $\Delta\mathcal{E}°$ for the proposed reduction of $Cu^{2+}(aq)$ by $I^-(aq$ is $(0.158) - (0.535) = -0.377$ V. The reaction of the two ions in their standard states to give $Cu^+(aq)$ and $I_2(s)$ in their standard states is therefore non-spontaneous. The answer is $\boxed{\text{no}}$.

(b) Recognize that $CuI(s)$ forms. Then the reduction half-reaction in part (a) is replaced by another half-reaction with an $\mathcal{E}°$ of 0.86 V. Now, $\Delta\mathcal{E}° = (0.86) - (0.535) = 0.32$ V. The reaction giving $CuI(s)$ in its standard state from aqueous Cu^{2+} and I^- in their standard states is spontaneous. The answer is therefore $\boxed{\text{yes}}$. The formation of insoluble $CuI(s)$ drives a redox interaction that would not occur if the product ions formed in their standard states.

13-66 The $Fe^{3+}IFe^{2+}$ half-reaction has a standard reduction potential of 0.770 V. The model for the "insoluble Mn(III) and Mn(IV) compounds" is $MnO_2(s)$, which has a reduction potential of 1.208 V at pH 0. A reducing agent like $Br^-(aq)$, would reduce $MnO_2(s)$ at pH 0 without reducing the $Fe^{3+}(aq)$ because $\mathcal{E}°$ of the $Br_2(l)|Br^-(aq)$ half cell is 1.065 V. Deviations from standard conditions of course affect voltages. This fact and the different rates at which reactions might proceed in a restoration operation make the suggestion of Br^- very tentative.

13-68 (a) The half-cell potential is calculated using the Nernst equation at 25°C:

$$\mathcal{E} = \mathcal{E}° - \frac{0.0592}{n} \log Q_{\text{hc}} = 1.229 - \frac{0.0592}{4} \log \left(\frac{1}{P_{O_2}[H^+]^4} \right)$$

$$\mathcal{E} = 1.229 - \frac{0.0592}{4} \log \left(\frac{1}{(1.0)(1.0 \times 10^{-7})^4} \right) = \boxed{0.815 \text{ V}}$$

(b) Aeration of a solution puts $O_2(g)$ at a pressure of about 0.2 atm in contact with its contents. The reaction

$$\boxed{O_2(g) + 4\,H^+(aq) + 4\,I^-(aq) \longrightarrow 2\,H_2O(l) + 2\,I_2(s)}$$

has $\Delta\mathcal{E}° = 0.815 - 0.535 = 0.280$ V at pH 7. This means the oxidation of the iodide ion is spontaneous at pH 7 if P_{O_2} is 1 atm. The $\Delta\mathcal{E}$ is somewhat less if $P_{O_2} = 0.2$ atm, but the reaction is still spontaneous, as can be confirmed by use of the Nernst equation on the $O_2|H_2O$ couple:

$$\mathcal{E} = 0.280 - \frac{0.0592}{4} \log \frac{1}{P_{O_2}}$$

$$= 0.280 - \frac{0.0592}{4} \log \frac{1}{(0.2)} = 0.280 - 0.01 = 0.27 \text{ V}$$

(c) $\boxed{\text{No}}$. The standard reduction potentials of $Br_2(l)$ and $Cl_2(g)$ to $Br^-(aq)$ and $Cl^-(aq)$ are larger than the standard reduction potential of $I_2(s)$. Both exceed 0.815 V, so the $\Delta\mathcal{E}°$'s for reactions analogous to the aerial oxidation of $I^-(aq)$ are negative. The $Br^-(aq)$ and $Cl^-(aq)$ ions resist oxidation even better than $I^-(aq)$. Oxygen from the air is not a sufficiently strong oxidizing agent to take them to their elemental form.

(d) Decomposition of solutions of I^- by aeration is $\boxed{\text{favored}}$ by increasing the H^+ concentration. Increasing acidity makes the denominator of Q in the Nernst equation larger. The quantity being subtracted from the standard potential difference is then smaller, and $\Delta\mathcal{E}$ is larger.

13-70 (a) Both half-reactions involve species that are not in standard states. The standard reduction potentials for the two have to be corrected to the actual conditions by means of the Nernst equation. First, compute the reduction potential for the half-reaction at the anode:

$$\mathcal{E}(Pb^{2+}|Pb) = \mathcal{E}° - \frac{0.0592}{2} \log \frac{1}{[Pb^{2+}]}$$

$$\mathcal{E}(Pb^{2+}|Pb) = -0.1263 - \frac{0.0592}{2} \log \frac{1}{1.0 \times 10^{-2}} = -0.1855 \text{ V}$$

For the cell:

$$\Delta\mathcal{E} = 0.640 \text{ V} = \mathcal{E}(\text{cathode}) - \mathcal{E}(\text{anode}) = \mathcal{E}(\text{cathode}) - (-0.1855)$$

This means that the (non-standard) reduction potential for the $VO^{2+}|V^{3+}$ half-cell (which is at the cathode) is 0.4545 V. To compute the standard reduction potential, write the Nernst equation for the $VO^{2+}|V^{3+}$ half-cell:

$$\mathcal{E}° = \mathcal{E}° - \frac{0.0592}{n} \log \left(\frac{[V^{3+}]}{[VO^{2+}][H^+]^2} \right)$$

$$0.4545 = \mathcal{E}° - \frac{0.0592}{1} \log \left(\frac{(1.0 \times 10^{-5})}{(0.10)(0.10)^2} \right)$$

Solving gives $\mathcal{E}° = \boxed{0.336 \text{ V}}$.

(b) The chemical equation in this part is the combination of the two half-reactions of part (a). The standard voltage is

$$\Delta\mathcal{E}° = \mathcal{E}°(\text{cathode}) - \mathcal{E}°(\text{anode}) = 0.336 - (-0.1263) = 0.462 \text{ V}$$

As developed on text page 544, the relationship:

$$\log_{10} K = \frac{n}{0.0592 \text{ V}} \Delta\mathcal{E}°$$

obtains. Substitution of $\Delta\mathcal{E}^\circ$ and $n = 2$ gives $\boxed{K = 4 \times 10^{15}}$.

13-72 The equation is $2\,Cu^{2+}(aq) + 4\,I^-(aq) \longrightarrow 2\,CuI(s) + I_2(s)$. The $\Delta\mathcal{E}^\circ$ for this reaction is 0.325 V. Therefore

$$\log_{10} K = \frac{n}{0.0592}\Delta\mathcal{E}^\circ = \frac{2}{0.0592\ V}0.325\ V = 11.0 \qquad \boxed{K = 1 \times 10^{11}}$$

13-74 (a)

$$Zn(s) + 2\,OH^-(aq) \longrightarrow ZnO(s) + H_2O(l) + 2e^- \quad \text{oxidation}$$

$$2\,e^- + 2H_2O(l) + Ag_2O(s) \longrightarrow 2\,Ag(s) + 2\,OH^-(aq) \quad \text{reduction}$$

(b) $Zn(s) + Ag_2O(s) \longrightarrow ZnO(s) + 2\,Ag(s)$ for which $n = 2$

(c)

$$\Delta G^\circ = 2\underbrace{(0)}_{Ag(s)} + 1\underbrace{(-318.32)}_{ZnO(s)} - 1\underbrace{(0)}_{Zn(s)} - 1\underbrace{(-10.84)}_{Ag_2O(s)} = -307.48\ kJ$$

$$\Delta\mathcal{E}^\circ = \frac{-\Delta G^\circ}{n\mathcal{F}} = \frac{-(-307.48 \times 10^3\ J)}{(2\ mol)(96485.3\ C\ mol^{-1})} = \boxed{1.5934\ V}$$

13-76 The recharging reaction is

$$2\,PbSO_4(s) + 2\,H_2O(l) \longrightarrow PbO_2(s) + Pb(s) + 4H^+(aq) + 2\,SO_4^{2-}(aq)$$

After this reaction is complete, the following reaction starts

$$\boxed{H_2O(l) \longrightarrow H_2(g) + 1/2\,O_2(g)}$$

13-78 Compute ΔG° at 1200 K for the reaction as it is written and then determine $\Delta\mathcal{E}^\circ$ using the relationship $\Delta\mathcal{E}^\circ = -n\mathcal{F}\Delta G^\circ$. The ΔH° and ΔS° values for the reaction are derived from the data in Appendix D. They combine as follows:

$$\Delta G^\circ = \Delta H^\circ - T\Delta S^\circ$$

$$= (-282.99 \times 10^3\ J) - 1200\ K(-86.44 J\ K^{-1}) = \boxed{-179.26 \times 10^3\ J}$$

The computation of $\Delta\mathcal{E}^\circ$ now proceeds:

$$\Delta\mathcal{E}^\circ = -\frac{\Delta G^\circ}{n\mathcal{F}} = -\frac{(-179.26 \times 10^3\ J)}{(2\ mol)(96485\ C\ mol^{-1})} = \boxed{0.93\ V}$$

13-80 (a) The electrolysis of aqueous HCl to give $H_2(g)$ and $Cl_2(g)$ involves these half-reactions:

$$\text{cathode:} \quad H^+(aq) + e^- \longrightarrow 1/2\,H_2(g)$$

$$\text{anode:} \quad Cl^-(aq) \longrightarrow 1/2\,Cl_2(g) + e^-$$

In the molten mixture of salts, the half-reactions are:

$$\text{cathode:} \quad HCl(sol) + e^- \longrightarrow 1/2\,H_2(g) + Cl^-(sol)$$

$$\text{anode:} \quad HCl(sol) \longrightarrow 1/2\,Cl_2(g) + H^+(sol) + e^-$$

(b) The electrolysis operates at a lower voltage in the molten salt system (1 V instead of 1.34 V). In an electrolysis, the energy consumption equals the operating voltage multiplied by the number of coulombs of charge transferred. Producing one mole of Cl_2 requires 2 mol of electrons, which equals 1.9297×10^5 C. In the aqueous system, the energy equals this Q times 1.34 V; in the non-aqueous system, the energy equals this Q times only 1 V. The energy savings is per mole of Cl_2 is

$$(0.34 \text{ V})(1.9297 \times 10^5 \text{ C}) = 65600 \text{ V C} = \boxed{66 \text{ kJ}}$$

(c) One secondary advantage of the new process is that it avoids handling aqueous HCl, which is quite corrosive.

(d) The mixture of salts has a lower melting point that either pure salt. This reduces the cost of maintaining the molten salt.

Answers to Even-Numbered Problems, Chapter 14

14-2 If the $CO_2(g)$ behaves ideally, its concentration in the vessel at 11:00:00 am is 0.02125 mol L^{-1} (using $n/V = P/RT$). Exactly 30 minutes and 0 seconds later, its concentration is 0.02636 mol L^{-1}. The average change in concentration is the final concentration minus the initial concentration divided by the elapsed time. It is $\boxed{2.84 \times 10^{-6} \text{ mol L}^{-1} \text{ s}^{-1}}$.

14-4 In each case, the average rate is the change in the concentration divided by the elapsed time.

$$\text{average rate}_1 = \frac{1.0 \text{ mol}}{60 \text{ s}} = \boxed{1.7 \times 10^{-2} \text{ mol L}^{-1} \text{ s}^{-1}}$$

$$\text{average rate}_2 = \frac{1.5 \text{ mol}}{120 \text{ s}} = \boxed{1.2 \times 10^{-2} \text{ mol L}^{-1} \text{ s}^{-1}}$$

$$\text{average rate}_3 = \frac{0.75 \text{ mol}}{120 \text{ s}} = \boxed{6.25 \times 10^{-3} \text{ mol L}^{-1} \text{ s}^{-1}}$$

14-6 The rate of the reaction at $t = 100$ s is the slope of a straight line that is tangent to the blue curve in Figure 14-3 (text page 570) at the point that it cuts the vertical line labeled 100 s. When this straight line is sketched in, it is seen to pass through the points (0.0300,150) and (0.0125,0). The slope of the line is

$$\frac{(0.0300 - 0.0125)}{(150 - 0)} = \boxed{1.17 \times 10^{-4} \text{ mol L}^{-1} \text{ s}^{-1}}$$

14-8 The expressions are

$$\text{rate} = -\frac{1}{2}\frac{\Delta[\text{H}_2\text{CO}]}{\Delta t} = -\frac{\Delta[\text{O}_2]}{\Delta t} = +\frac{1}{2}\frac{\Delta[\text{CO}]}{\Delta t} = +\frac{1}{2}\frac{\Delta[\text{H}_2\text{O}]}{\Delta t}$$

14-10 (a) The rate of disappearance of $\text{O}_3(g)$ is 1/3 of the rate of disappearance of $\text{SO}_2(g)$ throughout the course of the reaction provided that no intermediates exists and that if they exist their concentrations are constant (in a steady state). The answer is therefore $\boxed{0.0333 \text{ mol L}^{-1}}$,

(b) The rate of appearance of the $\text{SO}_3(g)$ equals the rate of disappearance of the $\text{SO}_2(g)$, which is $\boxed{0.100 \text{ mol L}^{-1}}$.

14-12 (a) This reaction is $\boxed{3/2}$ order overall.   (b) This reaction is $\boxed{\text{zero}}$ order overall.

14-14 (a) The units of k for a one-half order reaction are $\boxed{\text{mol}^{1/2}\text{L}^{-1/2} \text{ s}^{-1}}$. When this unit is multiplied by $(\text{mol L}^{-1})^{1/2}$, the proper units for a rate result.

14-16 (a) The ratio is first order in $\text{C}_4\text{H}_8(g)$ since a reduction of the concentration of the $\text{C}_4\text{H}_8(g)$ causes a reduction in the observed rate in direct proportion. The rate law is therefore $\boxed{\text{rate} = k[\text{C}_4\text{H}_8]}$.

(b) Substitution of the first pair of data gives $k = \boxed{0.0036 \text{ s}^{-1}}$; the identical answer is obtained from the other two pairs.

14-18 (a) The reaction is second order in ClO_2. Tripling the concentration of this reactant therefore $\boxed{\text{increases the rate 9-fold}}$.

(b) Decreasing the pH from 12.0 to 11.0 increases the concentration of H^+ ion by a factor of 10, which causes a decrease in the OH^- concentration by a factor of 10. Since the reaction is first order in OH^-, this $\boxed{\text{decreases the rate 10-fold}}$.

14-20 (a) The rate equation is

$$\boxed{\text{rate} = k[SO_2][SO_3]^{-1/2} = k\frac{[SO_2]}{[SO_3]^{1/2}}}$$

The reaction is of order one-half overall, so the units of k are $\boxed{\text{mol}^{1/2}\text{L}^{-1/2}\text{s}^{-1}}$, as established in problem 14-14.

(b) The accelerating effect of the larger concentration of SO_2 exactly offsets the decelerating effect of the larger concentration of SO_3. $\boxed{\text{The rate is unchanged}}$.

14-22 (a) Increasing the concentration of Fe^{2+} by a factor of 1.6 between the first and second experiments (while the concentration of Ce^{4+} is constant) increases the rate by a factor of 1.6; the reaction is first-order in Fe^{2+}. Increasing the concentration of Ce^{4+} by a factor of 3.1 between the second and third experiments (while the concentration of Fe^{2+} stays constant) increases the rate by a factor of 3.1; the reaction is first-order in Ce^{4+}. The rate law is

$$\boxed{\text{rate} = k[Fe^{2+}][Ce^{4+}]}$$

(b) Substitute one of three sets of data into the rate law. The first set gives:

$$2.0 \times 10^{-7} \text{ mol L}^{-1}\text{s}^{-1} = k\,(1.8 \times 10^{-5} \text{ mol L}^{-1})(1.1 \times 10^{-5} \text{ mol L}^{-1})$$

From this,

$$k = \boxed{1.0 \times 10^3 \text{ L mol}^{-1} \text{ s}^{-1}}$$

The other two sets of data give the same k (to two significant figures).

(c) Substitute the given concentrations and the k from part (b) into the rate law. The initial rate should be $\boxed{3.4 \times 10^{-7} \text{ mol L}^{-1} \text{ s}^{-1}}$.

14-24 (a) Comparing the first and second lines of data shows that increasing the concentration of $C_2O_4^{2-}$ by a factor of $0.21/0.13 = 1.615$ while the concentration of $HgCl_2$ is kept constant increases the rate of reaction by the factor $5.5/2.1 = 2.62$. It is clear that 2.62 is the square of 1.615. Hence the reaction is second order in $C_2O_4^{2-}$. Similar reasoning shows that the reaction is first order in $HgCl_2$. Therefore the rate expression is

$$\boxed{\text{rate} = k[HgCl_2][C_2O_4^{2-}]}$$

(b) the rate constant is $\boxed{1.3 \times 10^{-4} \text{ L}^2 \text{ mol}^{-2} \text{ s}^{-1}}$. This same answer is obtained by substituting the data on each of the three lines of the table into the rate expression.

14-26 (a) The rate constant of a first-order process is $\ln 2$ (which is 0.6931) divided by the half-life. The answer is $\boxed{4.68 \times 10^{-4} \text{ s}^{-1}}$.

(b)

$$\frac{[FClO_2]}{[FClO_2]_0} = e^{-kt} = \exp(-4.68 \times 10^{-4} \text{ s}^{-1})(8280 \text{ s}) = 0.02075$$

This answer is the fraction that remains after the stated time; the fraction that has reacted is $\boxed{0.9792}$.

14-28 The concentration of the CH_3NC diminishes to 0.71 of its original value in 520 s in a first-order process. This means:

$$\ln(0.71) = -kt + \ln(1.00) = -k(520 \text{ s}) + 0$$

Solving gives $k = \boxed{6.6 \times 10^{-4} \text{ s}^{-1}}$.

14-30 (a) $\boxed{\text{False}}$. The compound B is a reactant. Its concentration *decreases* over time.

(b) $\boxed{\text{False}}$. The time needed for half of the reactant to be consumed in a second-order reaction depends on the initial concentrations of the reactants. In this case, half of the B will be consumed in 1.0 s only if the original concentration of B is 0.25 mol L^{-1}.

(c) $\boxed{\text{True}}$. The reaction is second order in B. Doubling the concentration of B therefore does double the rate of the reaction, in exact accord with the rate law.

(d) $\boxed{\text{False}}$. The rate of change of [B] falls off as [B] falls off.

(e) $\boxed{\text{True}}$. See text Figure 14-7.

14-32 The rate of disappearance of the HO_2 is second order in the concentration of HO_2 and second order overall. Hence

$$\frac{1}{[HO_2]} - \frac{1}{[HO_2]_0} = 2kt$$

The $[HO_2]_0$ and k are given, so $[HO_2]$ at $t = 1.0$ s is $\boxed{3.5 \times 10^{-10} \text{ mol L}^{-1}}$.

14-34 The following shows the concentrations of the product and reactant as a function of time in the second-order dimerization of C_2F_4:

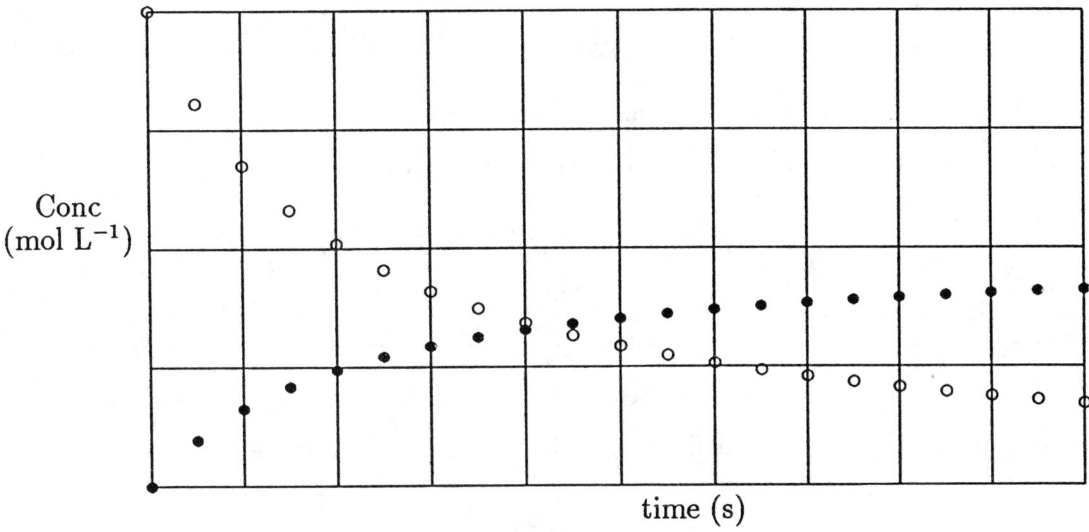

14-36 Each of the reactions is an elementary reaction. Therefore, the molecularity of each equals the number of species on the left of the arrow.
(a) unimolecular, rate $= k[BrONO_2]$.
(b) termolecular rate $= k[HO][NO_2][Ar]$.
(c) bimolecular rate $= k[O][H_2S]$.

14-38 (a) The mechanism has three steps, which are of the following types: $\boxed{\text{unimolecular}}$, rate $= k_1[NO_2Cl]$; $\boxed{\text{bimolecular}}$, rate $= k_2[Cl][H_2O]$; $\boxed{\text{termolecular}}$, rate $= k_3[OH][NO_2][N_2]$.

(b) The overall reaction is

$$\boxed{H_2O(g) + NO_2Cl(g) \longrightarrow HCl(g) + HNO_3(g)}$$

(c) $\boxed{\text{Cl, OH and NO}_2}$, which are generated during the course of the reaction but later consumed, are all intermediates. The N_2 is not a chemical intermediate in this reaction.

14-40 The equilibrium constant for the mer $\rightleftharpoons$ fac reaction is $K = [\text{fac}]/[\text{mer}]$. At equilibrium the rate of the mer $\longrightarrow$ fac reaction equals the rate of the fac $\longrightarrow$ mer reaction. Then $2.10 \text{ s}^{-1}[\text{fac}] = 2.33 \text{ s}^{-1}[\text{mer}]$. Solving for the ratio $[\text{fac}]/[\text{mer}]$ gives $K = \boxed{1.11}$.

14-42 (a) For $2\,A + 2\,B \longrightarrow E + G$ the rate law is:

$$\text{rate} = k_2[D][B] = K_1 k_2[A]^2[B]^2$$

according to the mechanism given.

(b) The reaction $A + B + D \longrightarrow G$ has the rate law:

$$\text{rate} = k_3[F] = k_3 K_2[C][D] = k_3 K_2 K_1[A][B][D]$$

according to the mechanism given.

14-44 The mechanism $\boxed{\text{(a)}}$, which starts with a rate-determining collision between $Cl_2(aq)$ and $H_2S(aq)$ $\boxed{\text{is consistent}}$ with the observed rate law:

$$\text{rate} = k_1[Cl_2][H_2S]$$

According to the second mechanism, the rate would be

$$\text{rate} = k_2[Cl_2][HS^-] = k_2 K_1[Cl_2]\frac{[H_2S]}{[H^+]}$$

Under this rate law, the reaction would still be first order in both $Cl_2(aq)$ and $H_2S(aq)$, but -1 order in $H^+(aq)$ and therefore only first order overall. This does not fit the experimental facts. According to the third mechanism, the reaction would be first order in both $Cl_2(aq)$ and $H_2S(aq)$, but -1 order in $H^+(aq)$ and also -1 order in $Cl^-(aq)$ and therefore zeroth order overall.

14-46 Both mechanisms $\boxed{\text{(a)}}$ and $\boxed{\text{(b)}}$ predict the experimental rate expression. Mechanism (c) predicts that the reaction would be second order in NO_2 and first order in O_3.

14-48 According to the Arrhenius equation, the logarithm of the rate constant is a linear function of the reciprocal of the absolute temperature with a slope of $-E_a/R$ and an intercept of $\ln A$.

(a) Most students use the Arrhenius equation with one or another of the six data pairs that are given. Their answers vary because there is some experimental scatter to the points. To use all of the data, perform a linear least-square fit of $\ln k$ versus $1/T$. The least-squares slope of the line is -4360 K, and the intercept is 23.962. This means that $E_a = -(-4360 \text{ K})(8.315 \text{ J K}^{-1} \text{ mol}^{-1}) = \boxed{36,200 \text{ J mol}^{-1}}$.

(b) The constant A always has the same units as k: $A = \boxed{2.6 \times 10^{10} \text{ s}^{-1}}$.

14-50 (a) The reaction is first order in $N_2O_4(g)$, so the partial pressure of N_2O_4 depends on time according to

$$\ln P = -kt + \ln P_0$$

With $P_0 = 0.100$ atm and $P = 0.099$ atm, at 30°C:

$$\ln\left(\frac{0.099}{0.100}\right) = -(5.1 \times 10^6 \text{ s}^{-1})t$$

Solving gives $t = \boxed{2 \times 10^{-9} \text{ s}}$. The reaction is very rapid. The back-reaction soon becomes important and the system reaches equilibrium in less than a second.

(b) The calculation is the same as in part (a), but a different rate constant is used (a larger one) because the temperature is higher. The new rate constant comes from the activation energy, the two temperatures (303 K and 333 K) and the old rate constant by use of the Arrhenius equation:

$$\ln\left(\frac{5.1 \times 10^6 \text{ s}^{-1}}{k_{333}}\right) = \frac{-54,000 \text{ J mol}^{-1}}{8.315 \text{ J K}^{-1} \text{ mol}^{-1}}\left(\frac{1}{303 \text{ K}} - \frac{1}{333 \text{ K}}\right)$$

Solving gives $k_{333} = 3.5 \times 10^7 \text{ s}^{-1}$. At the higher temperature of 333 K, it takes only $\boxed{3 \times 10^{-10} \text{ s}}$ for the partial pressure of $N_2O_4(g)$ to fall by 1.0%.

14-52 (a) The Arrhenius equation gives $\ln A$ in terms of k, T, and E_a, which are stated explicitly in the problem:

$$\ln A = \ln k + \frac{E_a}{RT} = \ln(6.1 \times 10^{-4}) + \frac{272,000 \text{ J mol}^{-1}}{(8.315 \text{ J K}^{-1} \text{ mol}^{-1})(773 \text{ K})} = 34.9$$

Hence $A = \boxed{1.5 \times 10^{15} \text{ s}^{-1}}$. Note that the units of A are always the same as the units of k. Also, the units in the E_a/RT term cancel out completely.

(b) Writing the Arrhenius equation at the two temperatures and taking the ratio (as in Example 14-9, text page 593) gives

$$\ln\left(\frac{k_{773}}{k_{298}}\right) = -\frac{272,000 \text{ J}}{8.134 \text{ J K}^{-1} \text{ mol}^{-1}}\left(\frac{1}{773 \text{ K}} - \frac{1}{298 \text{ K}}\right)$$

The ratio k_{773}/k_{298} equals 2×10^{29}, and k_{773} is given. The rate constant at 298 is easily calculated by substitution: $k_{298} = \boxed{3.1 \times 10^{-33} \text{ s}^{-1}}$. The reaction is very slow at room conditions.

14-54 The ΔE for the HOCl $\longrightarrow$ HClO conversion should equal the activation energy for the forward reaction minus the activation energy for the reverse reaction, or $311 - 31 = \boxed{280 \text{ kJ mol}^{-1}}$.

14-56 In homogeneous catalysis, the catalyst is present in the same phase as the reactants, as in the catalysis of the oxidation of $Tl^+(aq)$ by $Ag^+(aq)$. In heterogeneous catalysis, the catalyst is present as a distinct phase, as in the production of sulfuric acid using solid V_2O_5.

14-58 Both Jones and Smith err in trying to deduce the order of a reaction from balanced chemical equations that represent it. Except in the special case of an elementary reaction, in which the equation represents the actual interactions of the particles, the order of a reaction cannot be derived reliably from the balanced equation.

14-60 (a) The initial partial pressure (the pressure $t = 0$) of DTBP is 0.2362 atm. If it has decreased by x atm at time t, to a value of $0.2362 - x$ atm, then the total pressure of the gas mixture at time t must be $0.2362 + 2x$ because 3 atm of gas pressure are produced for every 1 atm consumed in this reaction. Write:

$$P_{\text{tot}} = 0.2362 + 2x \text{ atm} \quad \text{from which} \quad x = \frac{1}{2}P_{\text{tot}} - 0.1181 \text{ atm}$$

The expression for the partial pressure of DTBP is:

$$P_{\text{DTBP}} = 0.2362 - x = 0.2362 - \left(\frac{1}{2}P_{\text{tot}} - 0.1181\right) = \left(0.3543 - \frac{1}{2}P_{\text{tot}}\right) \text{ atm}$$

The following table shows the computed values for the pressure of DTBP based on the total pressure at the various times:

time (s)	P_{DTBP} (atm)	time (s)	P_{DTBP} (atm)
0	0.2362	26	0.1882
2	0.2310	30	0.1818
6	0.2236	34	0.1758
10	0.2158	38	0.1700
14	0.2088	40	0.1668
18	0.2018	42	0.1642
20	0.1982	46	0.1588
22	0.1949		

(b) The partial pressure of DTBP and its concentration are in direct proportion. If the reaction is first order in DTBP, then, over any set period of time, the natural logarithm of the DTBP pressure always changes by the same amount. If the reaction is second order in DTBP, then, instead of $\ln P_{DTBP}$, $1/P_{DTBP}$ exhibits a constant change over equal time intervals. Four minute intervals are common in the table. From 2 to 6 minutes $\Delta(1/P_{DTBP})$ is 0.1433; from 42 to 46 minutes, it is 0.2071. This is a substantial alteration. For the same pair of intervals $\Delta(\ln P_{DTBP}) = -0.0326$ and -0.0334. This is very little change. The reaction is $\boxed{\text{first order}}$ in DTBP. The same result is obtained by plotting the full set of data as in text Figure 14-8.

14-62 The forward rate of the reaction and the reverse rate of the reaction are

$$\text{rate}_f = k_f P_{CO_2} P_{NH_3}^2 \quad \text{and} \quad \text{rate}_r = k_r$$

The net rate is the forward rate minus the reverse rate. At equilibrium the net rate is zero. Hence

$$k_f P_{CO_2} P_{NH_3}^2 = k_r$$

and

$$\frac{k_f}{k_r} = \frac{1}{P_{CO_2} P_{NH_3}^2}$$

But the expression on the right is the equilibrium constant, so $K = k_f/k_r$. Substitution of the two k's gives $K = \boxed{1.49 \times 10^6}$.

14-64 (a) The two equations are:

$$Mg(ClO_4)_2(s) \longrightarrow MgCl_2(s) + 4\,O_2(g)$$

$$MgCl_2(s) + Mg(ClO_4)_2(s) \longrightarrow 2\,MgO(s) + 2\,Cl_2(g) + 3\,O_2(g).$$

(b) The $\boxed{\text{first reaction}}$ must be faster. If the second reaction were faster, it would consume $MgCl_2$ as fast as it was generated, and there would be no $MgCl_2$ residues when the $Mg(ClO_4)_2$ was all consumed.

14-66 The logarithm of the rate constant k is a linear function of the reciprocal of the absolute temperature. The slope of this line on an Arrhenius plot is $-E_a/R$. Converting the Celsius temperatures to Kelvin temperatures and performing a least-squares fit (the best way to use all four data points, see solution to problem 14-48) gives slope $= -7606.6$ K, and therefore $E_a = \boxed{63.2 \text{ kJ mol}^{-1}}$.

14-68 (a) The following table gives the half-lives

	1st half-life	2nd half-life	3rd half-life
zeroth order	227 s	114 s	57 s
first order	315 s	315 s	315 s
second order	227 s	455 s	680 s

The entries in the first two columns are read from the graph in the problem by noting that the concentration of A falls from 1.0 to 0.5 M during the first half-life and from 0.5 to 0.25 M during the second half-life. The entries in the third column come from extrapolation of the trends: for a zeroth-order reaction, each successive half-life is shorter by a factor of 2; for a first-order reaction, each half-life equals the previous one; for a second-order reaction each half-life is longer by a factor of 2. Some of the results can be confirmed by looking at the graph.

(b) The problem states that the rate constant is numerically the same for all three cases. It is easiest to estimate this value for the zeroth-order case because the rate of zeroth-order reactions remains constant from first to last. The rate equals the negative of the slope of the black line (the negative sign arises because A is a reactant). Using the line from time zero to 454 s gives

$$\text{rate} = -\frac{\Delta[A]}{\Delta t} = -\frac{(0-1) \text{ mol L}^{-1}}{(454-0) \text{ s}} = 0.0220 \text{ mol L}^{-1} \text{ s}^{-1}$$

For a zeroth-order reaction the rate is a constant—the rate constant: rate $= k$. Therefore,

$$k_{\text{zero}} = \boxed{0.0022 \text{ mol L}^{-1} \text{ s}^{-1}}$$

For the first-order and second-order cases, the units differ:

$$k_{\text{first}} = \boxed{0.0022 \text{ s}^{-1}}$$

$$k_{\text{second}} = \boxed{0.0022 \text{ L mol}^{-1} \text{ s}^{-1}}$$

(c) Substitute $t = 300$ s and the original concentration of A into the appropriate equation for the dependence of the concentrations on time:

zero $\quad$ $[A] = [A]_0 - kt = 1.0 \text{ mol L}^{-1} - (0.0022 \text{ mol L}^{-1} \text{ s}^{-1})(300 \text{ s})$

$$= \boxed{0.34 \text{ mol L}^{-1}}$$

first $\quad$ $[A] = [A]_0 e^{-kt} = (1.0 \text{ mol L}^{-1}) \exp\big((-0.0022 \text{s}^{-1})(300 \text{ s})\big)$

$$= \boxed{0.52 \text{ mol L}^{-1}}$$

second $\quad$ $\dfrac{1}{[A]} = 2kt + \dfrac{1}{[A]_0} = 2(0.0022 \text{ L mol}^{-1} \text{ s}^{-1})(300 \text{ s}) + 1.0 \text{ L mol}^{-1}$

$$= 2.32 \text{ L mol}^{-1} \qquad [A] = \boxed{0.43 \text{ mol L}^{-1}}$$

(d) The second-order reaction starts out fast because its rate depends on the square of the concentration of A. As A is used up, this value drops rapidly and the second-order reaction slows down rapidly. In contrast, the zeroth-order reaction continues at the same rate until the very last supply of A is used up.

14-70 (a) Let the ratio of the rates at 308 K and 298 K be x. Then, in the Arrhenius equation:

$$\ln x = \frac{-53,000 \text{ J mol}^{-1}}{8.315 \text{ J K}^{-1} \text{ mol}^{-1}} \left(\frac{1}{308 \text{ K}} - \frac{1}{298 \text{ K}}\right)$$

Solving gives $x = 2.00$.

(b) A similar substitution in the Arrhenius equation gives:

$$\ln x = \frac{-53,000 \text{ J mol}^{-1}}{8.315 \text{ J K}^{-1} \text{ mol}^{-1}} \left(\frac{1}{408 \text{ K}} - \frac{1}{398 \text{ K}}\right)$$

Solving gives $x = 1.48$, which is 1.5.

14-72 (a) The reactants are A_2, B, and CD, and the products are AC and BD. The balanced equation is

$$\boxed{A_2 + 2\,B + 2\,CD \longrightarrow 2\,AC + 2\,BD}$$

(b) Rate $= k_3[\text{AB}][\text{CD}] = k_3 K_2[\text{A}][\text{B}][\text{CD}] = k_3 K_2 K_1^{1/2}[\text{A}_2]^{1/2}[\text{B}][\text{CD}]$.

14-74 Cholinesterase is an enzyme. When it is inhibited by Soman, it no longer perform its function in the body. This function is crucial in the transmission of nervous impulses, which is why Soman is called a nerve gas.

14-76 (a)

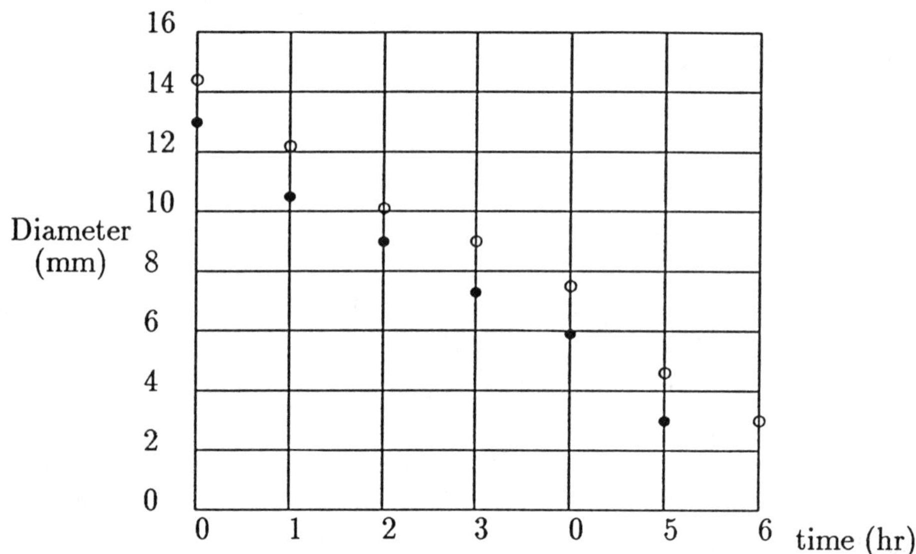

(b) The dissolution reaction is zeroth order under the conditions of the experiment. Note that the concentration of CO_2 in liquid CO_2 is a constant in any case. Also, the parallel lines on the graph show that the rate constant is the same in the two experiments.

(c) Because the reaction is zeroth-order, its rate constant just equals its rate. The rate of the reaction is the negative of the slope of the lines in the answer to part a. Thus:

$$k = \text{rate} = -\frac{\Delta(\text{diameter})}{\Delta t}$$

A least-squares fit to the data gives a slope of -1.86 mm hr^{-1} at 28 MPa and 1.87 at 35 MPa. Thus $k = \boxed{1.86 \text{ mm hr}^{-1}}$. Converting the rate constant to more familiar units, such as mol L^{-1} s^{-1}, requires knowledge of the density of the liquefied carbon dioxide.

Answers to Even-Numbered Problems, Chapter 15

15-2 The sums of the left superscripts must be equal on the two sides of the equation; the sums of the left subscripts must be equal on the two sides of the equation:

(a) $^{4}_{2}\text{He} + ^{235}_{99}\text{Es} \longrightarrow ^{237}_{101}\text{Md} + 2\,^{1}_{0}n$

(b) $^{249}_{98}\text{Cf} + ^{10}_{5}\text{B} \longrightarrow ^{257}_{103}\text{Lr} + 2\,^{1}_{0}m$

(c) $^{238}_{92}\text{U} + ^{12}_{6}\text{C} \longrightarrow ^{244}_{98}\text{Cf} + 6\,^{1}_{0}n$

15-4 The mass of each atom minus the sum of the masses of its constituent protons, neutrons, and electrons equals the mass change as the atom forms from its constituent particles. This difference in mass is negative for all atoms that actually exist. In computing such Δm's, it is both convenient and acceptable to use the mass of the ^1_1H atom as equivalent to the mass of a proton plus the mass of an electron.

The Δm's are converted to ΔE's using $\Delta E = c^2 \Delta m$ with SI units or, more conveniently, using the facts that 1 u is equivalent to 931.494 MeV, and 1 MeV equals 1.602177×10^{-13} J. Energies per mole equal energies per atom multiplied by Avogadro's number. Binding energies are the negatives of the ΔE's, and the binding energies per nucleon are the binding energies per atom divided by A, the number of protons plus neutrons.

(a) For the formation of $^{10}_{4}\text{Be}$: the Δm is -0.0697546 u;
$E_b = 64.9760$ MeV per atom $= 6.26923 \times 10^9$ kJ mol^{-1};
binding energy per nucleon $= 64.9760/10 = 6.49760$ MeV per nucleon.

(b) For the formation of $^{35}_{17}\text{Cl}$: Δm is -0.32014135 u;
$E_b = 298.210$ MeV per atom $= 28.7729 \times 10^9$ kJ mol^{-1};
binding energy per nucleon $= 298.210/35 = 8.52029$ MeV per nucleon.

(c) For the formation of $^{49}_{22}\text{Ti}$: Δm is -0.4582324 u;
$E_b = 426.841$ MeV per atom $= 41.1839 \times 10^9$ kJ mol^{-1};
binding energy per nucleon $= 426.841/49 = 8.71104$ MeV per nucleon.

15-6 Nobelium-257 decays spontaneously to give fermium-253 plus helium-4. The change in energy equals -8.45 MeV. This corresponds to a change in mass of:

$$-8.45 \text{ MeV} \times \left(\frac{1 \text{ u}}{931.494 \text{ MeV}}\right) = -0.00907 \text{ u}$$

This change equals the mass of the products minus the mass of the reactants. Hence:

$$\Delta m = m[^4_2\text{He}] + m[^{253}_{100}\text{Fm}] - m[^{257}_{102}\text{No}]$$

$$-0.00907 \text{ u} = 4.00260324 \text{ u} + m[^{253}_{100}\text{Fm}] - 257.096853 \text{ u}$$

$$m[^{253}_{100}\text{Fm}] = \boxed{253.08518 \text{ u}}$$

15-8 Compare the mass of two $^{16}_{8}\text{O}$ atoms to the mass of one $^{32}_{16}\text{S}$ atom. The latter atom has a smaller mass, so $\boxed{^{32}\text{S}}$ atom is more stable. Subtracting the mass of two O's from the mass of one S gives $\boxed{\Delta m = -0.017757 \text{ u}}$.

(b) The ^{32}S atoms *tend* to break down, but the activation energy for the change is enormous, and the actual rate is zero.

15-10

$$\text{(a) } ^{155}_{70}\text{Yb} \longrightarrow ^{151}_{68}\text{Er} + ^{4}_{2}\text{He} \qquad \text{(b) } ^{26}_{14}\text{Si} \longrightarrow ^{26}_{13}\text{Al}^- + ^{0}_{1}e^+ + \nu$$

$$\text{(c) } \left(65^{}_{30}\text{Zn}^+ + ^{0}_{-1}e^-\right) \longrightarrow ^{65}_{29}\text{Cu} + \nu \qquad \text{(d) } ^{100}_{41}\text{Nb} \longrightarrow ^{100}_{42}\text{Mo}^+ + ^{0}_{-1}e^- + \bar{\nu}$$

15-12 The first step is the absorption of a neutron: $^{209}_{83}\text{Bi} + ^{1}_{0}n \longrightarrow ^{210}_{83}\text{Bi}$.

Loss of a beta particle then gives the correct product: $^{210}_{83}\text{Bi} \longrightarrow ^{210}_{84}\text{Po}^+ + ^{0}_{-1}e^- + \bar{\nu}$.

15-14 The nuclear reaction, which also appears as an example on text page 620, is

$$^{14}_{6}\text{C} \longrightarrow ^{14}_{7}\text{N}^+ + ^{0}_{-1}e^- + \bar{\nu}$$

For this reaction the change in mass equals

$$\Delta m = m[^{14}_{7}\text{N}^+] + m[^{0}_{-1}e^-] - m[^{14}_{7}\text{C}]$$
$$= 14.00307400 - 14.00324198 = -0.00016798 \text{ u}$$

where the mass of the neutral N-14 atom (as found in text Table 15-1, text page 614) is used. This mass is essentially equal to the mass of the N^+ ion plus the mass of one electron. The energy change is the energy equivalent of this Δm:

$$\Delta E = \left(\frac{931.49 \text{ MeV}}{\text{u}}\right) \times (-0.00016798 \text{ u}) = -0.15647 \text{ MeV}$$

The maximum kinetic energy of the beta particle is $\boxed{+0.15647 \text{ MeV}}$.

15-16 When the emitted positron has its maximum kinetic energy, it carries the whole of the energy released by the reaction. The corresponding change in mass is:

$$-1.732 \text{ MeV} \times \left(\frac{1 \text{ u}}{931.494 \text{ MeV}}\right) = -0.001859 \text{ u}$$

This equals the mass of the products minus the mass of the reactants:

$$\Delta m = m[^{15}_{7}\text{N}^-] + m[^{0}_{1}e^+] - m[^{15}_{8}\text{O}]$$
$$-0.001859 \text{ u} = (15.0001089 + 0.00054856) + 0.00054856 - m[^{15}_{8}\text{O}]$$
$$m[^{15}_{8}\text{O}] = \boxed{15.003066 \text{ u}}$$

15-18 The change in mass equals the mass of the products minus the mass of the reactants:[8]

$$\Delta m = m[{}^{4}_{2}\text{He}] + m[{}^{227}_{89}\text{Ac}] - m[{}^{231}_{92}\text{U}]$$

$$= 4.00260324 \text{ u} + 227.0277470 \text{ u} - 231.036289 \text{ u} = -0.005939 \text{ u}$$

The energy change is:

$$\Delta E = -0.005939 \text{ u} \times \left(\frac{931.494 \text{ MeV}}{1 \text{ u}}\right) = \boxed{-5.532 \text{ MeV}}$$

15-20 Compute the half-life of ^{238}U in minutes:

$$t_{1/2} = 4.47 \times 10^9 \text{ yr} \times \left(\frac{365.25 \text{ day}}{1 \text{ yr}}\right) \times \left(\frac{1440 \text{ min}}{1 \text{ day}}\right) = 2.351 \times 10^{15} \text{ min}$$

The activity A of the sample is desired:

$$A = kN = \left(\frac{\ln 2}{t_{1/2}}\right) N$$

$$= \left(\frac{0.6931}{2.351 \times 10^{15} \text{ min}}\right) \left(\frac{0.0010 \text{ g U}}{238 \text{ g mol}^{-1}} \times \frac{6.022 \times 10^{23} \text{ atom U}}{1 \text{ mol U}}\right)$$

$$= 7.46 \times 10^2 \frac{\text{atom U}}{\text{min}} = 12.4 \text{ atom U s}^{-1} = \boxed{12.4 \text{ Bq}}$$

15-22 (a) The decay constant of the S-35 equals $\ln 2$ divided by the half-life:

$$(0.6931/87.1 \text{ day}) = 0.007958 \text{ day}^{-1}$$

The activity A is 370 Bq which is the same as 370 s^{-1}. Then:

$$N = \frac{A}{k} = \frac{370 \text{ s}^{-1}}{0.00758 \text{ day}^{-1}} \times \left(\frac{86,400 \text{ day}^{-1}}{1 \text{ s}^{-1}}\right) = 4.217 \times 10^9$$

This is the number of atoms of ^{35}S. Convert it to grams as follows:

$$4.217 \times 10^9 \text{ atoms} \times \left(\frac{1 \text{ mol } {}^{35}\text{S}}{6.022 \times 10^{23} \text{ atoms}}\right) \times \left(\frac{35 \text{ g } {}^{35}\text{S}}{1 \text{ mol } {}^{35}\text{S}}\right) = \boxed{2.5 \times 10^{-13} \text{ g } {}^{35}\text{S}}$$

---

[8]In some copies of the textbook, the nuclidic mass of Ac-227 is incorrectly given as 221.015576 u which is the mass of Ac-221. The correct nuclidic mass for Ac-227 is used here.

(b) The decay constant of the radioactive isotope and its original amount are known. The decay proceeds for 365 days. At that time:

$$\ln\left(\frac{N}{N_i}\right) = -kt$$

$$\ln\left(\frac{N}{4.017 \times 10^9 \text{ atoms}}\right) = -(0.007958 \text{ day}^{-1})(365 \text{ day})$$

Solving for N gives 2.200×10^8 atoms. This number of atoms of ^{35}S has a very small mass, only $\boxed{1.3 \times 10^{-14} \text{ g}}$.

15-24 The activity A of the ^{59}Fe in the mixture is related to the decay constant k of the ^{59}Fe and the number of atoms N of that radioactive nuclide:

$$N = \frac{A}{k} = A\left(\frac{t_{1/2}}{\ln 2}\right) = 3.0 \times 10^7 \text{ s}^{-1}\left(\frac{3.853 \times 10^6 \text{ s}}{\ln 2}\right) = 1.67 \times 10^{14}$$

Note that $t_{1/2}$ was converted from days to seconds, and that the equivalent unit s^{-1} was substituted for Bq. The mass of this number of ^{59}Fe atoms is

$$1.67 \times 10^{14} \text{ atoms}\left(\frac{1 \text{ mol } ^{59}\text{Fe}}{6.022 \times 10^{23} \text{ atoms}}\right)\left(\frac{59.0 \text{ g } ^{59}\text{Fe}}{1 \text{ mol } ^{59}\text{Fe}}\right) = 1.64 \times 10^{-8} \text{ g } ^{59}\text{Fe}$$

This mass is $\boxed{1.3 \times 10^{-5}\%}$ of the 125.3 mg sample.

15-26 The activity of the gallium citrate decays according to first-order kinetics: $A = A_i e^{-kt}$. From this equation and the definition of the half-life, it follows that

$$-kt = -\left(\frac{\ln 2}{t_{1/2}}\right)t = \ln\left(\frac{A}{A_i}\right)$$

The ratio A/A_i equals 0.050. Substitution of this value and the half-life of 77.9 h allows solution for t, which equals $\boxed{337 \text{ h}}$.

15-28 The 1.0 μg of ^{99m}Tc contains 6.08×10^{15} ^{99m}Tc atoms (obtained by dividing the 1.0×10^{-6} g by the molar mass of 99 g mol^{-1} and multiplying by Avogadro's number). The 6.0 h half-life equals 2.16×10^4 s. The activity is:

$$A = kN = \left(\frac{\ln 2}{2.16 \times 10^4 \text{ s}}\right)(6.08 \times 10^{15}) = 2.0 \times 10^{11} \text{ s}^{-1} = \boxed{2.0 \times 10^{11} \text{ Bq}}$$

15-30 Evidently, 1 out of every 121 atoms of ^{87}Rb has decayed since the rock was formed. This means that 120/121 atoms remain. Also, the decay constant of the ^{76}Rb is 1.414×10^{-11} yr^{-1}. Therefore

$$\ln\left(\frac{N}{N_i}\right) = \ln\left(\frac{120}{121}\right) = -kt = -(1.414 \times 10^{-11} \text{ yr}^{-1})t$$

Solving gives $\boxed{t = 5.9 \times 10^8 \text{ yr}}$.

15-32 The activity of the charcoal has fallen to 1/8 times its original activity. This means that 3 half-lives have elapsed. The half-life of ^{14}C is 5730 years, so the residues are about $\boxed{17{,}190 \text{ years old}}$.

15-34 The specific activity of the carbon from the wooden bow has decayed to a value as low as 3×10^{-4} or as high as 7×10^{-4} Bq g^{-1}. The original specific activity of the carbon is assumed to equal 0.255 Bq g^{-1}. Assume that the low end of the range given for the specific activity of the wood in the bow is correct:

$$\ln\left(\frac{A}{A_i}\right) = \ln\left(\frac{3 \times 10^{-4} \text{ Bq g}^{-1}}{0.255 \text{ Bq g}^{-1}}\right) = -kt = -\frac{\ln 2}{5730 \text{ yr}}t$$

Solving gives a t of 5.6×10^5 years. Duplicating the calculation using 7×10^{-4} Bq g^{-1}, the value at the high end of the range, gives a t of 4.9×10^4 years. The wood in the bow is $\boxed{49{,}000 \text{ to } 56{,}000 \text{ years old}}$.

15-36 The decays must occur by positron emission:
$$^{13}_{7}\text{N} \longrightarrow \, ^{13}_{6}\text{C}^- + \, ^{0}_{1}e^+ + \nu \quad \text{and} \quad ^{18}_{9}\text{F} \longrightarrow \, ^{18}_{8}\text{O}^- + \, ^{0}_{1}e^+ + \nu.$$
The positrons are annihilated in collisions with electrons in ordinary matter.

15-38 The ^{226}Ra emits more energetic particles faster. The emitted particles are moreover absorbed better. The $\boxed{^{226}\text{Ra is more dangerous}}$ than the ^{14}C.

15-40 (a) First compute the decay constant of the ^{239}Pu in s^{-1} from its half-life in years:

$$k = \frac{\ln 2}{t_{1/2}} = \frac{\ln 2}{(2.411 \times 10^4) \times 365.25 \times 24 \times 3600} = 9.110 \times 10^{-13} \text{ s}^{-1}$$

Then determine the activity of the little piece of plutonium:

$$A = kN$$

$$= (9.110 \times 10^{-13} \text{ s}^{-1})\left(5.0 \times 10^{-6} \text{ g Pu} \times \frac{1 \text{ mol Pu}}{239 \text{ g Pu}} \times \frac{6.022 \times 10^{23} \text{ atoms}}{1 \text{ mol Pu}}\right)$$

$$= \boxed{5.74 \times 10^4 \text{ Bq}}$$

(b) Compute the energy emitted annually by the radioactive Pu lodged in the worker's lung. Recall that a Bq is a decay event per second, and note that the activity of the radioactive Pu does not decay perceptibly during the year:

$$5.74 \times 10^4 \text{ Bq} \times \left(\frac{5.24 \text{ MeV}}{1 \text{ event}}\right) \times (365.25 \times 24 \times 3600 \text{ s})$$

$$\times \left(\frac{1.602 \times 10^{-13} \text{ J}}{1 \text{ MeV}}\right) = 1.52 \text{ J} = 152 \text{ cJ}$$

At a body mass of 60 kg the worker absorbs 2.53 cJ per kg of body mass per year. This dose equals 2.53 rad, or $\boxed{2530 \text{ millirad}}$, in the first year.

(c) Even after multiplying by 10 to account for the extra hazard of alpha radiation, the dose is less than 10% of the LD_{50}. By this reckoning, the dose is $\boxed{\text{not lethal}}$. The localization of the radiation damage in a lung will however cause worse short-term effects than getting an uniform dose throughout the body and will almost certainly lead to ill-health over the longer term.

15-42 (a) The alpha decay of ^{239}Pu is represented $\boxed{^{239}_{94}\text{Pu} \longrightarrow {}^{235}_{92}\text{U} + {}^{4}_{2}\text{He}}$.

(b) The Δm in this reaction is -0.0056267 u. Hence, $\Delta E = -5.24$ MeV, and the energy released is $\boxed{+5.24 \text{ MeV}}$.

(c) The 1.00 g of ^{239}Pu contains 2.519×10^{21} atoms of ^{239}Pu. The decay constant is ln 2 divided by the half-life, or 2.876×10^{-5} yr^{-1}. The product of these two numbers is the activity: $A = 7.245 \times 10^{16}$ yr^{-1}. The activity is requested in becquerels (Bq), which are the same as reciprocal seconds. Dividing the activity in yr^{-1} by the number of seconds in a year gives $\boxed{A = 2.30 \times 10^9 \text{ Bq}}$.

(d) Use the equation $A = A_i e^{-kt}$ with $A_i = 2.30 \times 10^9$ s^{-1}, $t = 100,000$ yr, and $k = 2.876 \times 10^{-5}$ yr^{-1} to find $\boxed{A = 1.30 \times 10^8 \text{ s}^{-1}}$. The passage of 100,000 years reduces the activity of the plutonium to approximately 5.7% of its initial value.

15-44 (a) $^{6}_{3}\text{Li} + {}^{1}_{0}n \longrightarrow {}^{3}_{1}\text{H} + {}^{4}_{2}\text{He}$.

(b) The molar mass of the lithium from which the ^{6}Li has been depleted will be $\boxed{\text{greater}}$ because ^{6}Li has a smaller atomic mass than the tabulated atomic mass of lithium.

15-46 For the fusion reaction:

$$2\,{}^{2}_{1}\text{H} \longrightarrow {}^{4}_{2}\text{He} \qquad \Delta m = -0.0256125 \text{ u}$$

a result computed from the nuclidic masses in text Table 15-1, Compute the change of mass in the reaction of exactly one gram of $^2_1$H:

$$\Delta m = 1 \text{ g } ^2_1\text{H} \times \left(\frac{1 \text{ mol } ^2_1\text{H}}{2.0141079 \text{ g } ^2_1\text{H}}\right) \times \left(\frac{6.022 \times 10^{23} \text{ atoms}}{1 \text{ mol } ^2_1\text{H}}\right)$$

$$\times \left(\frac{-0.0256125 \text{ u}}{2 \, ^2_1\text{H atoms}}\right) \times \left(\frac{1.00 \times 10^{-3} \text{ kg}}{6.022 \times 10^{23} \text{ u}}\right) = -6.358 \times 10^{-6} \text{ kg}$$

This Δm has been expressed in kilograms so that it can be multiplied by c^2 in m^2 s^{-2} to furnish an energy in joules; the ΔE is -5.715×10^{11} J. The energy released in the nuclear reaction is 5.715×10^{11} joules per gram of deuterium, which equals $\boxed{5.715 \times 10^8 \text{ kJ g}^{-1}}$. Fission of ^{235}U (problem 15-45) gives 7.59×10^7 kJ g^{-1}, which is about 13.3% of the energy released in the fusion of a gram of deuterium.

15-48 Since 1 u is 931.49 MeV, the mass equivalent of 76 eV (76×10^{-6} MeV) is a mere $\boxed{8.2 \times 10^{-8} \text{ u}}$. This is $\boxed{1.4 \times 10^{-34} \text{ kg}}$.

15-50 The nuclear reaction of Cockcroft and Walton was

$$^7_3\text{Li} + ^1_1\text{H} \longrightarrow 2 \, ^4_2\text{He}$$

for which

$$\Delta m = 2(4.0026033) - 7.016003 - 1.00782505 = -0.0186234 \text{ u}$$

This Δm also equals -3.09256×10^{-29} kg. The change in energy of the system equals the original E (the 0.70 MeV supplied by the incoming proton) minus the final E, which is the kinetic energy carried off by the two alpha particles:

$$\Delta E = 0.70 - 2(8.5) = -16.3 \text{ MeV} = -2.61 \times 10^{-12} \text{ J}$$

The speed of light based on these results is

$$c = \sqrt{\frac{\Delta E}{\Delta m}} = \sqrt{\frac{-2.61 \times 10^{-12} \text{ J}}{-3.09256 \times 10^{-29} \text{ kg}}} = \boxed{2.90 \times 10^8 \text{ m s}^{-1}}$$

This result is about 3% below the current accepted value of c.

15-52 The daughter nuclide must be $^{64}_{30}$Zn because beta emission always raises Z by one. The nuclear reaction is then:

$$^{64}_{29}\text{Cu} \longrightarrow ^{64}_{30}\text{Zn}^+ + \, ^{\,0}_{-1}e^- + \bar{\nu}$$

The change in mass in this reaction equals

$$\Delta m = -0.573 \text{ MeV} \times \left(\frac{1 \text{ u}}{931.494 \text{ MeV}}\right) = -0.000615141 \text{ u}$$

As shown on text page 622, in positron emission, the change in mass equals the mass of the neutral daughter nuclide minus the mass of the neutral parent nuclide. Hence:

$$m[^{64}_{30}\text{Zn}] = m[^{64}_{29}\text{Cu}] + \Delta m$$
$$= 63.92976 + (-0.000615141) = \boxed{63.929145 \text{ u}}$$

15-54 The decay constant for the double beta decay is 1.980×10^{-28} s^{-1}, and $N = 6.022 \times 10^{23}$ atoms. The product of these two values is the activity of the sample. It is $\boxed{A = 1.19 \times 10^{-4} \text{ s}^{-1}}$. Only about 10 atoms of ^{82}Se in this large sample decay per day!

15-56 (a) According to problem 15-55, the activity of ^{14}C equals 0.255 Bq per gram of carbon in the biosphere. The overall activity of the biosphere is 1.1×10^{19} Bq. The total mass of C in the biosphere therefore is

$$1.1 \times 10^{19} \text{ s}^{-1} \times \left(\frac{1 \text{ g C}}{0.255 \text{ s}^{-1}}\right) = \boxed{4.3 \times 10^{19} \text{ g C}}$$

(b) The carbon in the earth's crust amounts to

$$\left(\frac{250 \text{ g C}}{10^{-6} \text{ g crust}}\right) \times 2.9 \times 10^{25} \text{ g crust} = 7.25 \times 10^{21} \text{ g C}$$

The amount of carbon in the biosphere is only about $\boxed{0.006}$ of this or about 0.6%. Most of the carbon in the crust is tied up in rocks.

15-58 Write decay equations for both the ^{238}U and ^{235}U:

$$N_{238} = N_{i,238}\, e^{-k_{238}t} \qquad\qquad N_{235} = N_{i,235}\, e^{-k_{235}t}$$

where the subscripts distinguish the two isotopes. Dividing the first equation by the second cancels out the N_i terms because the original amounts of the two isotopes are equal. Taking the natural logarithm then gives

$$\ln\left(\frac{N_{238}}{N_{235}}\right) = (-k_{238}t) - (-k_{235}t) = t(k_{235} - k_{238})$$

The ratio of the present-day abundances is 137.1 to 1. Also, each k equals $\ln 2$ divided by the isotope's half-life. Therefore

$$\ln(137.1) = t \left(\frac{\ln 2}{7.04 \times 10^8 \text{ yr}} - \frac{\ln 2}{4.468 \times 10^9 \text{ yr}} \right)$$

Solving gives $\boxed{t = 5.9 \times 10^9 \text{ yr}}$.

15-62 Alpha decay decreases Z by 2 and A by 4; beta decay increases Z by 1 and leaves A unchanged. The atomic number of thorium is 90, the atomic number of uranium is 92, and the atomic number of lead is 82. The multi-step process of decay U $\longrightarrow$ Pb must emit more alpha particles than the process Th $\longrightarrow$ Pb in order to make A lower for the lead that derives from it.

15-60 Because $A = kN$,

$$k = \frac{A}{N} = \frac{2.5 \text{ s}^{-1}}{10 \times 10^9} = 2.5 \times 10^{-10} \text{ s}^{-1}$$

$$t = \frac{\ln 2}{k} = \frac{0.69315}{2.5 \times 10^{-10} \text{ s}^{-1}} \times \left(\frac{1 \text{ yr}}{3.15 \times 10^7 \text{ s}} \right) = \boxed{88 \text{ yr}}$$

15-64 The problem can be solved by a series of unit-conversions starting from the known power of the reactor. Note the use of the efficiency of the process as a conversion factor:

$$\left(\frac{1000 \times 10^6 \text{ J output}}{1 \text{ s}} \right) \times \left(\frac{100 \text{ J released}}{40 \text{ J output}} \right) \times \left(\frac{1 \text{ MeV}}{1.602 \times 10^{-13} \text{ J released}} \right)$$

$$\times \left(\frac{1 \text{ atom } ^{235}\text{U}}{200 \text{ MeV}} \right) \times \left(\frac{1 \text{ mol } ^{235}\text{U}}{6.022 \times 10^{23} \text{ atoms } ^{235}\text{U}} \right) \times \left(\frac{235 \text{ g } ^{235}\text{U}}{1 \text{ mol } ^{235}\text{U}} \right)$$

$$\times \left(\frac{3.156 \times 10^7 \text{ s}}{1 \text{ yr}} \right) = \boxed{9.6 \times 10^5 \frac{\text{g } ^{235}\text{U}}{\text{yr}}}$$

15-66 The important points are covered on text pages 633-640.

Answers to Even-Numbered Problems, Chapter 16

16-2 The wavelength of this chemical wave is 1.2 cm; its frequency is $1/42$ s^{-1}. The speed of propagation is the product of these two values, $\boxed{0.029 \text{ cm s}^{-1}}$.

16-4 Assume that the gamma rays are propagating through a vacuum. The wavelength is then

$$\lambda = \frac{c}{\nu} = \frac{2.9979 \times 10^8 \text{ m s}^{-1}}{2.83 \times 10^{20} \text{ s}^{-1}} = \boxed{1.06 \times 10^{-12} \text{ m}}$$

This equals $\boxed{0.0106 \text{ Å}}$.

16-6 (a) For electromagnetic waves, $c = \lambda\nu$. Therefore

$$\nu = \frac{c}{\lambda} = \frac{2.9979 \times 10^8 \text{ m s}^{-1}}{488 \times 10^{-9} \text{ m}} = \boxed{6.15 \times 10^{14} \text{ s}^{-1}}$$

Again, c is taken as its value in a vacuum in the absence of any information to the contrary.

(b) Divide twice the distance from the earth to the moon by the speed of light. The answer is $\boxed{2.5 \text{ s}}$.

16-8 The wavelength of the ultrasonic wave is its speed of propagation divided by its frequency. It comes out to 0.030 m, or $\boxed{3.0 \text{ cm}}$. The resolution in the sonic image cannot exceed the wavelength of sound used. If lower frequency ($\nu = 8000 \text{ s}^{-1}$) is used, then the wavelength is longer (18.7 cm) and all detail in the image of the fetus is lost.

16-10 The answer is $\boxed{\text{nothing}}$. The problem should be considered in conjunction with problem 16-9. The red light has lower frequency and hence lower energy per photon than the green light. If photons of green light eject no electrons from the copper surface, then photons of red light also eject no electrons from the copper surface.

16-12 A wavelength of 454 nm is in the blue region of the spectrum. The flame is $\boxed{\text{blue}}$.

16-14 The energy lost by the potassium atom is carried away by a photon. The wavelength of the photon is

$$\lambda = \frac{hc}{E} = \frac{(6.626 \times 10^{-34}\text{J s})(3.00 \times 10^8 \text{ m s}^{-1})}{4.9 \times 10^{-19} \text{ J}} = 4.1 \times 10^{-7} \text{ m} = \boxed{410 \text{ nm}}$$

This wavelength is in the $\boxed{\text{violet}}$ region of the spectrum.

16-16 (a)

$$E = h\frac{c}{\lambda} = \frac{(6.626 \times 10^{-34}\text{J s})(2.9979 \times 10^8 \text{ m s}^{-1})}{5.2 \times 10^{-7} \text{ m}} = \boxed{3.8 \times 10^{-19} \text{ J}}$$

(b) The power of the laser equals 10 W, which is 10 J s^{-1}. Hence:

$$\frac{10 \text{ J}}{1 \text{ s}} \times \left(\frac{1 \text{ photon}}{3.82 \times 10^{-19} \text{ J}} \right) = \boxed{2.6 \times 10^{19} \frac{\text{photon}}{\text{s}}}$$

16-18 The maximum wavelength is the wavelength of photons that are just energetic enough to overcome the work function of the surface:

$$\lambda = \frac{c}{\nu} = \frac{hc}{E} = \frac{(6.626 \times 10^{-34} \text{J s})(2.9979 \times 10^8 \text{ m s}^{-1})}{4.41 \times 10^{-19} \text{ J}} = \boxed{4.50 \times 10^{-7} \text{ m}}$$

16-20 Use the equation governing the photoelectric effect:

$$E_{\text{k}}(\text{max.}) = 1/2 m_e v^2 = h\nu - \Phi = \frac{hc}{\lambda} - \Phi$$

Radiation that barely causes detachment of electrons from the surface of the tungsten imparts a maximum kinetic energy of zero, so that

$$\frac{hc}{\lambda} = \Phi \qquad \text{or} \qquad \lambda = \frac{hc}{\Phi}$$

The Φ for tungsten is 7.29×10^{-19} J. Inserting this value and the two constants gives $\boxed{\lambda = 272 \text{ nm}}$.

If the ejected electrons are to have a maximum speed of 2.00×10^6 m s^{-1}, then:

$$\frac{hc}{\lambda} = 1/2 m_e v_{\text{max}}^2 + \Phi$$

$$= \frac{1}{2}(9.109 \times 10^{-31} \text{ kg})(2.00 \times 10^6 \text{ m s}^{-1})^2 + 7.29 \times 10^{-19} \text{ J}$$

$$= 2.55 \times 10^{-18} \text{ J}$$

Solving for λ gives a wavelength of $\boxed{77.9 \text{ nm}}$.

16-22 For all one-electron atoms and ions:

$$E_n = -\frac{Z^2}{n^2} \text{Ry} \qquad \text{and} \qquad r_n = \frac{n^2}{Z} a_0$$

(a) For He$^+$ ion, $Z = 2$. Substituting in turn the values $n = 2, 4, 7$ along with $Z = 2$ gives energies of:

$$E_2 = -\frac{2^2}{2^2}\text{Ry} = -1 \text{ Ry} = \boxed{-2.17987 \times 10^{-18} \text{ J}}$$

$$E_4 = -\frac{2^2}{4^2}\text{Ry} = -\frac{1}{4} \text{ Ry} = \boxed{-0.544967 \times 10^{-18} \text{ J}}$$

$$E_7 = -\frac{2^2}{7^2}\text{Ry} = -\frac{4}{49} \text{ Ry} = \boxed{-0.177949 \times 10^{-18} \text{ J}}$$

(b) For He^+ ion, $Z = 2$. Substituting in turn the values $n = 2, 4, 7$ along with $Z = 2$ gives the radii:

$$r_2 = \frac{2^2}{2}a_0 = 2a_0 = \boxed{1.058354 \times 10^{-10} \text{ m}}$$

$$r_4 = \frac{4^2}{2}a_0 = 8a_0 = \boxed{4.233416 \times 10^{-10} \text{ m}}$$

$$r_7 = \frac{7^2}{2}a_0 = 24.5a_0 = \boxed{12.964837 \times 10^{-10} \text{ m}}$$

16-24 For Li^{2+}, a one-electron ion,

$$E_n = -\frac{Z^2}{n^2}\text{Ry} = -\frac{3^2}{n^2}\text{Ry}$$

For a change between any two energy states of Li^{2+} ion:

$$\Delta E = -3^2\left(\frac{1}{n_{\text{final}}^2} - \frac{1}{n_{\text{initial}}^2}\right)\text{Ry}$$

This is the energy change in the system consisting of the lithium ion. The energy *lost* by the ion is the negative of this ΔE,

(a)   Energy lost $= +3^2\left(\frac{1}{1^2} - \frac{1}{3^2}\right)\text{Ry} = \boxed{8\text{ Ry}} = \boxed{17.44 \times 10^{-18}\text{ J}}$

(b)   Energy lost $= +3^2\left(\frac{1}{2^2} - \frac{1}{6^2}\right)\text{Ry} = \boxed{2\text{ Ry}} = \boxed{4.36 \times 10^{-18}\text{ J}}$

(c)   Energy lost $= +3^2\left(\frac{1}{4^2} - \frac{1}{3^2}\right)\text{Ry} = \boxed{-0.4375\text{ Ry}} = \boxed{-0.954 \times 10^{-18}\text{ J}}$

The transition in part (c) is an absorption as indicated by the negative loss of energy (the ion gains energy). The energies were converted to joules using the fact that $1\text{ Ry} = 2.17987 \times 10^{-18}\text{ J}$.

16-26 The wavelength of the single photon is related to the energy change by

$$\lambda = \frac{hc}{\Delta E} = \frac{(6.626 \times 10^{-34}\text{ J s})(2.9979 \times 10^8\text{ m s}^{-1})}{\Delta E}$$

Substitution of ΔE gives the wavelength. The negative loss of energy in part (c) means that the photon is absorbed, not emitted.

(a) $\lambda = \boxed{1.14 \times 10^{-8}\text{ m}}$ (b) $\lambda = \boxed{4.56 \times 10^{-8}\text{ m}}$ (c) $\lambda = \boxed{0.208 \times 10^{-8}\text{ m}}$

16-28 According to the Bohr model, the radius and energy of a one-electron atom or ion are

$$r_n = \frac{n^2}{Z}(5.29 \times 10^{-11} \text{ m}) \quad \text{and} \quad E_n = -\frac{Z^2}{n^2}(2.18 \times 10^{-18} \text{ J})$$

where Z is the atomic number of the nucleus of the atom or ion and n is the quantum number. Substitution of $Z = 2$ for helium and $n = 5$ gives $r_5 = 6.61 \times 10^{-10}$ m and $E_5 = -3.49 \times 10^{-19}$ J.

Removing the electron means changing the energy of the atom to $E = 0$. The change in energy of one atom is this final value minus the initial value, or $+3.49 \times 10^{-19}$ J. For a mole of atoms the energy change is Avogadro's number times larger or $\boxed{210 \text{ kJ}}$.

For He^+ ion $E_3 = -9.69 \times 10^{-19}$ J. The change in energy of the ion in the $5 \longrightarrow 3$ transition is E_3 (the final energy) minus E_5 (the initial energy). Hence $\Delta E = -6.20 \times 10^{-19}$ J. The transition gives off energy, as shown by its negative ΔE. The frequency of the photon that possesses this amount of energy is

$$\nu = \frac{E}{h} = \frac{6.20 \times 10^{-19} \text{ J}}{6.626 \times 10^{-34} \text{ J s}} = \boxed{9.36 \times 10^{14} \text{ s}^{-1}}$$

The corresponding wavelength is $\boxed{3.20 \times 10^{-7} \text{ m}}$.

16-30 In the emission spectrum of Be^{3+} ion (for which $Z = 4$) lines in the series analogous to the Lyman series of atomic hydrogen have wavelengths severally equal to 1/16 those of the lines in the Lyman series. The frequencies in this series are 16 times the frequencies in the Lyman series. Similar scaling occurs with the series analogous to the Balmer series. These conclusions come from the dependence of the energy of the hydrogen-like ions on Z^2. Thus, for the frequencies in the Lyman-like series of Be^{3+}

$$\nu = Z^2 \left(\frac{1}{1^2} - \frac{1}{n_{\text{initial}}^2} \right) \frac{\text{Ry}}{h} = 4^2 \left(\frac{1}{1^2} - \frac{1}{n_{\text{initial}}^2} \right) 3.29 \times 10^{15} \text{ s}^{-1}$$

The ν's and λ's for the first three lines in the series are

	$2 \longrightarrow 1$	$3 \longrightarrow 1$	$4 \longrightarrow 1$
ν	3.95×10^{16} s^{-1}	4.68×10^{16} s^{-1}	4.93×10^{16} s^{-1}
λ	7.59 nm	6.41 nm	6.07 nm

For the frequencies in the Balmer-like series of Be^{3+}

$$\nu = Z^2 \left(\frac{1}{2^2} - \frac{1}{n_{\text{initial}}^2} \right) \frac{\text{Ry}}{h} = 4^2 \left(\frac{1}{2^2} - \frac{1}{n_{\text{initial}}^2} \right) 3.29 \times 10^{15} \text{ s}^{-1}$$

	$3 \longrightarrow 2$	$4 \longrightarrow 2$	$5 \longrightarrow 2$
ν	7.31×10^{15} s^{-1}	9.87×10^{15} s^{-1}	1.11×10^{16} s^{-1}
λ	41.0 nm	30.4 nm	27.1 nm

16-32 (a) In the ground state of the standing wave, there is one half-wavelength along the bond, L. Hence $\boxed{\lambda_1 = 2.0 \text{ Å}}$. In the first excited state, $n = 2$; there are two half-wavelengths along the bond, so that $\boxed{\lambda_2 = L = 1.0 \text{ Å}}$.

(b) The number of nodes is one less than the quantum number describing the standing wave. Hence, there is $\boxed{1 \text{ node}}$.

16-34 (a) The mass of the electron is known so that the speed of the electron can be calculated from its kinetic energy. The kinetic energy is converted from the collective basis given to an individual basis by dividing 1.20×10^7 J mol^{-1} by N_0. It equals 1.993×10^{-17} J. Then:

$$v = \sqrt{\frac{2E_k}{m_e}} = \sqrt{\frac{2(1.993 \times 10^{-17} \text{ J})}{9.109 \times 10^{-31} \text{ kg}}} = 6.614 \times 10^6 \text{ m s}^{-1}$$

$$\lambda = \frac{h}{m_e v} = \frac{6.626 \times 10^{-34} \text{ J s}}{(9.109 \times 10^{-31} \text{ kg})(6.614 \times 10^6 \text{ m s}^{-1})} = \boxed{1.10 \times 10^{-10} \text{ m}}$$

(b) By a similar calculation, the He atom moving at 353 m s^{-1} has a wavelength of $\boxed{2.82 \times 10^{-10} \text{ m}}$.

(c) The krypton atom moving at 299 m s^{-1} has a wavelength of 0.159×10^{-10} m.

16-36 (a) Use the Heisenberg indeterminacy principle:

$$\Delta x \cdot \Delta(mv) \geq h/4\pi$$

with $m = 9.109 \times 10^{-31}$ kg and $v = 3.0 \times 10^8$ m s^{-1}. The result is $\Delta x \geq 1.93 \times 10^{-13}$ m. This means that if nothing is known about the speed of an electron except that it is less than the speed of light, its location can be known to at best $\boxed{\pm 0.002 \text{ Å}}$.

(b) Because the helium atom is much more massive than an electron its Δx is much smaller. By a similar computation it is $\boxed{2.64 \times 10^{-17} \text{ m}}$.

16-38 (a) $3d$ (b) $7g$ (c) $5p$.

16-40 (a) A $3d$-orbital has 0 radial and 2 angular nodes; (b) a $7g$-orbital has 2 radial and 4 angular nodes; (c) a $5p$-orbital has 3 radial nodes and 1 angular node.

16-42 An electron in a $3d_{x^2-y^2}$ orbital is most likely to be found in the xy-plane and along the directions of the x- and y-axes. The probability of finding it is zero at its nodes, which are two planes perpendicular to the xy-plane at 45° angles to the x and y axes and intersecting at the z-axis.

16-44 An atomic orbital is a wave function that is a solution of the Schrödinger wave equation. The square of an orbital evaluated at a point is proportional to the probability of finding an electron at that point. Orbitals are not trajectories, but the Bohr planetary orbits are.

16-46 The number of all kinds of nodes equals $n - 1$ and the number of angular nodes equals ℓ:

		No. of Nodes	
n	ℓ	Radial	Angular
1	0	0	0
2	0	1	0
5	1	3	1
5	3	1	3

The four orbitals are labeled $1s$, $2s$, $5p$, and $5f$ respectively.

16-48 (a) This combination is $\boxed{\text{not allowed}}$ because m_s is never equal to zero. If m_s were $\pm 1/2$, this combination would be allowed.

(b) This combination is $\boxed{\text{allowed}}$. It specifies a $2s$-electron.

(c) This combination is $\boxed{\text{allowed}}$. It specifies a $7d$-electron.

(d) This combination is $\boxed{\text{not allowed}}$. The quantum number l is never negative.

16-50 The time it takes for electromagnetic waves to arrive from Cygnus A is the distance divided by the speed of light. This is 3×10^{24} m divided by 3.00×10^8 m s^{-1} or 10^{16} s, which equals $\boxed{\text{300 million years}}$. It is thus said that Cygnus A is 300 million light-years away.

The frequency of the radio wave is c divided by λ, or $\boxed{3.0 \times 10^7 \text{ s}^{-1}}$.

16-52 The several paradoxical experiments are discussed in text section 16-2 (text page 652).

16-54 The energy of the photon is sufficient to overcome the work function of the nickel surface and eject an electron and also to impart a kinetic energy as large as 7.04×10^{-19} J to the ejected electron. It is known that

$$h\nu = \frac{hc}{\lambda} = \Phi + \frac{1}{2}mv^2$$

Hence:

$$\frac{(6.626 \times 10^{-34} \text{ J s})(2.9979 \times 10^8 \text{ m s}^{-1})}{131 \times 10^{-9} \text{ m}} = \Phi + 7.04 \times 10^{-19} \text{ J}$$

Solving gives $\boxed{\Phi = 8.1 \times 10^{-19} \text{ J}}$.

16-56 Green light has wavelengths ranging 500 nm to 570 nm. The upper end of the energy range of green light is therefore

$$E = \frac{hc}{\lambda} = \frac{(6.626 \times 10^{-34} \text{ J s})(2.9979 \times 10^8 \text{ m s}^{-1})}{500 \times 10^{-9} \text{ m}}$$

$$= 3.973 \times 10^{-19} \text{ J} \times \left(\frac{1 \text{ Ry}}{2.17987 \times 10^{-18} \text{ J}}\right) = 0.182 \text{ Ry}$$

The lower end of the energy range for green photons is computed similarly. It equals 0.160 Ry. The C^{5+} ion is a hydrogen-like ion with $Z = 6$. All transitions in its emission spectrum have energies that fit the following formula with $Z = 6$ and integral n's:

$$E = Z^2 \left(\frac{1}{n_{\text{final}}^2} - \frac{1}{n_{\text{initial}}^2}\right) \text{Ry}$$

Insert $Z = 6$ and the two energies at the ends of the range of green light into this formula. Using energies in rydbergs causes all units to cancel away to give

$$\left(\frac{1}{n_{\text{initial}}^2} - \frac{1}{n_{\text{final}}^2}\right) = \frac{0.182}{36} \text{ to } \frac{0.160}{36} = 0.005055 \text{ to } 0.00444$$

Now, systematically try out integers to find a combination that gives a result in the range 0.00444 to 0.005055. Note that n_{final} must be less than n_{initial}. The first combination that works is $\boxed{n_{\text{initial}} = 8; \; n_{\text{final}} = 7}$ for which

$$\left(\frac{1}{8^2} - \frac{1}{7^2}\right) = 0.004783$$

16-58 Obtain the frequency of the light that must be absorbed to excite hydrogen atoms ($Z = 1$) from $n = 2$ to $n = 5$ by substitution in the formula

$$\nu = Z^2 \left(\frac{1}{n_{\text{initial}}^2} - \frac{1}{n_{\text{final}}^2} \right) (3.29 \times 10^{15} \text{ s}^{-1})$$

The result is

$$\nu = 1^2 \left(\frac{1}{2^2} - \frac{1}{5^2} \right) 3.29 \times 10^{15} \text{ s}^{-1} = 0.691 \times 10^{15} \text{ s}^{-1}$$

This is the required output frequency from the doubling crystal. The input frequency is half of this, or 0.345×10^{15} s^{-1}. The wavelength corresponding to the input frequency is obtained by substituting in $\lambda = c/\nu$. It is $\boxed{868 \text{ nm}}$.

16-60 (a) The loss in intensity of the tone is due to destructive interference between the two loudspeakers. Destructive interference occurs when one wave-train lags the other by an integral number of half-wavelengths. The half-wavelength of the tone is thus 0.16 (or 0.080 or 0.0533...) m, and λ is 0.32 m. Accept $\boxed{0.16 \text{ m}}$ because loss of intensity of the tone would presumably have been reported at movement distances less than 0.16 m if the wavelength were shorter.

(b) The frequency of the sound is its speed divided by its wavelength. It is $\boxed{1.1 \times 10^3 \text{ s}^{-1}}$.

16-62 (a) Write the equation $\lambda = h/p = h/mv$ and set $h = 1$ J s. The wavelength of the 0.170 kg baseball moving at 30 m s^{-1} is then $\boxed{0.20 \text{ m}}$.

(b) From the Heisenberg indeterminacy principle write: $\Delta x \cdot m \Delta v \geq h/4\pi$. Substitute $m = 0.170$ kg, $h = 1$ J s, and $\Delta v = 2$ m s^{-1} to establish that $\boxed{\Delta x \geq 0.23 \text{ m}}$. Even at its smallest, this Δx is too large to permit the moving baseball to be caught reliably.

16-64 The amplitude of a wave function in a hydrogen atom, represented ψ, is the size of the wave disturbance. The average electron density is proportional to ψ^2. Its variation from point to point in the region of the nucleus tells how the probability of finding the electron varies from point to point.

Answers to Even-Numbered Problems, Chapter 17

17-2 It is best to cite subshells in order of increasing n and subshells with the same n in order of increasing l. (a) N: [He]$2s^2 2p^3$ (b) Mn: [Ar]$3d^5 4s^2$ (c) Ho: [Xe]$4f^7 6s^2$

17-4   The Li^- ion ($1s^2 2s^2$) is diamagnetic in the ground state; the B^+ ion ($1s^2 2s^2$), the F^- ion ($1s^2 2s^2 2p^6$) and the Al^{3+} ion ($1s^2 2s^2 2p^6$) are also diamagnetic in the ground state. All of the preceding have an even number of electrons and all electrons paired.

The S^- ion ($1s^2 2s^2 2p^6 3s^2 3p^5$) is paramagnetic in the ground state. The Ar^+ ion has the same ground-state configuration as S^- and is therefore also paramagnetic. The Br^+ ion ($[Ar]3d^{10}4s^2 4p^4$) and Te^- ion ($[Kr]4d^{10}5s^2 5p^5$) are paramagnetic. A species with an odd number of electrons must be paramagnetic; a species with an even number of electrons may be diamagnetic or paramagnetic.

17-6   (a) The species is an atom and is therefore electrically neutral. It has 54 electrons. The element with $Z = 54$ is $\boxed{\text{osmium}}$.   (b) The ion is $\boxed{O^{2-}}$.   (c) The ion is $\boxed{Rh^{3+}}$.

17-8   The alkaline earth elements have Z's of $4, 12, 20, 38, 56, 88$. The successive differences between these numbers are $8, 8, 18, 18, 32$. The pattern of these "first differences" suggests a gap of 32 elements before the next alkaline earth element. This element would follow the completion of the $n = 7$ row in the periodic table and would have $Z = 120$.

Finding $Z = 138$ for the eight period alkaline earth element implies the interpolation of an additional 18 elements in the seventh period. This departure from the expected aufbau pattern is explainable as the result of the intermediate filling of nine $5g$ orbitals by 18 electrons before the closing of the seventh row of the periodic table.

17-10   The three noble-gas atoms would have $Z = \boxed{3}$ (corresponding to "$1s^3$"), $Z = \boxed{15}$ ("$1s^3 2s^3 2p^9$ "), and $Z = \boxed{27}$ ("$1s^3 2s^3 2p^9 3s^3 3p^9$ ").

17-12   (a) $\boxed{Xe}$ should have a higher IE_1 than Bi. Its IE_1 should exceed that of Sb; the IE_1 of Sb should exceed that of Bi by application of periodic trends. See text Figure 17-9.

(b) $\boxed{Se}$ should have a higher IE_1 than Te; the two are in the same group and Se is higher up the periodic table.

(c) $\boxed{Y}$ should have a higher IE_1 than Rb. The two are in the same row and Y is farther to the right.

(d) $\boxed{Ne}$ should have a higher IE_1 than K since its IE_1 exceeds that of Ar, which definitely has a higher IE_1 than K.

17-14 The reasoning uses the patterns in the periodic table as in problem 17-12: (a) Rb (b) I (c) Te (d) Cl.

17-16 $(IE_1$ versus $Z)$

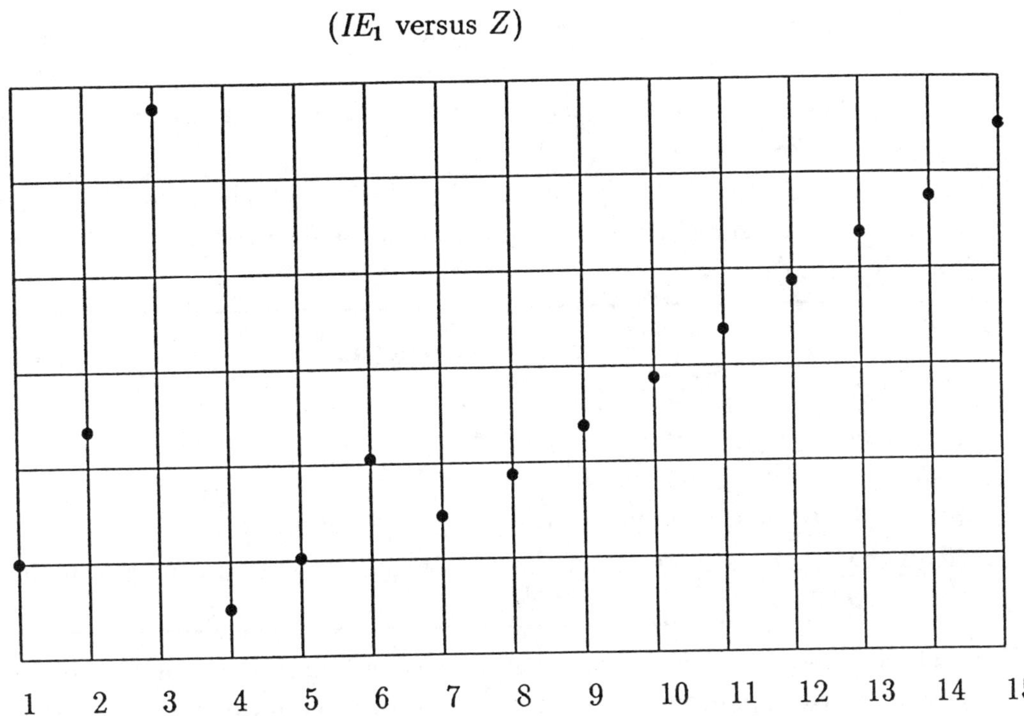

This problem uses the same hypothetical universe as problem 17-10. IE_1 has relative maxima when noble-gas electron configurations are attained. The graph rises to a maximum at $Z = 3$ (corresponding to the "$1s^3$" noble-gas configuration), falls sharply from $Z = 3$ to $Z = 4$, rises again to a lesser relative maximum at $Z = 6$ (the "$1s^3 2s^3$" configuration), falls somewhat at $Z = 7$, and resumes increasing to a relative maximum at $Z = 15$, as the $2p$ subshell is filled with nine electrons.

17-18 Calcium does have a positive affinity for an electron, but the quoted number is very small. Convert it from kJ mol^{-1} to J per atom and then use $E = hc/\lambda$ to compute the wavelength. The answer is $\boxed{2.9 \times 10^{-2} \text{ mm}}$. $\boxed{\text{Infrared}}$ radiation is energetic enough to remove the extra electron from Ca$^-$ (see text Figure 16-5, text page 651).

17-20 The positive charge on the nucleus increases steadily across the period. Meanwhile, additional electrons are added to the same shell and are roughly the same distance from the nucleus. They do not screen each other very effectively from the greater nuclear charge so the atom contracts.

17-22 (a) Mn has more electrons but the same nucleus as Mn^{3+} ion. Therefore $\boxed{Mn}$ is larger.

(b) Ca uses the $n = 3$ shell while Mg uses only the $n = 2$ shell. Therefore $\boxed{Ca}$ is larger.

(c) I^- ion and the Xe atom are isoelectronic, but I^- has a smaller nuclear charge. $\boxed{I^-}$ is accordingly larger.

(d) Ge is just to the left of As in the fourth period. Therefore $\boxed{Ge}$ is larger.

(e) Rb and Sr^+ are isoelectronic and Rb has the smaller nuclear charge. Therefore $\boxed{Rb}$ is larger.

17-24 (a) The Cl^- and S^{2-} ions are isoelectronic, and the latter has a lesser nuclear charge; the $\boxed{S^{2-} \text{ ion}}$ is larger.

(b) The $\boxed{Tl^+ \text{ ion}}$ is larger. Loss of two electrons to give Tl^{3+} reduces electron-electron repulsions and so allows contraction.

(c) The $\boxed{Ce^{3+} \text{ ion}}$ is larger, considering the lanthanide contraction.

(d) The $\boxed{I^- \text{ ion}}$ is larger than the S^- ion; its outer electrons are in the $n = 5$ shell.

17-26 The bond enthalpy increases with increasing bond order, and the bond length decreases with increasing bond order. None of the relationships among the three are linear, however.

17-28 The order of bond length is $FCl < BrCl < IBr$. The smaller bond enthalpy should be for the longest bond, the one in $\boxed{IBr}$.

17-30 (a) $\boxed{MgBr_2}$ is more ionic than PBr_3 and should have a higher boiling point.

(b) $\boxed{SrO}$ should boil higher than OsO_4 for the same reason. (c) $\boxed{Al_2O_3}$ should boil higher than Cl_2O.

17-32 (a) The ΔE of the first reaction is the ionization energy of one mole of $Na(g)$ added to the electron attachment energy of one mole of $I(g)$. These values are the IE_1 of Na multiplied by 1 mol and the negative of the EA (electron affinity) of I (also multiplied by 1 mol). Adding them gives

$$\Delta E = 495.8 + (-295.2) = \boxed{200.6 \text{ kJ}}$$

The ΔE of the second reaction is the IE of 1 mol of I added to the electron attachment energy of 1 mol of Na:

$$\Delta E = 1008.4 + (-52.867) = \boxed{955.5 \text{ kJ}}$$

(b) Similar combination of the ionization energies and electron attachment energies of K and Cl gives ΔE of the first reaction as $\boxed{69.8 \text{ kJ}}$ and ΔE of the second reaction as $\boxed{1202.7 \text{ kJ}}$. Even if Na^-I^+ or K^-Cl^+ were to form, the reactions transferring electrons to form Na^+I^- or K^+Cl^- would be strongly favored energetically and would occur quickly.

17-34 The third and fourth columns of the following table present the comparison. Note that the absolute value of Δ must be used in the equation.

Compound	Δ	$16\Delta + 3.5\Delta^2$	Expt. Ionic Character
ClF	0.82	15	11%
BrF	1.02	20	15%
BrCl	0.20	3.3	5.6%
ICl	0.50	8.9	5.8%
IBr	0.30	5.1	10%

17-36 (a) The polarity of the bonds follows the difference Δ in the electronegativities of the two atoms. These differences are given in parentheses in the following ranking

$$\boxed{\text{IF } (1.32) > \text{ICl } (0.50) > \text{ClF } (0.82) > \text{BrCl } (0.15) > \text{ClCl } (0.0)}$$

(b) The more electropositive element in a two-center partial covalent bond carries the fractional positive charge. In the preceding ranking, the more electronegative element is the one given first in the formula (except in the case of Cl_2 where the two atoms are equal in electronegativity).

17-38 The predicted maximum oxidation states are W (+6), At (+7), Xe (+8), Fr (+1). Research[9] suggests strongly that Cs(III) exists. A better answer for Fr might therefore be +3.

17-40 The problem asks in effect for an explanation of the tendency (identified on text page 708) for the heavier transition elements to form stable compounds in higher oxidation states. The $5d$ wave-functions in osmium, which have two radial and two angular nodes, are diffuse and well-screened from the nucleus. The $5d$ electrons are therefore not as tightly held, even in the +2 ion created by the removal of the $6s$ electrons.

---

[9]Angew. Chem. Intl. Ed. Engl. **28** 1676 (1989).

17-42 Tin(IV) chloride is a liquid at ordinary conditions, but tin(II) chloride is a solid. The bonding in tin(IV) chloride is more covalent than in tin(II) chloride, according to the trends discussed on text page 706. This makes for weaker cohesive forces between the particles in solid tin(IV) chloride and a lower melting point.

17-44 By comparison to other oxoacids of the general formula $XO(OH)_2$, the oxoacid $CO(OH)_2$ should have K_a equal to about 2×10^{-2}. Apparently most CO_2 does not convert to $CO(OH)_2$ when dissolved in water. The low measured K_a for "carbonic acid" is then justified as the product of the theoretical value and a small K for the reaction $CO_2(aq) + H_2O(l) \rightleftharpoons H_2CO_3(aq)$. This point is also made in the marginal note on text page 359.

17-46 The strength of the acids increases from left to right:

$$H_3PO_4 \approx H_4P_2O_6 < H_4P_2O_7 << H_5P_3O_{10} < H_3P_3O_9$$
$$OP(OH)_3 \approx O_2P_2(OH)_4 < O_3P_2(OH)_4 << O_5P_3(OH)_5 < O_6P_3(OH)_3$$

In the second line the formulas have been re-written to highlight the ratio of the number of lone oxygen atoms to phosphorus atoms.

17-48 The oxoacid H_2MnO_4 is analogous to H_2SO_4. From Table 17-5 (text page 710) the K_a is about 10^{-2}. By similar analogies, K_a for H_4ZrO_4 is about 10^{-14}; for $HMnO_4$ about 10^3; for H_3MnO_4 about 10^{-6}.

17-50 All that is necessary is to write the ground-state electron configurations. The species that are spherically symmetric are F^- (filled shell), Na (half-filled s-subshell, S^{2-} (filled shell), Cu (half-filled s-subshell, filled d-subshell), Mo (half-filled s subshell, half-filled d-subshell), Sb (filled s-subshell, half-filled p-subshell), Au (half-filled s-subshell, filled d-subshell).

17-52 The third ionization energy of Li is ΔE for the process $Li^{2+}(g) \longrightarrow Li^{3+}(g) + e^-$. The ejection of a $1s$-electron from a lithium atom on the other hand is represented $Li(g) \longrightarrow Li^{+*}(g) + e^-$. where the star indicates that the Li^+ ion is in an excited state. In both cases a $1s$-electron is removed. The IE_3 is larger because the $1s$-electron is removed against the full attraction of the $Z = 3$ nucleus whereas the removal of the $1s$-electron in the photoelectron spectroscopy experiment is against the attraction of a $Z = 3$ nucleus screened somewhat by the other two electrons.

17-54 The ions N^{3-}, O^{2-}, and F^- all have the ground-state electron configuration $[He]2s^2 2p^6$. The size of the ion drops sharply (from 1.71 to 1.40 Å) going from nitrogen to oxygen, but only slightly (from 1.40 to 1.36 Å) going from oxygen to fluorine. This occurs because contraction of the electron cloud in the second step must overcome stronger electron-to-electron repulsions than in the first step (the 10 electrons are closer together).

17-56 In general, the longer the bond is, the weaker it is. The C—O bond in $Ni(CO)_4$ is weaker than the C—O bond in free carbon monoxide.

17-58 The difference in electronegativity between Rb and Cs is small, only 0.03, but the difference in electronegativity between Au and Cs is large, 1.75. This is slightly larger than ΔEN in NaI. The compound between Cs and Au can reasonably be formulated $Cs^+ \cdots Au^-$ (ionic).

17-60 (a) The Lewis structures are:

$$: \ddot{N} = \ddot{S} - \ddot{F}: \quad \text{and} \quad : \ddot{S} = \ddot{N} - \ddot{F}: \quad \text{and} \quad : \ddot{N} = \ddot{F} - \ddot{S}:$$

In the first, the formal charges are (from left to right) -1, $+1$, and 0. In the second, all three atoms have zero formal charge. In the third, the formal charges are -1, $+2$, and -1.

(b) The structure with the least separation of formal charge has a central N. This is not consistent with the observation of a central S atom.

(c) The electronegativity of N exceeds that of S; that is, N has a greater tendency than S to accept electrons in a chemical bond. This helps explain why the observed structure corresponds to a formal build-up of negative charge on the N and formal build-up of positive charge on the S. Also, the two most electronegative atoms (N and F) are separated, reducing electron-electron repulsions. Finally, the resonance structure $: N \equiv \ddot{S} - \ddot{F}:$ probably contribute, In this structure all formal charges equal zero but the sulfur has an expanded octet (it sees 10 electrons).

17-62 The greatest electronegativity should be listed for the $\boxed{\text{the highest}}$ oxidation state.

17-64 The acidity constant is in the range for oxoacids of the $XO(OH)_m$ type (see text Table 17-5 on text page 710). The problem is that the molecule has three hydrogens, not just one. Two hydrogen atoms must be bound directly to the phosphorus atom. The P also has an OH group and a lone oxygen atom attached to it: $\boxed{(\text{HO})\text{OPH}_2}$. There is a single O—H bond in this molecule, which is a $\boxed{\text{monoprotic}}$ acid.

Answers to Even-Numbered Problems, Chapter 18

18-2

Species	Ground-State Configuration	Bond Order
Li_2^-	$(\sigma_{2s})^2(\sigma_{2s}^*)^1$	1/2
Be_2^-	$(\sigma_{2s})^2(\sigma_{2s}^*)^2(\pi_{2p})^1$	1/2
B_2^-	$(\sigma_{2s})^2(\sigma_{2s}^*)^2(\pi_{2p})^3$	3/2
C_2^-	$(\sigma_{2s})^2(\sigma_{2s}^*)^2(\pi_{2p})^4(\sigma_{2p_z})^1$	5/2
N_2^-	$(\sigma_{2s})^2(\sigma_{2s}^*)^2(\pi_{2p})^4(\sigma_{2p_z})^2(\pi_{2p}^*)^1$	5/2
O_2^-	$(\sigma_{2s})^2(\sigma_{2s}^*)^2(\sigma_{2p_z})^2(\pi_{2p})^4(\pi_{2p}^*)^3$	3/2
F_2^-	$(\sigma_{2s})^2(\sigma_{2s}^*)^2(\sigma_{2p_z})^2(\pi_{2p})^4(\pi_{2p}^*)^4(\sigma_{2p_z}^*)^1$	1/2

18-4 (a) The correlation diagram for O_2^- is identical to the diagram for O_2 given in Figure 18-5 (text page 723) except that an additional electron is added to either the $\pi_{2p_x}^*$ or $\pi_{2p_y}^*$ MO (the two are equivalent).

(b) The ground-state electron configuration of O_2 is listed in Table 18-2 on text page 725. The configurations of the ions in the problem are derived by removal or addition of electrons from the highest occupied and lowest unoccupied molecular orbitals:

O_2^+ $\quad (\sigma_{2s})^2(\sigma_{2s}^*)^2(\sigma_{2p_z})^2(\pi_{2p})^4(\pi_{2p}^*)^1$

O_2 $\quad (\sigma_{2s})^2(\sigma_{2s}^*)^2(\sigma_{2p_z})^2(\pi_{2p})^4(\pi_{2p}^*)^2$

O_2^- $\quad (\sigma_{2s})^2(\sigma_{2s}^*)^2(\sigma_{2p_z})^2(\pi_{2p})^4(\pi_{2p}^*)^3$

O_2^{2-} $\quad (\sigma_{2s})^2(\sigma_{2s}^*)^2(\sigma_{2p_z})^2(\pi_{2p})^4(\pi_{2p}^*)^4$

(c) The bond orders of the species are $\boxed{5/2}$ (for O_2^+), $\boxed{2}$ (for O_2), $\boxed{3/2}$ (for O_2^-) and $\boxed{1}$ (for O_2^{2-}).

(d) All of the species except O_2^{2-} should be paramagnetic. A species with an odd number of electrons is automatically paramagnetic; the reason for the paramagnetism of ordinary O_2 is discussed on text page 725.

(e) In the series given in (c), each additional electron occupies a π^* antibonding orbital. The bond dissociation energy decreases along the series.

18-6 It is not necessary to show any configuration for core electrons, which play little part in bonding. Then:

I_2: $(\sigma_{5s})^2(\sigma_{5s}^*)^2(\sigma_{5p_z})^2(\pi_{5p})^4(\pi_{5p}^*)^4$ This substance should be diamagnetic.

18-8 The point of the problem is to count the valence electrons and deal with the charge: (a) oxygen   (b) boron   (c) fluorine

---

18-10 (a) bond order $= 2$ (b) bond order $= 3/2$ (c) bond order $= 2$

18-12 (a) paramagnetic   (b) paramagnetic   (c) paramagnetic

18-14 The molecule of BeN has 7 valence electrons and a bond order of 3/2. The substance should be paramagnetic.

18-16 The nitrosyl molecular ion forms from nitrogen monoxide by the loss of an electron from the highest occupied molecular orbital. The electron configuration of NO is $(\sigma_{2s})^2(\sigma_{2s}^*)^2(\pi_{2p})^4(\sigma_{2p_z})^2(\pi_{2p}^*)^1$ so the electron comes from a π_{2p}^* orbital. Because this is an antibonding orbital, the bonding in NO$^+$ should be stronger , and the bond should be shorter than in NO. Moreover, NO$^+$ should be diamagnetic because of its electrons are paired.

18-18 The HeH$^+$ ion has the electron configuration $(\sigma_{1s})^2$. Its bond order is 1, and it is diamagnetic.

The lower energy state is reached by the reaction HeH$^+$ $\longrightarrow$ He $+$ H$^+$. In this set of products two electrons are both close to the $+2$ charge of a helium nucleus and the same distance from it. In He$^+$ one electron is under the influence of a $Z = 2$ nucleus, but the other experiences only a $Z = 1$ central field. The energy of the second state is therefore less negative. Experiment confirms this conclusion. The first ionization energy of He is 2372 J, but the first ionization energy of H is only 1312 J. Hence, for

$$\text{He}(g) + \text{H}^+(g) \longrightarrow \text{He}^+(g) + \text{H}(g) \quad \Delta E = 2372 - 1312 = 1060 \text{ J}$$

18-20 The central O in the H$_3$O$^+$ ion has $SN\,4$. This means there is sp^3 hybridization on the O. The molecular ion will be pyramidal.

18-22 The Lewis structures are

(a) $:\ddot{F}:$

$:\ddot{F}—B$
$\quad\ \ |$
$\quad :F:$

(b) $\begin{array}{c} H \\ | \\ H—B—H \\ | \\ H \end{array}\ ^{-}$

(c) $\begin{array}{c} H \\ | \\ H—P: \\ | \\ H \end{array}$

(d) $\ddot{S}{=}C{=}\ddot{S}$

(e) $H—C\begin{array}{c} {}^H \\ {}_H \end{array}\ ^{+}$

(a) The central B in BF_3 has $SN\ 3$. It is sp^2 hybridized, and the molecule is trigonal-planar.

(b) The central B in BH_4^- has $SN\ 4$. It is sp^3 hybridized, and the molecular ion is tetrahedral.

(c) The central P in PH_3 has $SN\ 4$. It is sp^3 hybridized, and the molecule is pyramidal.

(d) The central C in CS_2 has $SN\ 2$. It is sp hybridized, and the molecule is linear.

(e) The central C in CH_3^+ has $SN\ 3$. It is sp^2 hybridized, and the molecular ion is trigonal-planar.

18-24 The tetrahedral ClO_4^- ion uses sp^3 hybrid orbitals on the central Cl (which has $SN\ 4$). The pyramidal ClO_3^- ion also uses sp^3 hybrid orbitals on the central Cl, which has $SN\ 4$.

18-26 The carbon atom in $N{\equiv}C—Cl$ has a steric number of 2. It is $\boxed{sp\ \text{hybridized}}$, and the molecule should be $\boxed{\text{linear}}$.

18-28 (a) N_2 has 1 σ-bond and 2 π-bonds.

(b) Formaldehyde has 3 σ-bonds and 1 π-bond.

(c) Boron trifluoride has 3 σ-bonds.

(d) Acetylene has 3 σ-bonds and 2 π-bonds.

18-30 The molecule of acetic acid has the structure

$$H—\overset{\displaystyle H}{\underset{\displaystyle H}{C}}—C\begin{array}{c} {\nearrow}^O \\ {\searrow}_{OH} \end{array}$$

The left carbon atom is sp^3 hybridized, and the right is sp^2 hybridized. A p orbital on the right carbon atom and a p orbital on one oxygen atom combine to form a π and π^* orbital. Two electrons occupy the π orbital, accounting for the double bond between the C and the O. The six angles at the sp^3 hybrid C are all very close to 109.5°. The three angles at the sp^2 C are close to 120°.

18-32 The structure of acrylonitrile is:

$$: N \equiv C - C \cdots$$

The molecule is planar with the six bond angles at the two double-bonded carbon atoms (both sp^2-hybridized) all close to 120° and the bond angle at triple-bonded C atom (sp-hybridized) equal to 180°. The p-orbitals perpendicular to the molecular plane form a π-bonding system with 4 electrons resembling that in butadiene (text page 734). The other two p-orbitals on the sp-hybridized C atom and the N atom form an additional doubly occupied π-orbital.

18-34 (a) There are three isomers of C_2Cl_2BrF. In one, both Cl's are on the same carbon. In the other two, the Cl's are on different C's and disposed *cis* and *trans* to each other, respectively. The three are:

(b) Refer to part (a). In each of the three structures there are two different Cl atoms that can be replaced by an H atom to give $C_2HBrClF$. Hence there are six structures. Another approach works if the problem is not considered in the context of part (a). Focus on the H as a substituent. Suppose it is on the same carbon as the Br. There are two possible isomers, one in which the F is *cis* to the H and the other in which the Cl is *cis* to the H. Now suppose that the H is on the same carbon as the Cl. There are again two isomers. Finally, suppose the H is on the same carbon as the F. There are another two isomers. This makes a total of six isomers:

18-36 The nitrate ion has 24 valence electrons. Three resonance structures are needed to represent the equivalence of the three N to O bonds:

$$:\ddot{O}-N\diagdown\diagup\ddot{O}:\diagdown\ddot{O} \quad\longleftrightarrow\quad \ddot{O}=N\diagdown\diagup\ddot{O}:\diagdown\ddot{O}: \quad\longleftrightarrow\quad :\ddot{O}-N\diagdown\diagup\ddot{O}:\diagdown\ddot{O}:$$

The *SN* of the central nitrogen is 3, which in VSEPR theory predicts trigonal-planar geometry about the central nitrogen. This corresponds to sp^2 hybridization of the valence orbitals of the central N. Six valence electrons are shared between the N and the O's in these orbitals. The remaining $2p$-orbital of the nitrogen combines with one $2p$-orbital on each of the three oxygen atoms to form a set of four π-molecular orbitals. The lowest lying of these, a bonding orbital, is occupied by a pair of electrons. The net effect is that the four atoms are bonded by eight electrons: each of the three links between N and O has bond order 4/3.

18-38 Absorption of x-rays excites core electrons up to high-energy outer orbitals or completely away from the molecule. The wavelengths of absorption give information about the bonding.

18-40 The electron lost in the ionization of ground-state C_2H_4 comes from a bonding π-orbital. The π system in C_2H_4 is less than half filled, which assures that all of the π-electrons are bonding electrons in the ground state. The bond order between the two carbons in $C_2H_4^+$ $\boxed{\text{decreases}}$ to 3/2.

18-42 Crystal violet is purple, which implies that it absorbs the color that is the complement of purple, which is yellow. This means it has a maximum in its absorption $\boxed{\text{near 600 nm}}$ (see text Figure 20-22).

18-44 The π-electrons in naphthalene are more delocalized in both the ground state and the excited states than the π-electrons in benzene, but the energy of the orbitals are lowered more in the excited state than in the ground state. This shifts the maximum absorption of light to a wavelength $\boxed{\text{longer than 255 nm}}$.

18-46 The energy 330 kJ mol^{-1} of bonds is equivalent to 5.48×10^{-19} J per bond. The maximum wavelength that can supply this amount of energy is

$$\lambda_{\text{max}} = \frac{hc}{E} = \frac{(6.626 \times 10^{-34} \text{J s})(3.00 \times 10^8 \text{ m s}^{-1})}{5.48 \times 10^{-19} \text{ J}} = 3.62 \times 10^{-7} \text{ m} = 362 \text{ nm}$$

18-48 Carbon dioxide and sulfur dioxide have similar formulas, but quite different molecular structures. Carbon dioxide has a linear molecule. In it, the central C forms σ-bonds to the two oxygen atoms using sp hybrid orbitals. These bonds are joined by π bonds from the overlap of C-atom $2p$ and O-atom $2p$ orbitals. The central atom in CO_2 has no lone pairs. The molecule of SO_2 is a bent molecule. The central S is sp^2 hybridized: two of these orbitals overlap with orbitals of the O's in σ bonds and the third accommodates a lone pair. Thus, the *SN* of the central S is 3. The molecule is bent with an O—S—O angle near 120°. One $3p$ orbital of the S mixes with one $2p$ orbital from each of the O's in a π-system.

18-50 The carbide ion has 10 valence electrons. Its ground-state configuration is $\boxed{(\sigma_{2s})^2(\sigma_{2s}^*)^2(\pi_{2p})^4(\sigma_{2p_z})^2}$, the same as the configuration of N_2, and its bond order is $\boxed{3}$.

18-52 The ground-state electron configuration of Be_2 is $\boxed{(\sigma_{2s})^2(\sigma_{2s}^*)^2}$. The predicted bond order of the molecule is $\boxed{\text{zero}}$: according to simple theory it should not exist. The observed bond length and dissociation enthalpy of Be_2 are respectively high and quite low: the molecule indeed almost does not exist.

18-54 The cyclic H_3^+ ion is a perfect equilateral triangle. It has two electrons. The electrons are in a low-energy bonding orbital $\boxed{\text{strongly localized in the center}}$ of the triangle so as best to nullify internuclear repulsions.

18-56 The molecule of O_3 has resonance Lewis structures like SO_2 (discussed on text page 736). Like SO_2, it has sp^2 hybridization on the central atom, σ bonds between the central O and the terminal O's, and one pair of electrons in a π bonding orbital spread out over all three atoms. The added electron to convert neutral O_3 to O_3^- ion would go into the lowest unoccupied molecular orbital, which is a π^* orbital. The second additional electron required to give O_3^{2-} would complete the occupancy of this antibonding orbital. The central O has a *SN* of 3 in all three species so that the O—O—O angle would be about $\boxed{120°}$ in all three. $\boxed{\text{Neutral } O_3}$ would have the shortest bond distances because the bond order in it is largest: 3/2 in O_3, compared to 5/4 in O_3^- ion and 1 in O_3^{2-} ion.

18-58 *Trans*-tetrazene and *cis*-tetrazene have the structures:

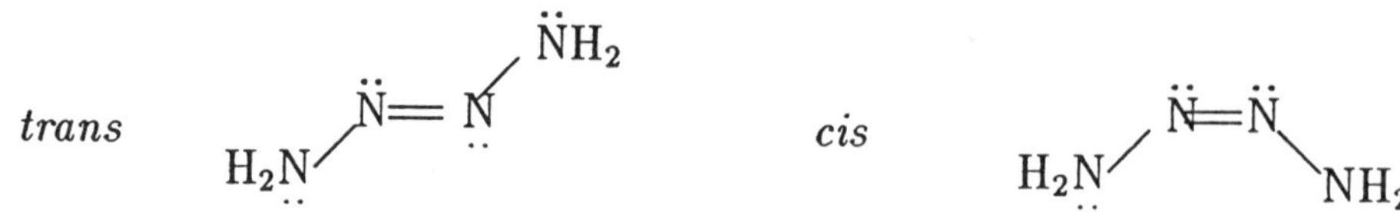

The *SN*'s of the inner nitrogen atoms in both structures are 3, and the *SN*'s of the outer N's are both 4. There is sp^2 hybridization at the inner N's and sp^3 at the outer N's. Both molecules are planar, except for the terminal hydrogen atoms, because of the π system.

18-60 (a) The ΔH° of the reaction $6\,C(g) + 6\,H(g) \longrightarrow C_6H_6(g)$ equals the ΔH_f°'s of the products minus the ΔH_f°'s of the reactants:

$$\Delta H^\circ = 1(82.93) - 6(716.682) - 6(217.96) = \boxed{-5524.92 \text{ kJ}}$$

This is an experimental result because it is based on experimental ΔH_f°'s and Hess's law.

(b) The formation of 1 mol of non-resonance benzene requires the formation of 6 mol of C—H bonds, 3 mol of C=C bonds, and 3 mol of C—C bonds. The negative of the sum of the bond dissociation enthalpies of these bonds is -5367 kJ, which is less negative than -5523 kJ mol^{-1}. Non-resonance benzene is $\boxed{\text{less stable}}$ than regular benzene. (c) The resonance stabilization of 1 mol of benzene equals $(-5367) - (-5524.92) = \boxed{158 \text{ kJ}}$.

18-62 The combustion of $C_{60}(s)$ is represented:

$$C_{60}(s) + 60\,O_2(g) \longrightarrow 60\,CO_2(g)$$

The ΔH° of this reaction is the difference between the ΔH_f°'s of the products and the reactants. It is

$$60 \underbrace{(-393.51)}_{CO_2(g)} - 1 \underbrace{\Delta H_f^\circ}_{C_{60}(s)} - 60 \underbrace{(0.000)}_{O_2(g)} = -25891 \text{ kJ}$$

Solving gives $\boxed{\Delta H_f^\circ = +2280.4 \text{ kJmol}^{-1}}$.

18-64 The equilibrium bond distance in the excited state of C_2H_4 will be $\boxed{\text{longer}}$ because the excited electron goes to a π^*-orbital, in which it actively opposes bonding.

18-66 (a) Excitation of the electron in H_2^+ to an antibonding orbital will lead to the dissociation of the molecule. The effect of an antibonding electron is to oppose the continued association of the nuclei.

(b) If even more energy is fed into a H_2^+ molecule, the electron can be brought into a higher-level bonding molecular orbital. The molecule should persist for a while in this bound state, but it will eventually dissociate to H and H^+.

18-68 Overall equations are:

$$S(s) + 3/2\,O_2(g) + H_2O(l) \longrightarrow H_2SO_4(aq)$$

$$N_2(g) + 5/2\,O_2(g) + H_2O(l) \longrightarrow 2\,HNO_3(aq)$$

Answers to Even-Numbered Problems, Chapter 19

19-2 (a) $[Cr(CO)_4]^0$ (b) $[CoCl_2(C_2O_4)_2]^{3-}$ (c) $[Rh(NH_3)_3Br_3]^-$.

19-4 The glycinate ion (H_2N—CH_2—COO^-) has two donor sites. They are the nitrogen, which has one lone pair of electrons, and the carboxylate oxygen, which has three lone pairs. The ligand can bind to a metal ion by donating electron pairs from either of these sites or both.

19-6 The dissolution of tetraamminegold(III) phosphate is not a simple dissociation into solvated $[Au(NH_3)_4]^{3+}$ ion and solvated PO_4^{3-} ion. The PO_4^{3-} reacts extensively with water (hydrolyzes) to give mainly HPO_4^{2-} but also $H_2PO_4^{2-}$ and H_3PO_4. These reactions remove PO_4^{3-} and thereby increase the solubility. The cation may also react with water to form H_3O^+ and $[Au(NH_3)_3(NH_2]^{2+}$ ion. The desired equation should include this fact too:

$$\boxed{[Au(NH_3)_4]PO_4(s) \longrightarrow [Au(NH_3)_3NH_2]^{2+}(aq) + HPO_4^{2-}(aq)}$$

19-8 The oxidation state of Mn in $Mn_2(CO)_{10}$ is $\boxed{\text{zero}}$. Assigning this oxidation number relies on recognizing CO (carbon monoxide) as a neutral ligand. The oxidation state of Re in $[Re_3Br_{12}]^{3-}$ is $\boxed{+3}$; the oxidation state of the Fe in $[Fe(H_2O)_4(OH)_2]^+$ is $\boxed{+3}$; the oxidation state of the Co in $[Co(NH_3)_4Cl_2]^+$ is $\boxed{+3}$.

19-10 (a) $Ag_4[Fe(CN)_6]$ (b) $K_2[Co(NCS)_4]$ (c) $Na_3[VF_6]$ (d) $K_3[Cr(C_2O_4)_3]$

19-12 Use the rules on text page 766: (a) tetraaquadihydroxonickel(II)

(b) chloroiodomercury(II)

(c) potassium hexacyanoosmate(II)

(d) bromochlorobis(ethylenediamine)iron(III) chloride

19-14 Such a complex suffers ligand exchange at only a very slow rate (the meaning of "inert"). It however does tend to decompose spontaneously in some fashion (the meaning of "thermodynamically unstable").

19-16 Combine ΔG_f°'s to determine the ΔG° of the reaction as it is given in the problem:

$$\underbrace{1(-45.6)}_{1\,Ni^{2+}(aq)} + \underbrace{4(172.4)}_{4\,CN^-(aq)} - \underbrace{1(+472.1)}_{1\,[NiCN]_4^{2-}(aq)} = 171.9 \text{ kJ}$$

A positive ΔG° means that the complex is thermodynamically stable with respect to decomposition to aquated metal ion and ligands. Still, ligands exchange readily. The equilibrium constant is found from:

$$\ln K = \frac{-\Delta G^\circ}{RT} = \frac{-171.9 \times 10^3 \text{ J mol}^{-1}}{(8.315 \text{ J K}^{-1} \text{ mol}^{-1})(298.15 \text{ K})} = -69.34$$

$$K = \exp(-69.34) = \boxed{7.7 \times 10^{-31}}$$

19-18 Conductivity increases in the order

$$[Fe(NH_3)_3Cl_3] < [Cr(NH_3)_4Cl_2]Cl < BaCl_2 < K_4[Fe(CN)_6]$$

19-20 (a) The square-planar complex bromochloro(ethylenediamine)platinum(II) has only one isomer because the $NH_2CH_2CH_2NH_2$ ligand does not span *trans* positions. The structure is:

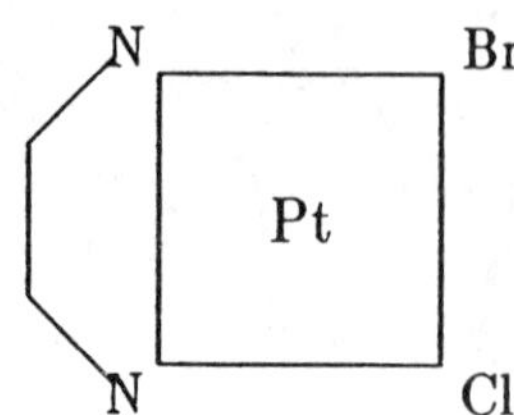

(b) The complex ion has *cis* and *trans* isomers just like those of text Figure 19-8 (text page 771) except with a central iron(III) ion.

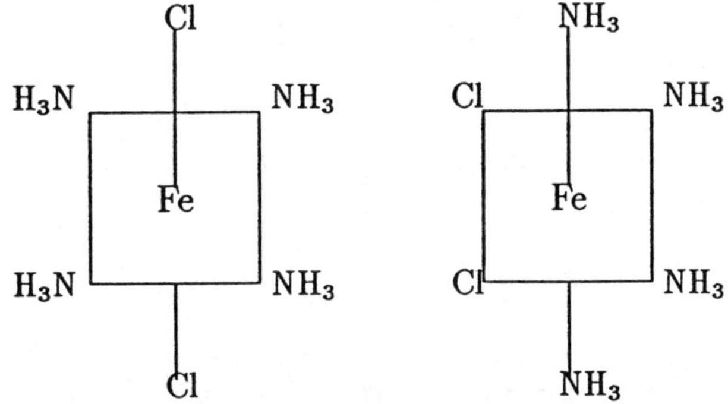

(c) The amminechlorobis(ethylenediamine)iron(III) ion, a +1 ion, has two *cis* forms and one *trans* form. Both of the *cis* forms (center and right in the following) are chiral, that is, non-superimposable upon their mirror images. This makes a total of five isomers, of which three are:

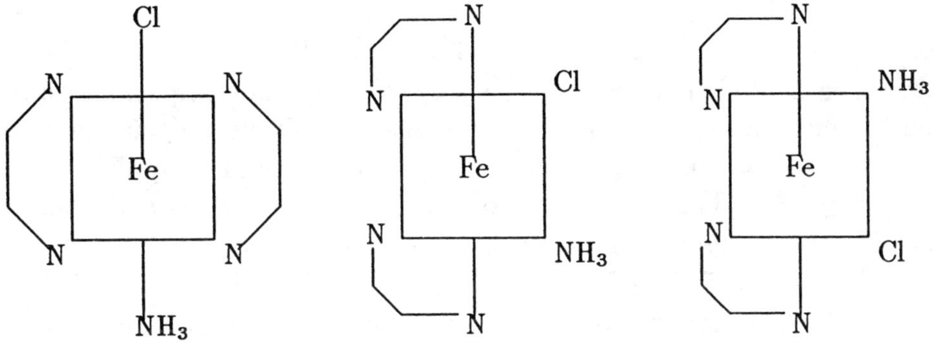

19-22 There are two each of the three ligands NH_3, Cl^-, and F^-. The members of each pair may be either *cis* or *trans* about the central Pt(IV) in octahedral coordination. This would seem to mean that there are eight isomers because there are eight possible triple combinations of *cis* and *trans* (*cis-cis-cis*, *cis-cis-trans*, *cis-trans-cis*, and so forth). But, placing two pairs *trans* forces the third pair to be *trans*, also. This eliminates *cis-trans-trans*, *trans-cis-trans*, *and cis-trans-trans* from the list of eight. Only five *cis-trans* isomers are therefore possible:

Isomer No:	1	2	3	4	5
NH_3	*cis*	*cis*	*cis*	*trans*	*trans*
Cl	*cis*	*cis*	*trans*	*cis*	*trans*
F	*cis*	*trans*	*cis*	*cis*	*trans*

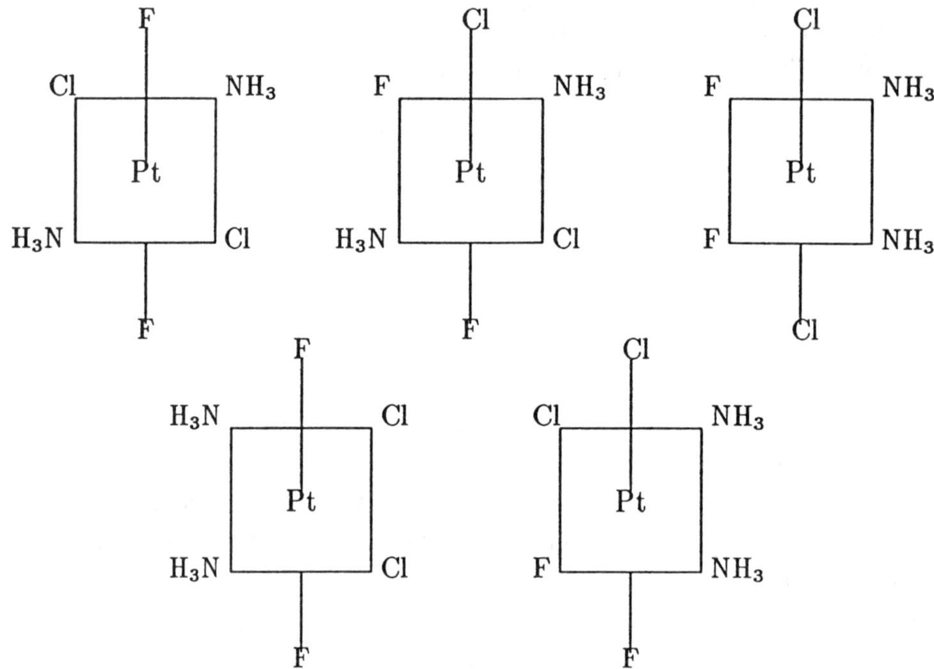

Each of these isomers is now checked to see if it is superimposable on its mirror image. Any isomer that has even one *trans* disposition of ligands has a plane of symmetry and *is* superimposable on its mirror image. The applies to four of the above five cases. The *cis-cis-cis* geometry (bottom right) is *not* superimposable on its mirror image and exists as a pair of enantiomers.

19-24 Strong-field octahedral complexes have a large splitting between the t_{2g} and e_g sets of orbitals; weak-field complexes have a small splitting between the t_{2g} and e_g orbitals. When the energy by which the levels are split exceeds the pairing energy of the electrons, electrons pair up in the t_{2g} level and fill it completely before occupying the e_g level. Otherwise, electrons remain unpaired as long as possible. The number of unpaired electrons is shown in the following occupancy diagrams in which the weak-field case is at the left.

(a) Cr^{2+}: $[Ar]3d^4$ 4 unpaired e^- in a weak field, 2 unpaired e^- in a strong field:

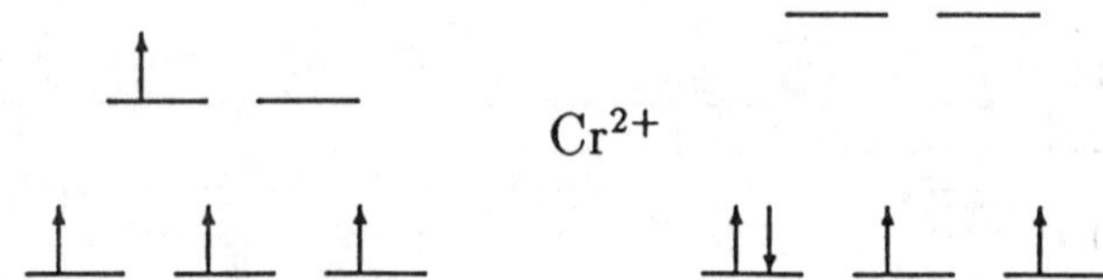

(b) V^{3+}: $[Ar]3d^2$ 2 unpaired e^- in a weak field, 2 unpaired e^- in a strong field.

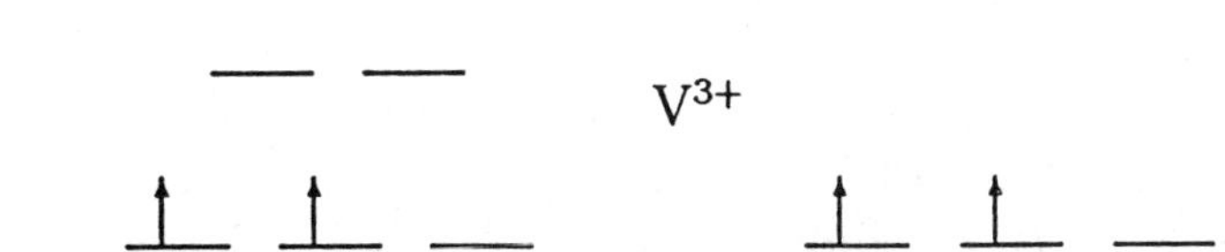

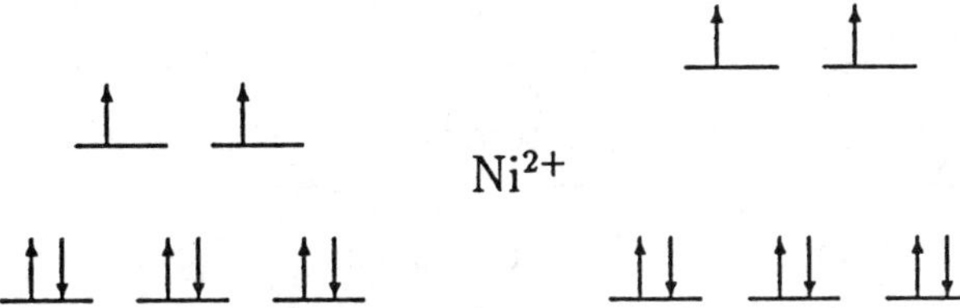

(c) Ni^{2+}: $[Ar]3d^8$ 2 unpaired e^- in a weak field, 2 unpaired e^- in a strong field.

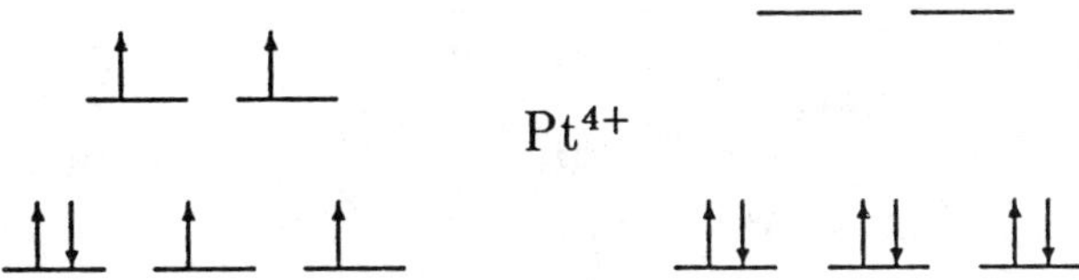

(d) Pt^{4+}: $[Xe]3d^6$ 4 unpaired e^- in a weak field, 0 unpaired e^- in a strong field.

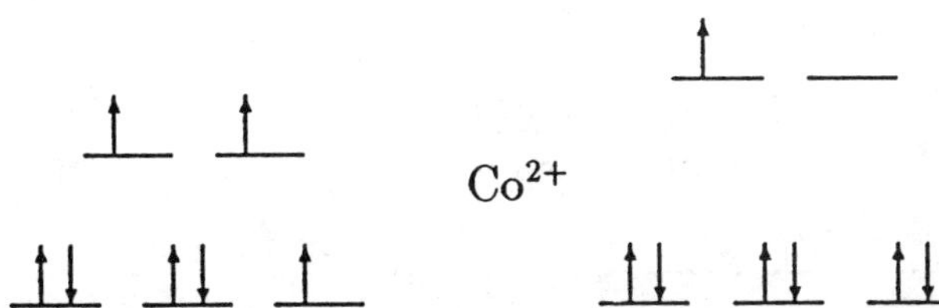

(e) Co^{2+}: $[Ar]3d^7$ 3 unpaired e^- in a weak field, 1 unpaired e^- in a strong field.

19-26 The Mn(III) in these two complexes has four d-electrons. When coordinated by Cl^- in the $[MnCl_6]^{3-}$ ion, the result is a high-spin complex—all $\boxed{\text{four}}$ electrons are unpaired. With a stronger-field ligand in the $[Mn(CN)_6]^{3-}$, no electrons are in the higher-energy e_g-levels. Putting four electrons into the t_{2g} level requires the formation of one electron pair, but $\boxed{\text{two}}$ electrons can remain unpaired. The two electron configurations are $\boxed{(t_{2g}^3\, e_g^1)}$ and $\boxed{(t_{2g}^4\, e_g^0)}$.

19-28 The problem concerns the standard potentials for reduction of $Mn^{3+}(aq)$, $Fe^{3+}(aq)$ and $Co^{3+}(aq)$ to the corresponding 2+ ions. All six of the ions are octahedrally coordinated by H_2O molecules in acidic aqueous solution. The metals in the 3+ ions are d^4, d^5, and d^6 species; the metals in the 2+ ions are d^5, d^6 and d^7 species, respectively. These hexaaqua complexes are all high-spin complexes because water is a weak-field ligand. The reduction potential of $Fe^{3+}(aq)$ is less than the reduction potential of its neighbors in the periodic table, which means that it is *harder* to reduce to the +2 state than the neighbors. The $Fe^{3+}(aq)$ ion has the high-spin $t_{2g}^3 e_g^2$ configuration. This is a relatively stable electron configuration relative to configurations with four or six d-electrons (note that both of the levels are half-filled). As a consequence, the $Fe^{3+}(aq)$ ion resists reduction better.

19-30 The ion $[PtI_6]^{2-}$ absorbs light all across the visible spectrum.

19-32 A solution of hexaaquanickel(II) ion is green. It therefore absorbs the complement of green, which is red (see Figure 18-23, text page 742). Use text Figure 16-5 (text page 651) to estimate the wavelength of the absorption in the red as $\boxed{670 \text{ nm}}$. The corresponding energy is

$$E = \frac{hc}{\lambda} = \frac{(6.626 \times 10^{-34} \text{ J s})(3.00 \times 10^8 \text{ m s})}{670 \times 10^{-9} \text{ m}} = \boxed{2.97 \times 10^{-19} \text{ J}}$$

The absorption occurs with the excitation of an electron from a t_{2g} to an e_g level so this energy is Δ_o. Multiplying Δ_o by Avogadro's number puts it on a molar basis. It is about $\boxed{180 \text{ kJ mol}^{-1}}$.

19-34 (a) The complement of yellow is violet; the hexaamminecobalt(III) ion therefore absorbs in the $\boxed{\text{violet}}$.

(b) Estimate the wavelength from the wavelength of the complement of the observed color. It is probably near $\boxed{430 \text{ nm}}$ as a maximum (see text Figure 16-5).

19-36 (a) The complex $[Mn(H_2O)_6]^{2+}$ results from the dissolution of $Mn(NO_3)_2$ in water. It is a high-spin d^5 complex. The excitation of an electron requires the reversal of a spin; such processes are forbidden. Their rarity makes the color of the ion faint. The $[Mn(CN)_6]^{4-}$ ion on the other hand is a low-spin complex. The $t_{2g} \longrightarrow e_g$ transitions are not spin-forbidden and are more intense.

(b) The $Zn(NO_3)_2$, $CdSO_4$ and $AgClO_3$ should be colorless in aqueous solution based on the full occupancy of the d-orbitals of the metal ion when coordinated by the solvent. The other three compounds should be colored.

19-38 The text points out (text page 780) that high-spin d^5 complexes exhibit only very weak absorption of light because processes in which a spin is reversed occur rarely. A tetrahedral d^5 complex is virtually certain to be a high-spin because Δ_t is always much less than Δ_o for a given ligand: few ligands exert a large enough crystal field to force a low-spin tetrahedral complex. Any slight absorption that does take place in the d^5 case will occur at low energies (because Δ_t is low). This is toward at the red end of the spectrum. The complex is therefore likely to be $\boxed{\text{very faint blue}}$.

19-40 The mass of iron in one mole of hemoglobin is 0.33% of the mass of one mole of hemoglobin, which is given as 6.8×10^4 g. This comes out to 224 g of iron. Dividing this mass of iron by the molar mass of iron establishes that there are four moles of iron in one mole of hemoglobin. Hence there are $\boxed{\text{four}}$ atoms of iron per hemoglobin molecule.

19-42 The reaction is:

$$2\,[\mathrm{Cr(en)_2(NCS)_2}]\mathrm{SCN}(s) \longrightarrow [\mathrm{Cr(en)_2(NCS)_2}][\mathrm{Cr(en)(NCS)_4}](s) + \mathrm{en}(g)$$

The oxidation number of the chromium does not change from $\boxed{+3}$ in this reaction. Rather, two NCS^- anions replace an ethylenediamine in the coordination sphere of one of the complexes.

19-44 The point of the problem is that coordination alters the usual reactions of both ligands and central metals.

$$\mathrm{CN}^-(aq) + \mathrm{HCl}(aq) \longrightarrow \mathrm{HCN}(g) + \mathrm{Cl}^-(aq)$$

$$\mathrm{Fe}^{2+}(aq) + 2\,\mathrm{OH}^-(aq) \longrightarrow \mathrm{Fe(OH)_2}(s)$$

$$6\,\mathrm{CN}^-(aq) + \mathrm{Fe}^{2+}(aq) \longrightarrow [\mathrm{Fe(CN)_6}]^{4-}(aq)$$

19-46 (a) $[\mathrm{Fe(H_2O)_5Cl}]\mathrm{SO_4}$ is most likely to have the same electrical conductivity per mole as $\mathrm{ZnSO_4}$, which is also a 2 to 2 ionic compound.
(b) $[\mathrm{Mn(H_2O)_6}]\mathrm{Cl_3}$ matches best to $\mathrm{LaCl_3}$, which also has Cl^- as the anion and is a 1 to 3 ionic compound.
(c) $[\mathrm{Zn(H_2O)_3(OH)}]\mathrm{ClO_4}$ matches best to NaCl.
(d) $[\mathrm{Fe(NH_3)_6}]_2(\mathrm{SO_4})_3$ matches best to $\mathrm{Fe_2(SO_4)_3}$.
(e) $[\mathrm{Cr(NH_3)_3Br_3}]$ matches best to HCN; both are nonelectrolytes.

19-48 A planar hexagonal structure for $[\mathrm{Co(NH_3)_4Cl_2}]^+$ would have three isomers: one with the two Cl's next to each other, one with the two Cl's separated by one $\mathrm{NH_3}$, and one with the two Cl's separated by two $\mathrm{NH_3}$'s around the hexagonal ring. A trigonal prismatic structure would have three isomers as well: one with both Cl's on the same triangular end of the prism, one with the two Cl's on opposite triangular ends and lying on the same edge, and one with the two Cl's on opposite ends and lying on different edges. The last of these would be chiral.

19-50 In a crystal field of octahedral symmetry all three p orbitals have the same energy. There is no splitting. Electrons in the p-orbitals have their greatest probability density along a coordinate axis and differ among themselves only in which axis. The axes are indistinguishable in the octahedral case; each has one ligand on each end. A square-planar field is derived from an octahedral field by removal of ligands from one axis, say, the z. If so, the p_z orbital would be split to lower energy (since it would not encounter ligands directly) while the p_y and p_x orbitals would remain unsplit.

19-52 The compound $[Co(NH_3)_6]Cl_2$ has an odd number of electrons. It follows that this compound is $\boxed{\text{paramagnetic}}$.

19-54 The two compounds have the $\boxed{\text{same number of unpaired electrons}}$, and both have Mn^{2+}, a d^5-metal ion, as the central metal ion. The octahedral complex is a weak-field, high-spin complex and so is the tetrahedral complex.

19-56 (a) The compound $(NH_4)_2[Fe(H_2O)F_5]$ has an odd number of valence electrons and is $\boxed{\text{paramagnetic}}$ regardless on the strength of the ligands. Because the ligands are weak-field ligands, the octahedral field is weak and the five d-electrons of the Fe(III) should remain unpaired. This makes the compound paramagnetic to the extent of $\boxed{\text{five}}$ unpaired electrons.

(b) The anion is a high-spin d^5 complex like $[Fe(H_2O)_6]^{3+}$ so the complex is probably $\boxed{\text{pale violet}}$. Note that the cation (the ammonium ion) is colorless.

(c) The ground-state configuration of the d-electrons of the iron is $\boxed{t_{2g}^3 e_g^2}$.

(d) The compound is named $\boxed{\text{ammonium aquapentafluoroferrate(III)}}$.

19-58 The answers are given in text Section 19-5.

Answers to Even-Numbered Problems, Chapter 20

20-2 The problem gives 2θ, λ, and n ready for substitution in the (rearranged)

Bragg law:

$$d = \frac{n\lambda}{2\sin\theta} = \frac{2(1.237\ \text{Å})}{2\sin(35.58/2)} = \boxed{4.049\ \text{Å}}$$

20-4 Rearrange the Bragg law and substitute:

$$\sin\theta = \frac{n\lambda}{2d} = \frac{2(1.539\ \text{Å})}{2(4.28\ \text{Å})} = 0.35958$$

The inverse sine of 0.3596 is θ; $2\theta = \boxed{\pm 42.1°}$.

20-6 (a) Write the Bragg law as $\sin\theta = n\lambda/2d$. Substitute the given values of λ and d, which are 0.6500 Å and 12.55 Å respectively. The result is $\sin\theta = 0.025896n$. Diffracted beams appear at 2θ's corresponding to integral values of n. The problem requests the answers for n equal to 1 through 5:

n	$\sin\theta$	$\pm 2\theta$	n	$\sin\theta$	$\pm 2\theta$
1	0.025896	2.967°	2	0.051928	5.938
3	0.077689	8.911	4	0.10356	11.89
5	0.129492	14.88			

(b) If $n > 38$, then $\sin\theta > 1.00$, and θ is not defined. Hence, exactly $\boxed{76}$ diffracted beams are possible from this set of planes. This counts the plus and minus values separately.

21-8 The diffraction pattern of a liquid resembles that of a microcrystalline solid in consisting of a series of concentric rings. The rings in the pattern from a liquid are diffuse and few in number because there is only local, short-range structure in liquids. The rings in the pattern from a polycrystalline solid are sharply defined and numerous because there is long-range order within the individual crystalline grains.

21-10 (a) A cereal box does *not* have four-fold rotational symmetry since no two perpendicular edges are equal in length.
(b) A stop sign (octagonal in shape) does have a four-fold rotational axis of symmetry (perpendicular to the plane of the sign).
(c) A tetrahedron (solid with equivalent equilateral triangles as its sides) does *not* have a four-fold axis; it does have four different three-fold axes.
(d) A cube has three four-fold rotational axes (passing through the centers of its three pairs of parallel faces).
(e) In general, a rectangular solid does not have a four-fold axes of rotation.

21-12 A baseball (counting the seam) has two mirror planes perpendicular to each other and each bisecting the two pieces of leather than are sewn together to form the surface of the ball[10]. The letters "bood" have a vertical mirror line; the digits "6009" have no mirror line (but do have a 2-fold axis centered between the two 0's. A closed textbook has two mirror planes of symmetry if the writing and pictures on the cover are not considered: The first relates the top to the bottom, and the second relates the front to the back.

20-14 According to Table 20-1 (text page 802), crystals in the orthorhombic system always have three mutually perpendicular 2-fold axes. Hence, a $\boxed{\text{2-fold axis}}$ must be present in the diffraction pattern.

20-16 The formula for the volume of a general parallelepiped is given on text page 807. It is

$$V = abc\sqrt{(1 - \cos^2\alpha - \cos^2\beta - \cos^2\gamma + 2\cos\alpha\cos\beta\cos\gamma}$$

The crystal is monoclinic, so $\alpha = \gamma = 90°$. The cosine of $90°$ is zero, hence, using the trigonometric identity $\sin^2\beta + \cos^2\beta = 1$:

$$V = abc\sqrt{1 - \cos^2\beta} = abc\sin\beta$$

Substitution gives $\boxed{V = 589 \text{ Å}^3}$.

20-18 (a) In a cubic unit cell, all three angles are $90°$, and the three edges are equal. The cell volume is just the cube of the length of the cell edge. In the case of gold, $V = \boxed{67.847 \text{ Å}^3}$.

(b) The density of bulk gold is the same as the density of a unit cell of gold. The density of a unit cell is the mass of the contents of the cell divided by the volume. The four gold atoms as a group have a mass of 787.868 u, so the density is $\boxed{11.612 \text{ u Å}^{-3}}$.

(c) This part is a straightforward conversion of units

$$\frac{11.612 \text{ u}}{1\text{Å}^3} \times \left(\frac{1.000 \text{ g}}{6.0221 \times 10^{23} \text{ u}}\right) \times \left(\frac{10^{24} \text{ Å}^3}{1 \text{ cm}^3}\right) = \boxed{19.282 \text{ g cm}^{-3}}$$

The measured density of gold at room temperature equals 19.32 g cm^{-3}.

---

[10]The overall symmetry is D_{2d}

20-20 Suppose a sample of iron has a volume of exactly 1 cm^3 in volume. It weighs 7.86 g. The volume of this iron can also be expressed in cubic ångströms: 10^{24} Å^3. How many unit cells are in this volume? Each unit cell has a volume of $(2.866)^3$ Å$^3 = 23.541$ Å^3. It follows that 4.24787×10^{22} unit cells are present in 1 cm^3 of iron. There are two lattice points per unit cell, because the cells are body-centered. Counting one iron atom at every lattice point gives a total of 8.49574×10^{22} atoms of iron in the specimen. Each Fe atom weighs 55.847 u (the relative atomic mass of Fe expressed in u). The 1 cm^3 of iron therefore has a mass of 4.7446×10^{24} u. But the ratio of the mass of a object in u to its mass in grams is Avogadro's number. Dividing 4.7248×10^{24} by 7.86 gives $\boxed{6.04 \times 10^{23}}$ as Avogadro's number.

20-22 The density of the NiO is the mass of the contents of the unit cell divided by the volume of the unit cell. The unit cell is cubic so its volume is the length of the cell edge cubed, 72.877 Å^3. The contents of the unit cell are four NiO units, which weigh $4 \times (58.69 + 15.9994) = 298.758$ u. The density of the solid is therefore 4.099 u Å^{-3}. Change to more usual units of density (g cm^{-3}) by the unit-factor method employing the facts that there are 10^{24} Å^3 per cm^3 and 6.022×10^{23} u per gram. The answer is $\boxed{6.808 \text{ g cm}^{-3}}$.

20-24 The volume of the unit cell of strontium chloride hexahydrate is determined by substituting in the general formula for the volume of a unit cell with the understanding that in a trigonal cell $a = b = c$ and $\alpha = \beta = \gamma$. Substitution of the a and α give $V = 678.413$ Å$^3 = 678.413 \times 10^{-24}$ cm^3. Meanwhile, the molar mass of $SrCl_2 \cdot 6H_2O$ is 266.617 g mol^{-1}. A mole of unit cells of this crystal therefore weighs $266.617 \times 3 = 799.851$ g because there are three formula units per cell. The volume of a mole of unit cells is the volume of a single cell multiplied by 6.022×10^{23} (Avogadro's number) and equals 408.54 cm^3. The density of the substance is the mass of the mole of unit cells divided by its volume. It is $\boxed{1.958 \text{ g cm}^{-3}}$. The measured density of $SrCl_2 \cdot 6H_2O$ is 1.93 g cm^{-3}.

20-26 Verify the volume of the unit cell of turquoise by substitution of $\alpha = 68.61°$, $\beta = 69.71°$, $\gamma = 65.08°$, $a = 7.424$ Å, $b = 7.629$ Å, and $c = 9.910$ Å into the formula:

$$V = abc\sqrt{(1 - \cos^2 \alpha - \cos^2 \beta - \cos^2 \gamma + 2 \cos \alpha \cos \beta \cos \gamma}$$

Now, suppose a particular turquoise contains one mole of unit cells. The volume of this stone is

$$461.40 \text{Å}^3 \times \left(\frac{1 \text{ cm}^3}{10^{24} \text{ Å}^3} \right) \times 6.022 \times 10^{23} = 277.855 \text{ cm}^3$$

This mass of the stone is its volume multiplied by its density (2.927 g cm^{-3}) or 813.28 g. This is the mass of a mole of unit cells of turquoise. The formula mass of turquoise is obtained by adding up the contributions of the several elements in the formula given in the problem. It is 813.44 g mol^{-1}. Since $813.28 \approx 813.44$, every unit cell contains one formula unit. There is one Cu atom per formula unit, so every unit cell contains $\boxed{\text{one}}$ Cu atom.

20-28 (a) Despite the huge size of the unit cell of this protein, the unit cell is still primitive; it has $\boxed{\text{one}}$ lattice point.

(b) The cell angles are all 90°, as required by the orthorhombic crystal system. The volume of the cell is then just the product of the lengths of the three edges. It is $\boxed{5.675 \times 10^6 \text{ Å}^3}$.

(c) The volume of the box-shaped crystal is also the product of the lengths of the three edges. It is 3 mm^3—small, but still easily visible with the unaided eye. We convert the unit cell volume to mm^3 by multiplying by 10^{-21} mm^3/Å^3. The result is 5.675×10^{-15} mm^3. The ratio of the volume of the crystal to the volume of the unit cell gives the number of units cells in the crystal. It is $\boxed{5 \times 10^{14}}$.

20-30 (a) Rb is metallic.　　(b) C_5H_{12} is molecular.　　(c) B is covalent.　　(d) Na_2HPO_4 is ionic.

20-32 If the boiling point of pentane is slightly less than the melting point of rubidium, then the melting point of pentane is certainly less than the melting point of rubidium. Moreover, rubidium must be a low-melting metal because the boiling point of a molecular liquid like pentane is low. Boron, a covalent substance, has the highest melting point, and the sodium hydrogen phosphate therefore has the second highest melting point.

20-34 Nitrogen is a gas at room conditions and a very poor conductor. Germanium is a semi-conductor, and chromium is a metallic conductor.

20-36 The nearest neighbors of a Na^+ ion are 6 Cl^- ions. The second nearest neighbors are a set of 12 Na^+ ions. (Next comes a set of 8 Cl^-'s.)

20-38 In a face-centered cubic lattice, the face of the unit cell is a square. Consider an Al atom at a face center. It has 4 equidistant neighbors in the plane of the face, 4 more equidistant neighbors in a second plane perpendicular to this face, and a final 4 equidistant neighbors in a second perpendicular plane. There are $\boxed{12}$ nearest neighbors, all at a distance of $\sqrt{2}/2$ times the edge of the unit cell. There are $\boxed{6}$ second-nearest neighbors. This group of neighbors all lie at a distance equal to the edge of the unit cell.

20-40 If a thallium atom is in the center of the unit cube, then eight bromine atoms lie "in contact" with it at the eight corners. See Fig. 20-21 on text page 812. The distance from the center of the thallium to the center of a chlorine is $\sqrt{3}/2$ times the edge of the cell, or $\boxed{3.43\text{Å}}$.

20-42 The presence of Schottky defects (vacancies) means that the measured density of a real crystal will inevitably be less than density of a hypothetical ideal crystal having no vacancies. This will make estimates of Avogadro's number by the method of problem 20-20 $\boxed{\text{high}}$ because the computation requires division by the density. This can illustrated by replacing the density of iron in problem 20-20 by a somewhat smaller number and repeating the computation.

20-44 (a) The empirical formula of the nickel oxide is $\boxed{\text{Ni}_{0.9796}\text{O}}$. (The answer $\text{NiO}_{1.0208}$ is equivalent.) The formula is determined by comparing the relative number of moles of nickel and oxygen in a sample of any arbitrary size. Note that the mass percentage of oxygen is 21.77%.

(b) The set-up of the problem is just like the set-up in Example 20-9 (text page 821). The key equation is $3y + 2(0.9796 - y) = 2$ where y is the number of moles of nickel(III) per 1.000 mol of O. The value of y is 0.0408 mol, so the fraction of nickel in the +3 oxidation states is $0.0408/0.9796 = \boxed{0.0416}$.

20-46 The entropy of $\boxed{\text{the isotropic liquid phase}}$ exceeds the entropy in the smectic liquid crystal phase of a substance. Compare the degrees of order apparent in text Figure 20-34a and 20-34c. Note also that heating the smectic phase converts it to the isotropic phase.

01-48 The Bragg law is written for each of the two experiments, the first with x-rays of wavelength $\lambda_1 = 1.54$ Å and the second with x-rays of unknown wavelength λ_2. The ratio of the second to the first is:

$$\frac{n_1 \lambda_1}{n_2 \lambda_2} = \frac{d_1 \sin \theta_1}{d_2 \sin \theta_2}$$

But $n_1 = n_2$, $d_1 = d_2$, and λ_1, θ_1, and θ_2 and given. Substitution gives $\lambda_2 = 1.64$ Å.

20-50 In x-ray diffraction, the greatest possible scattering angle 2θ is 180°, which corresponds to direct back-scattering. When 2θ is 180°, θ is 90° and $\sin\theta$ is 1.00. Substitute into the Bragg law with $n = 1$ and $d = 4.20$ Å. The value of λ is $\boxed{8.40 \text{ Å}}$. If the wavelength of the probing x-rays exceeds the dimension of the feature being probed (the distance between the parallel planes) by more than a factor of two, no diffraction can occur.

21-52 The problem illustrates the nature and number of symmetry operations.

(a) The golden cube can be placed into the case with any of its six square faces down. Once a given face is down, the cube can be rotated into four different positions. There are therefore $\boxed{24}$ ways to put it in the case.

(b) The tetrahedron has four triangular faces. It can be placed into a form-fitting niche with any of the four faces up. With a given face up, the tetrahedron can be rotated to three different positions. There are therefore $\boxed{12}$ ways to put it in the case.

(c) The diagrammed shape is a rectangular prism. Its sides are different from its ends and top and bottom. The shape is in its symmetry like the shape of an ordinary ruler. Once it is fit into a niche it could be to reversed end for end and turned over front to back, but would not fit otherwise. There are therefore only $\boxed{\text{four}}$ ways to put it in the case.

20-54 Text Figure 20-30 (text page 818) shows the outline of the unit cell of graphite. The eight atoms at the corners contribute a total of 1 atom to the cell. The four atoms lying at the midpoints of the vertical edges contribute a total of 1 atom to the cell. The two atoms lying in the top and bottom faces (and bonded to corner atoms) contribute a total of 1 atom to the cell. Finally, an atom in the middle layer is entirely enclosed in the outline of the unit cell, and so contributes 1 atom to the cell. The total is $\boxed{4}$.

20-56 Let the radius of the hard spheres equal r. Then the body diagonal of the cubical unit cell equals $4r$ because the spheres are in contact along the body diagonal. Then, the edge of the cell equals $4r/\sqrt{3}$. The volume of the unit cell equals the edge cubed:

$$V_{\text{cell}} = a^3 = \left(\frac{4r}{\sqrt{3}}\right)^3 = \frac{4^3}{3\sqrt{3}}r^3$$

The unit cell contains a total of 2 atoms, so the volume of actual atoms in a cell equals

$$V_{atoms} = 2\left(\frac{4}{3}\pi r^3\right)$$

The fraction of the unit cell that contains atoms is the second of these divided by the first:

$$f_{occupied} = \frac{V_{atoms}}{V_{cell}} \frac{2\left(\frac{4}{3}\pi r^3\right)}{\frac{4^3}{3\sqrt{3}}r^3} = \frac{\sqrt{3}\pi}{8} = 0.680$$

The fraction of the cell that is void is therefore $\boxed{0.319}$.

20-58 The problem tells us the number of atoms of each kind in the unit cell. The unit cell is the repeating motif of the crystalline substance, so that its contents have the same ratio of atoms as the substance. There is 1 O atom per unit cell (1/8 atom per corner times 8 corners), 1 Cl atom (entirely within the cell), and there are 3 Na atoms (1/2 atom per face times 6 faces). The formula is $\boxed{Na_3ClO}$.[11]

20-60 Define one of the corners of the tetragonal unit cell of white tin as the origin of a Cartesian coordinate system. An Sn atom occupies this origin. Another tin atom occupies a site on the *bc* face of the cell at a distance of *b*/2 (which is 5.8315 Å/2) along the *b* axis and *a*/4 (which is 3.1813 Å/4) along the *c* axis. Because the *b* and *c* axes are at right angles, the Pythagorean theorem can be used to compute the distance between the origin and specified location:

$$d = \sqrt{(5.8315\text{ Å}/2)^2 + (3.1813\text{ Å}/4)^2} = \boxed{3.0223\text{ Å}}$$

In some copies of the text, the location of the additional tin atom is described as "one fourth of the distance along the face diagonal of the *bc* face." Nothing occupies this site in the actual unit cell of white tin. Still, the distance between the site and the nearest tin atom can be calculated by a similar application of the Pythagorean theorem. The answer is 1.6607 Å.

20-62 (a) Substitute the given data into the Bragg law and compute $d = \boxed{3.344\text{ Å}}$.

(b) The volume of the unit cell of polonium is d^3 because the distance between the parallel faces of the unit cell is the cell edge. The contents of the unit cell are a single Po atom because there is only one lattice point per cell, and there is an atom of Po for every lattice point. Also, the mass of a Po atom is 209 u.

---

[11]This substance actually exists; the answer is not a mix-up with sodium chlorate.

The density of Po is therefore $209 \text{ u}/(3.344 \text{ Å})^3 = 5.59 \text{ u Å}^{-3}$. This converts (as in problem 20-18c) to $\boxed{9.28 \text{ g cm}^{-3}}$.

20-64 Nonmetals like the chalcogens and the halogens form molecular crystals in their solid states; metals like the transition metals and the alkali metals form metallic crystals. Elements at the center of the periodic table (in Group IV) such as carbon and silicon form covalent crystals.

20-66 Solid CO_2 and solid I_2 are both molecular crystals. The molecules in each are maintained in position in the crystal by van der Waals forces. Both molecules are non-polar, so the van der Waals forces are weak ones (dispersion forces, see text page 225). The forces are so weak that the two solids have significant vapor pressures at room temperature. In other molecular crystals, the intermolecular forces are usually stronger and the room temperature vapor pressure is less. The lower vapor presure of solid iodine implies stronger dispersion forces between its molecules. This is in accord with the presence of more electrons in iodine molecules.

20-68 The volume of the NaCl crystal with the vacancies is the same as the volume of the perfect NaCl crystal but its mass is only 0.9985 as large because 0.15% of the NaCl that is present in a perfect crystal is missing. Therefore the density is

$$0.9985 \times 2.165 \text{ g cm}^{-3} = \boxed{2.162 \text{ g cm}^{-3}}$$

20-70 (a) covalent   (b) covalent within the chains, but molecular between chains (c) covalent along long chains of alternating atoms ($—Si—O—Si—$). The chains are themselves negatively charged and bonded ionically to positive ions.   (d) metallic.

20-72 The molecule of methane is more globular than stick-like in shape, so a plastic crystal is more likely than a liquid crystal. It should be relatively difficult to form an amorphous solid from methane; the small symmetrical molecules will easily organize into a crystalline solid.

Answers to Even-Numbered Problems, Chapter 21

21-2 In each case the maximum wavelength is calculated by inserting the band-gap energy in the equation $\lambda = hc/E$. For GaAs, it is $\boxed{867 \text{ nm}}$; for CdS, it is $\boxed{512 \text{ nm}}$. The former is in the infrared region of the spectrum; this latter is near the middle of the visible region of the spectrum, which explains its use as a light-sensing material in cameras.

21-4 The equation for the number of electrons excited to the conduction band per cubic centimeter in a semiconductor is

$$n_e = \left(4.8 \times 10^{15} \text{ cm}^{-3}\text{K}^{-3/2}\right) T^{3/2} e^{-E_g/2RT}$$

Substitution of the band gap energy with due regard for the cancellation of units gives $n_e = \boxed{4.3 \times 10^{13} \text{ cm}^{-3}}$ for germanium.

21-6 (a) Ge doped with In is a *p*-type semiconductor (b) CdS doped with As is a *p*-type semiconductor if As substitutes for S. If As substitutes at random for Cd or S then it raises the number of electrons in the product and makes it *n*-type.

21-8 Substitute the band-gap energy of the LED into the equation $\lambda = hc/E$ to compute the required wavelength. It is $\boxed{585 \text{ nm}}$, in the yellow region of the visible spectrum.

21-10 The decrease in the band-gap energy occasioned by the conversion of cinnabar to metacinnabar causes a shift in the minimum wavelength required to excite electrons across the band gap. The minimum wavelength rises from 621 nm to 764 nm. In cinnabar, red light ($\lambda > 621$ nm) is transmitted and the pigment appears red; in metacinnabar, all colors of visible light are absorbed, and the sample appears black. Thus, the $\boxed{\text{color faces}}$.

21-12 The Lewis dot structure of the cyclosilicate ion displays a total of 144 valence electrons. It consists of 6 Si and 6 O atoms alternating in a ring. Each Si atom has 2 other O's linked to it as side-groups on the ring. All bonds are single bonds.

21-14 (a) Tremolite consists of $\boxed{\text{infinite double chains}}$. All O's are in the -2 oxidation state, and all Si's in the $+4$ state. The Mg and Ca are both $+2$ and the hydrogen is $+1$. The assignment to a structural type is made on the basis of the ratio of silicate oxygen to silicon, which is 2.75 to 1.

(b) Gillespite consists of $\boxed{\text{infinite sheets}}$. The Si and O are in the $+4$ and -2 oxidation states respectively (as they are in all the silicate minerals in this problem) and the Ba and Fe are both $+2$.

(c) Uvarovite contains $\boxed{\text{silicate tetrahedra}}$. The Ca is in the $+2$ state, and the Cr is in the $+3$ state.

(d) Barysilate contains $\boxed{\text{pairs of silicate tetrahedra}}$. The Mn and Pb are both in the +2 oxidation state.

21-16 (a) The aluminosilicate mineral amesite derives formally from a silicate with a 5 to 2 oxygen-to-silicon ratio. Based on this ratio it contains $\boxed{\text{infinite sheets}}$ of aluminosilicate units. The aluminum is in the +3 oxidation state, the Si and O are in the +4 and −2 states, respectively. This is true in all the aluminosilicates in this problem. The Mg is in the +2 state and the "other aluminum" (that which is not part of the silicate framework) is +3 aluminum. The hydroxide hydrogen and hydroxide oxygen are +1 and −2 respectively.

(b) Phlogopite contains $\boxed{\text{infinite sheets}}$ of aluminosilicate units. The K is +1 and the Mg is +2. Other atoms are as before.

(c) Thomsonite contains an $\boxed{\text{infinite network}}$ of aluminosilicate units. The Na and Ca are +1 and +2 respectively.

21-18 (a) The reaction of tremolite with calcite and quartz is

$$Ca_2Mg_5Si_8O_{22}(OH)_2(s) + 3\,CaCO_3(s) + 2\,SiO_2(s) \longrightarrow$$

$$5\,CaMgSi_2O_6(s) + H_2O(g) + 3\,CO_2(g)$$

(b) The $\Delta H°$ and $\Delta S°$ for this reaction is computed in the usual way, taking additional data from Appendix D:

$$\Delta H° = 5(-3206.2) + 1(-241.82) + 3(-393.51)$$

$$- 1(-12360) - 3(1206.92) - 2(-910.94) = \boxed{349\ \text{kJ}}$$

$$\Delta S° = 5(142.93) + 1(188.72) + 3(213.63)$$

$$- 1(548.9) - 3(92.9) - 2(41.84) = \boxed{633\ \text{J K}^{-1}}$$

(c) The approximate temperature at which $\Delta G°$ is zero is given by

$$T = \frac{\Delta H°}{\Delta S°} = \frac{349 \times 10^3\ \text{J}}{633\ \text{J K}^{-1}} = \boxed{550\ \text{K}}$$

21-20 Metals are ductile, but ceramics are brittle. Ceramics tend to fail brittlely when stressed by rapid heating or on mechanical impact. Metals are also often altered structurally by rapid heating, but gross failure is rare. Slow heating gives the grain structure of a ceramic a chance to rearrange itself, and the ceramic is less likely to fail.

21-22 A metal consists of a three-dimensional lattice of positive ions whose mutual repulsions are countered by a sea of mobile electrons. This bonding is delocalized and non-directional. Elastic or plastic deformation results when a metal is stressed. A ceramic has a structure in which directed covalent bonds have a major role. The localization of electrons in bonds gives rigidity to the structures of ceramics, but reduces their tolerance for stresses.

21-24

$$3\ \underbrace{Al_2Si_4O_{10}(OH)_2(s)}_{\text{pyrophyllite}} \longrightarrow \underbrace{Al_6Si_2O_{13}(s)}_{\text{mullite}} + 10\ \underbrace{SiO_2(s)}_{\text{cristobalite}} + 3\ H_2O(g)$$

21-26 The dehydration of kaolinite ($Al_2Si_2O_5(OH)_4$) produces 2 mol of water per mol of the mineral. The molar mass of kaolinite is 258.16 g mol^{-1}. The chemical amount of kaolinite is its mass (4.0×10^3 g) divided by its molar mass. Doubling this gives the chemical amount of steam, 31.0 mol. Then:

$$V = \frac{nRT}{P} = \frac{(31.0\ \text{mol})(0.08206\ \text{L atm mol}^{-1}\ \text{K}^{-1})(873.15\ \text{K})}{1.00\ \text{atm}} = \boxed{2.2 \times 10^3\ \text{L}}$$

21-28 Imagine a 100-g sample of portland cement and compute the masses of the several oxides. From these obtain the chemical amounts of the oxides by dividing each mass by the proper molar mass. The formulas of the oxides then allow computation of the chemical amounts of the six elements and of course of the total chemical amount of oxygen. Put this on a basis of 1.00 mol of O. The answer can be summarized as

$$\boxed{Ca_{0.514}Si_{0.158}Al_{0.052}Fe_{0.016}Mg_{0.028}S_{0.010}Na_{0.020}O_{1.00}}$$

21-30 The slaking of lime is approximately represented:

$$\boxed{CaO(s) + H_2O(l) \longrightarrow Ca(OH)_2(s)}$$

The $\Delta H°$ for this reaction is

$$\Delta H° = 1(-986.09) - 1(-635.09) - 1(-285.83) = -65.17\ \text{kJ}$$

The slaking of 1 mol (56.08 g) of CaO(s) is exothermic to this extent. The slaking of 1.00 kg releases more heat in the proportion 1000/56.08. This comes out to $\boxed{1162\ \text{kJ}}$. This is only an estimate because the actual product is not entirely solid $Ca(OH)_2$ when lime is slaked under ordinary conditions. Dissolution of some of the $Ca(OH)_2(s)$ releases more heat.

21-32 (a) $BCl_3(g) + NH_3(g) \longrightarrow BN(s) + 3\,HCl(g)$

(b) For this reaction:

$$\Delta H^\circ = 1(-254.4) + 3(-92.31) - 1(-403.76) - 1(-46.11) = -81.46 \text{ kJ}$$

Thus, the ΔH° of $BN(s)$ is $\boxed{-81.46 \text{ kJ mol}^{-1}}$.

(c) Boron nitride has extremely low electrical conductivity and a very high melting point. It is nearly as hard as diamond.

21-34 The oxidation of $B_4C(s)$ is represented: $B_4C(s) + 4\,O_2(g) \longrightarrow 2\,B_2O_3(s) + CO_2(g)$. The ΔG° of this reaction at 298.15 K is computed from ΔG_f°'s in the usual way:

$$\Delta G^\circ = 2(-1193.70) + 1(-394.36) - 4(0) - 1(-71) = \boxed{-2711 \text{ kJ}}$$

Boron carbide is thermodynamically unstable in oxygen at standard conditions and certainly thermodynamically unstable in air as well. It reacts only slowly, however.

21-36 The sum of the oxidation states of the single atoms must equal zero. The oxidation states of Tl, Ca, Ba and O are +3, +2, +2 and −2 respectively. Let the oxidation state of Cu be y. Then:

$$(2 \times 3) + (2 \times 2) + (2 \times 2) + (3y) + (10.5 \times -2) = 0 \quad \text{and} \quad y = \boxed{7/3}$$

21-38 The band gap $\boxed{\text{gets larger}}$ going from a metal to a semiconductor to an insulator.

21-40 The conductivity of the Sb-doped Si (an n-type semiconductor) should decrease as Ga is added to a minimum when the ratio of the dopants is 1 to 1. At this point the doubly-doped Si is electronically equivalent to pure silicon itself. More Ga beyond this point should make the conductivity increase, as the semiconductor becomes p-type

---

21-42 (a) Apophyllite contains $\boxed{\text{infinite sheets}}$ of silicate units. The Si and O are in the +4 and −2 oxidation states respectively. The F is −1, the K is +1 and the Ca is +2. The water of hydration in the mineral has H in the +1 and O in the −2 states.

(b) Rhodonite contains $\boxed{\text{infinite single chains}}$ of SiO_4 units. The Ca and Mn are both +2.

(c) Margarite is an aluminosilicate. It contains ⌈infinite sheets⌋ ofaluminosilicate groups. The Ca and non-infinite-sheet Al are in the +2 and +3 oxidation states, respectively. The Si, Al and O in the aluminosilicate framework are in the +4, +3, and −2 states; the hydroxide H and O are +1 and −2 respectively.

21-44 Let x equal the oxidation number of the Fe and write an equation using the oxidation numbers to express the electrical neutrality of the compound:

$$\underbrace{2.36(2)}_{\text{Mg}} + \underbrace{0.48x}_{\text{Fe}} + \underbrace{0.16(3)}_{\text{Al}} + \underbrace{2.72(4)}_{\text{Si}} + \underbrace{1.28(3)}_{\text{Al}} + \underbrace{10(-2)}_{\text{O}} + \underbrace{2(-1)}_{\text{OH}} + \underbrace{0.32(2)}_{\text{Mg}} = 0$$

From which $\boxed{x = 3}$.

21-46 The Na^+ ion and Ca^{2+} ion have nearly the same radius (0.98 and 0.99 Å). They substitute freely in each other's sites in solid solutions of albite and anorthite. The radius of the K^+ ion is greater (1.33 Å) and its substitution in sites in albite and anorthite is restricted.

21-48 The given quantities define the zeolite $K_2O \cdot Al_2O_3 \cdot 4(SiO_2) \cdot 6(H_2O)$. This compound is ⌈9.91% Al⌋ by mass.

21-50 Soda-lime glass is an amorphous solid of approximate composition 73% SiO_2, 17% Na_2O, 5% CaO, and 5 % MgO. It contains little aluminum. Fired kaolinite contains a great deal of aluminum, having the approximate composition $Al_6Si_2O_{13}$. Soda-lime glass contains a random, three-dimensional ionic network $(SiO_3^{2-})_n$ the charge of which is locally neutralized by the Na^+, Ca^{2+} and Mg^{2+} ions. Fired kaolinite includes Al as part of structural aluminosilicate chains. Fired kaolinite has undergone an irreversible chemical reaction; its physical shape in practically unalterable without breakage. Glass can be easily melted and reformed.

21-52 The CaF_2 would lower the melting point of the oxide mixture and save energy because lower temperatures would suffice in processing.

21-54 Magnesia (MgO) is an O^{2-} donor and is a basic refractory. Silica is an O^{2-} acceptor and is an acidic refractory. The two react: $MgO + SiO_2 \longrightarrow MgSiO_3$.

21-56 The likeliest problem with thoria ceramics would be their ⌈radioactivity⌋. All isotopes of Th are radioactive.

21-58 The oxide ceramics are in general thermodynamically stable with respect to their constituent elements. The nonoxide ceramics are often *not* thermodynamically stable with respect to reaction to form oxides (as in an atmosphere of air).

Answers to Even-Numbered Problems, Chapter 22

22-2 By-products are not the main product of economic value of a process, but nevertheless have some value. Waste materials have zero value or negative value (if they present a disposal problem).

22-4 Biochemical processes can be mapped without difficulty onto a process diagram of the type shown in Figure 22-1. The biochemical production of sulfur, for example, has methane, carbon dioxide, and calcium sulfate as starting materials that undergo a chemical transformation by the intermediacy of marine bacteria to give the "desired product" sulfur and the respective by-product and waste product calcium carbonate and water (see the equation on text page 877). The methane itself (in natural gas) arises from the decay of the remains of marine organisms (along with petroleum). This too should be shown in the process diagram. Biochemical processes differ from processes in the chemical industry (with rare exceptions) in taking place at room pressure and near to room temperature, in employing enzymes instead of catalysts, in requiring an aqueous medium. Industrial processes tend to use gas-phase reactions or heterogeneous reactions, and to employ extremes of T and P.

22-6 (a) The only important useful source of phosphorus is phosphate rock.

(b) Silicon is widely available in the rocks of the earth's crust; a useful source is common sand.

(c) Magnesium comes from sea-water, see diagram, text page 515.

22-8 The coating of corrosion products that protects mild steel against the attack of motionless sulfuric acid is removed by friction when the sulfuric acid is set in motion.

22-10 The main reaction in the Claus process is the oxidation of sulfur dioxide by hydrogen sulfide to produce elemental sulfur and water

$$SO_2(g) + 2\,H_2S(s) \longrightarrow 3\,S(s) + 2\,H_2O(g)$$

Determine $\Delta H°$ and $\Delta S°$ based on data from Appendix D:

$$\Delta H° = 3(0) + 2(-241.82) - 1(-296.83) - 2(-20.63) = \boxed{-145.55 \text{ kJ}}$$

$$\Delta S° = 3(31.80) + 2(188.72) - 1(248.11) - 2(205.68) = \boxed{-186.63 \text{ J K}^{-1}}$$

The reaction is exothermic and involves a decrease in entropy of the reacting system. To obtain a high equilibrium yield of the products, run the reaction at the lowest temperature consistent with an acceptable rate. Also, force the formation of products with high pressure.

22-12 The combustion of sulfur gives mainly $SO_2(g)$ although some other oxides of sulfur form in very small amounts. Collecting the gas from the combustion in hydrogen peroxide solution oxidizes the SO_2 to $H_2SO_4(aq)$:

$$SO_2(g) + H_2O_2(aq) \longrightarrow H_2SO_4(aq)$$

The amount of H^+ in this dilute aqueous sulfuric acid can be quite accurately determined by an acid-base titration. Alternatively, the sulfate ion could be precipitated with $Ba^{2+}(aq)$ and weighed. This is probably a better way to determine the sulfur unless it is known for certain that no other acidic oxides are released by burning the sample.

22-14 The overall reaction is: $H_2S(g) + 2\,O_2(g) \longrightarrow H_2SO_4(l)$.

22-16 The oxidation of sulfur by sulfuric acid is represented:

$$\boxed{2\,H_2SO_4(aq) + S(s) \longrightarrow 3\,SO_2(g) + 2\,H_2O(l)}$$

It is easy to balance this equation by inspection (text Chapter 2) but the formal method of Chapter 12 can also be used. Note that the sulfur from the sulfuric acid is reduced from the $+6$ to the $+4$ oxidation state and the elemental sulfur is oxidized from the 0 to the $+4$ oxidation state. This assures that the coefficient of H_2SO_4 in the equation is twice the coefficient of S.

22-18 The copper will be oxidized to Cu(II)—the most common oxidation state of this transition metal. Three answers are possible depending on whether SO_2 or S or H_2S is the main S-containing product:

$$Cu(s) + 2\,H_2SO_4(aq) \longrightarrow CuSO_4(aq) + 2\,H_2O(l) + SO_2(g)$$

$$3\,Cu(s) + 4\,H_2SO_4(aq) \longrightarrow 3\,CuSO_4(aq) + 4\,H_2O(l) + S(s$$

$$4\,Cu(s) + 5\,H_2SO_4(aq) \longrightarrow 4\,CuSO_4(aq) + 4\,H_2O(l) + H_2S(g)$$

22-20 Phosphoric acid is converted to NaH_2PO_4 in one run and to Na_2HPO_4 in another by controlled neutralization with NaOH. "Controlled neutralization" means that amounts of NaOH sufficient to neutralize one mole hydrogen ions in one batch and two moles of hydrogen ions in another batch are added. The sodium hydrogen phosphate and sodium dihydrogen phosphate are then reacted:

$$NaH_2PO_4(s) + 2\,Na_2HPO_4(s) \longrightarrow Na_5P_3O_{10}(s) + 2\,H_2O(g)$$

The mixture has to be heated to at least 300°C, well above the boiling point of water.

22-22 If *pairs* of phosphoric acid molecules react and split out water, the reaction must be $2\,H_3PO_4(l) \longrightarrow H_4P_2O_7(aq) + H_2O(l)$. The structure of $H_4P_2O_7$ is:

$$
\begin{array}{ccc}
O\!-\!H & & O\!-\!H \\
| & & | \\
O\!=\!P\!-\!O\!-\!P\!=\!O \\
| & & | \\
O\!-\!H & & O\!-\!H
\end{array}
$$

22-24 Multiply the number of molecules of NO created per flash by the number of flashes per year. Division by Avogadro's number gives the chemical amount of NO produced by lighting per year. This is also the chemical amount of N that is fixed per year. Multiplying this amount by the molar mass of N gives the mass of N that is fixed each year. The number of flashes per year is 3.15×10^9, so 3.15×10^{36} molecules of NO are produced. This is 5.24×10^{12} mol of NO, which contains 5.24×10^{12} mol of N. This amounts to 7.33×10^{13} g of nitrogen, or $\boxed{7 \times 10^7 \text{ metric tons}}$.

22-26 If the calcium hydroxide and carbon dioxide are *not* recycled, then the overall equation for the cyanamide process is

$$CaCO_3(s) + 2\,C(s) + N_2(g) + 4\,H_2O(l)$$
$$\longrightarrow 2\,NH_3(g) + 2\,CO_2(g) + CO(g) + Ca(OH)_2(s)$$

The $\Delta H°$ corresponding to this equation is $+374.4$ kJ.

Recycling the $Ca(OH)_2$ means that the equation

$$Ca(OH)_2(s) + CO_2(g) \longrightarrow CaCO_3(s) + H_2O(l)$$

is added to the preceding. The result is

$$2\,C(s) + N_2(g) + 3\,H_2O(l) \longrightarrow 2\,NH_3(g) + CO_2(g) + CO(g)$$

The ΔH° of the second reaction is -113.15 kJ. Hence, ΔH° of the overall reaction is $+261.25$ kJ. The ΔH° per mole of $NH_3(g)$ is $\boxed{+130.62 \text{ kJ mol}^{-1}}$.

22-28 The equilibrium constant for the production of ammonia from its elements at 600 K is readily estimated using the van't Hoff equation as on text page 473. The answer is $\boxed{K_{600} \approx 4 \times 10^{-3}}$.

Such a catalyst would be useful indeed. It would allow the use of lower pressures to achieve the same equilibrium yields of ammonia. This would save considerably on the cost of the equipment for the process.

22-30 The answer requires the application of LeChatelier's principle. The three steps of the Ostwald process are given on text page 894-5. The first step, the oxidation of ammonia by oxygen, is favored by $\boxed{\text{low pressure}}$ because 10 moles of gas are formed from 9 moles of gas. The two subsequent steps are $\boxed{\text{both}}$ driven to the right by $\boxed{\text{high pressure}}$.

22-32 The balanced equations are:

$$3\,Ni(s) + 8\,HNO_3(aq) \longrightarrow 3\,Ni(NO_3)_2(aq) + 2\,NO(g) + 4\,H_2O(l)$$

$$Tl(s) + 4\,HNO_3(aq) \longrightarrow Tl(NO_3)_3(aq) + NO(g) + 2\,H_2O(l)$$

When rewritten as net ionic equations these become:

$$3\,Ni(s) + 2\,NO_3^-(aq) + 8\,H^+(aq) \longrightarrow 3\,Ni^{2+}(aq) + 2\,NO(g) + 4\,H_2O(l)$$

$$Tl(s) + 4\,NO_3^-(aq) + 4\,H^+(aq) \longrightarrow Tl^{3+}(aq) + NO(g) + 2\,H_2O(l)$$

22-34 (a) The Lewis structure of diimine is $\boxed{H\!:\!\ddot{N} = \ddot{N}\!:\!H}$.

(b) The conversion of hydrazine to diimine is an oxidation—the oxidation number of nitrogen increases from -2 to -1. An oxidizing agent is required.

22-36 (a) The oxidation number of nitrogen in NH_2OH is $\boxed{-1}$; this is computed taking oxygen as having a -2 oxidation number and hydrogen a $+1$ oxidation number.

(b) The disproportionation of hydroxylamine N_2 and N_2H_4 at pH 14 has the standard potential difference $\Delta\mathcal{E}^\circ = -0.73 - (-1.59) = 0.86$ V. The positive potential difference means the reaction is spontaneous under standard conditions. Hydroxylamine is $\boxed{\text{not stable}}$ with respect to disproportionation under the stated conditions.

22-38 The balanced equation is

$$\boxed{2\,N_2H_4(l) + N_2O_4(g) \longrightarrow 4\,H_2O(l) + 6\,N_2(g)}$$

$$\Delta H^\circ = 4(-285.83) + 3(0) - 2(50.63) - 1(9.16) = \boxed{-1253.74 \text{ kJ}}$$

22-40 A practical sequence to make N_2O would be to prepare ammonia from hydrogen and nitrogen by the Haber process, to oxidize some of the ammonia to nitric acid by the Ostwald process and then to react remaining ammonia (a base) with nitric acid to give ammonium nitrate. Heating ammonium nitrate gives N_2O in good yield.

22-42 In addition to its value as a propaganda ploy, this practice added a good reducing agent to the explosive and made the explosion more energetic. The reaction $2\,Al(s) + 3/2\,O_2(g) \longrightarrow Al_2O_3(s)$ has ΔH° equal to -1675.7 kJ. The aluminum served the same function as fuel oil does in its explosive mixtures with ammonium nitrate.

22-44 The course of the explosive decomposition of HNIW is approximated by the equation given on text page 894:

$$C_6H_6N_{12}O_{12}(l) \longrightarrow 3\,CO(g) + 3\,CO_2(g) + 6\,N_2(g) + 3\,H_2O(g)$$

The standard enthalpy change for this reaction is computed from the "per-gram" enthlapy change that is given for this reaction:

$$\Delta H^\circ = -5.3 \text{ kJ g}^{-1} \times \left(\frac{438.2 \text{ g HNIW}}{\text{mol}}\right) = -2322 \text{ kJ mol}^{-1}$$

The ΔH_f° of HNIW is not known; the ΔH_f°'s of all the other substances in the equation appear in Appendix D. Because the ΔH° of the explosion reaction equals the standard enthalpy of formation of its products minus the standard enthalpy of formation of the reactant, the following equation can be written:

$$-2322 \text{ kJ} = \underbrace{3\,(-110.52)}_{CO(g)}\underbrace{3\,(-393.51)}_{CO_2(g)} +\underbrace{6\,(0.00)}_{N_2(g)} +\underbrace{3\,(-241.82)}_{H_2O(g)} - \underbrace{\Delta H_f^\circ}_{HNIW(s)}$$

Solving for the unknown gives the molar enthalpy of formation of HNIW as $\boxed{84 \text{ kJ mol}^{-1}}$.

22-46 An igneous rock is the first result of the crystallization of the molten silicate fluid, the magma, from the earth's interior. Examples are aluminosilicates. Sedimentary rocks are formed by the weathering of other rocks. An example is bauxite (Al_2O_3. Metamorphic rocks result from the re-application of heat and pressure to sedimentary rocks. An example is marble (calcium carbonate).

22-48

$$63 \times 10^9 \text{ cans} \times \left(\frac{0.355 \text{ L}}{1 \text{ can}}\right) \times \left(\frac{0.15 \text{ mol } CO_2}{1 \text{ L}}\right) \times \left(\frac{44 \text{ g } CO_2}{1 \text{ mol } CO_2}\right) = \boxed{1.5 \times 10^{11} \text{ g } CO_2}$$

22-50 In the desired process, the SO_2 must be oxidized. Obviously, water will provide the hydrogen for the sulfuric acid. A possible first step is the reaction

$$SO_2(g) + NO_2(g) \longrightarrow SO_3(g) + NO(g)$$

The SO_3 would then be taken up in sulfuric acid as in the contact process and put into a sulfuric-acid production stream. The $NO(g)$ from this reaction and the portion originally in the flue gas could be oxidized

$$2\,NO(g) + O_2(g) \longrightarrow 2\,NO_2(g)$$

and the NO_2 then reacted with water to make nitric acid:

$$3\,NO_2(g) + H_2O(l) \longrightarrow 2\,H^+(aq) + 3\,NO_3^-(aq) + NO(g)$$

The $NO(g)$ produced in this last step would be recycled. Alternatively, the $NO(g)$ could be converted to completely innocuous products:

$$6\,NO(g) + 4\,NH_3(g) \longrightarrow 5\,N_2(g) + 6\,H_2O(g)$$

This approach would be costly because of the expense of the ammonia, which has better use in fertilizer. Such equations are only the barest skeleton of practical process for removing pollutants from flue gases. One real problem is the low concentrations of the pollutants as they are found in the flue gas.

22-52 In the Glover tower, the reaction:

$$2\,HNO_2(aq) + SO_2(g) \longrightarrow H_2SO_4(aq) + 2\,NO(g)$$

produces sulfuric acid. Nitrogen in the $+3$ state is reduced to $+2$ as sulfur is oxidized to the desired $+6$ state. In the lead-chamber process, the oxidizing agent was $NO_2(g)$, in which nitrogen is in the $+4$ oxidation state.

22-54 The problem is solved by a series of unit conversions:

$$1.30 \times 10^{11} \text{ kg H}_2\text{SO}_4 \times \left(\frac{32.07 \text{ kg S}}{98.08 \text{ kg H}_2\text{SO}_4} \right) \times \left(\frac{0.03 \text{ kg S lost}}{99.97 \text{ kg S made}} \right)$$

$$\times \left(\frac{64.06 \text{ kg SO}_2 \text{ emitted}}{32.07 \text{ kg S lost}} \right) = \boxed{2.54 \times 10^7 \text{ kg SO}_2 \text{ emitted}}$$

22-56 The metallic elements that form a part of the oxide scale on a piece of steel are $\boxed{\text{removed from the metal}}$ as soluble or suspended solid sulfates and carried away in the acid. This creates the problem of what to do with the spent acid, which is contaminated with compounds of iron and other metals and requires recycling or careful disposal.

22-58 The coke reduces the Na_2SO_4 according to the equation:

$$\text{Na}_2\text{SO}_4(s) + 2\,\text{C}(s) \longrightarrow \text{Na}_2\text{S}(s) + 2\,\text{CO}_2(g)$$

It is also possible that carbon monoxide forms as a product:

$$\text{Na}_2\text{SO}_4(s) + 4\,\text{C}(s) \longrightarrow \text{Na}_2\text{S}(s) + 4\,\text{CO}(g)$$

Combining the ΔH_f°'s of the reactants and products gives ΔH° of the first reaction as $+235.26$ kJ and ΔH° of the second as $+580.2$ kJ. Either way, the production of the sodium sulfide is endothermic. It $\boxed{\text{consumes heat}}$.

22-60 In the wet-acid process, phosphate rock (mainly fluorapatite) is reacted with sulfuric acid to make phosphoric acid. The production of triple superphosphate fertilizer consists of reacting phosphoric acid with phosphate rock. The balanced equations for these processes are given on text page 881. Multiply the first equation by 7 and the second by 3 and add them together. The H_3PO_4 terms cancel out, and the resultant equation, written with smallest whole-number coefficients, is:

$$2\,\text{Ca}_5(\text{PO}_4)_3\text{F}(s) + 7\,\text{H}_2\text{SO}_4(aq) + 17\,\text{H}_2\text{O}(l)$$
$$\longrightarrow 7\,\text{CaSO}_4{\cdot}2\text{H}_2\text{O}(s) + 2\,\text{HF}(g) + 3\,\text{Ca}(\text{H}_2\text{PO}_4)_2{\cdot}\text{H}_2\text{O}(s)$$

This equation shows that $\boxed{7/3 \text{ mol}}$ of gypsum is generated per mole of triple superphosphate.

22-62 (a) According to the second equation on text page 881, 5 mol of gypsum is produced for every 3 mol of phosphoric acid in the wet-acid process. Daily production in a fertilizer plant of 100 metric tons of phosphoric acid gives by-product according to:

$$100 \text{ tons} \times \left(\frac{10^6 \text{ g}}{1 \text{ ton}}\right) \times \left(\frac{1 \text{ mol H}_3\text{PO}_4}{98.0 \text{ g H}_3\text{PO}_4}\right) \times \left(\frac{5 \text{ mol CaSO}_4\cdot 2\text{H}_2\text{O}}{3 \text{ mol H}_3\text{PO}_4}\right)$$

$$\times \left(\frac{172.2 \text{ g CaSO}_4\cdot 2\text{H}_2\text{O}}{1 \text{ mol CaSO}_4\cdot 2\text{H}_2\text{O}}\right) = 2.93 \times 10^8 \text{ g CaSO}_4\cdot 2\text{H}_2\text{O}$$

This is the *daily* production of gypsum. If the plant operates 365 days a year, its annual production is 1.07×10^{11} g which is $\boxed{107{,}000 \text{ metric tons}}$.

(b) The volume of the gypsum is its mass divided by its density. Dividing the mass from part (a) by 2.32 g cm^{-3} gives a volume of 4.61×10^{10} cm^3, which equals $\boxed{4.61 \times 10^4 \text{ m}^3}$. This would be a cube of gypsum about 36 meters on an edge.

22-64 The platinum is catalyzing the oxidation of ammonia. Observation of brown fumes means that $NO_2(g)$ is formed. One possible reaction is

$$\boxed{4\,NH_3(g) + 7\,O_2(g) \longrightarrow 4\,NO_2(g) + 6\,H_2O(g)}$$

Also possible is a reaction between NH_3 and O_2 to give $NO(g)$, which then reacts with $O_2(g)$ in a separate step to generate the $NO_2(g)$. The platinum foil stays hot (red-hot, in fact) as long as the reaction continues, being continually heated by the burning of the ammonia at its surface.

22-66 Assume that the $NO(g)$ behaves ideally and use the ideal-gas equation to calculate the chemical amount of $NO(g)$ that is generated. The data are $V = 0.01146$ L, $P = 0.965$ atm, and $T = 291.95$ K, so $n_{NO} = 4.616 \times 10^{-4}$ mol. Students often confuse the volume of the gas and the volume of the solution. According to the balanced equation, 2 mol of nitrate reacts to form 2 mol of nitrogen monoxide. The concentration of the nitrate is the chemical amount of nitrate divided by the volume of the solution: 4.616×10^{-4} mol$/0.357$ L $=$ $\boxed{1.29 \times 10^{-3} \text{ M}}$.

22-68 The observation of a non-zero dipole moment for hydrazine definitely eliminates centrosymmetric structures such as:

$$
\begin{array}{ccc}
\text{H} & & \text{H} \\
& \searrow\;\;\;\;\nearrow & \\
& \text{N}\!-\!\text{N} & \\
& \nearrow\;\;\;\;\searrow & \\
\text{H} & & \text{H}
\end{array}
$$

22-70 The ingredients in gunpowder are KNO_3, C, and S in a *molar* ratio of 2 to 3 to 1, according to Table 22-1. These are in a *mass* ratio of 202.21 to 36.033 to 32.07. Out of every 270.31 g of gunpowder 202.21 are potassium nitrate. Hence, potassium nitrate is 74.8 g out of every 100 g of gunpowder. Similarly, carbon is 13.3 g out of every 100 g, and sulfur is 11.9 g out of every 100 g.

22-72 (a) The nitric acid is formed by the reaction

$$
NO_3^-(aq) + H^+(aq) \longrightarrow HNO_3(aq)
$$

The sulfuric acid supplies the hydrogen ion. The method works because nitric acid in solution is somewhat volatile. Heating the mixture drives the above reaction to the right by removing the product.

(b) As explained on text page 889, the nitric acid distills (at 1 atm pressure) at a maximum concentration of 69% by mass.

Answers to Even-Numbered Problems, Chapter 23

23-2 First calculate $\Delta G°$ of the reaction from the standard free energies of formation. Then compute K using $\Delta G° = -RT \ln K$. From the balanced equation and the data in Appendix D,

$$
\Delta G° = 1 \underbrace{(0)}_{Cl_2(g)} + 1 \underbrace{(-228.1)}_{Mn^{2+}(aq)} + 2 \underbrace{(-237.18)}_{H_2O(l)} - 2 \underbrace{(-131.23)}_{Cl^-(aq)}
$$

$$
- 4 \underbrace{(0)}_{H^+(aq)} - 1 \underbrace{(-465.17)}_{MnO_2(s)} = +25.2 \text{ kJ}
$$

$$
K = \exp\left(\frac{-\Delta G°}{RT}\right) = \exp\left(\frac{25.2 \times 10^3 \text{ J}}{(8.315 \text{ J K}^{-1}\text{ mol}^{-1})(298.15 \text{ K})}\right) = \boxed{3.9 \times 10^{-5}}
$$

23-4 The by-product of the Solvay process for sodium carbonate is calcium chloride, which is used to de-icing streets. The by-products of the Leblanc process are calcium sulfide, carbon dioxide and hydrogen chloride. The HCl can be converted to bleach. The calcium sulfide could be treated as waste, but also dealt with by reaction with CO_2 and water to give H_2S and $CaCO_3$. Oxidation of the hydrogen sulfide with oxygen would then generate elemental sulfur, a useful commodity, and water.

23-6 The net ionic equation for the Solvay process is

$$2\,Na^+(aq) + CaCO_3(s) \longrightarrow Na_2CO_3(s) + Ca^{2+}(aq)$$

The standard enthalpy change of this reaction is

$$\Delta H° = 1\underbrace{(-1130.68)}_{NaCO_3(s)} + 1\underbrace{(-542.83)}_{Ca^{2+}(aq)} - 1\underbrace{(-1206.92)}_{CaCO_3(s)} - 2\underbrace{(-240.12)}_{Na^+(aq)} = \boxed{+13.6\ \text{kJ}}$$

Overall, the Solvay process is slightly endothermic.

23-8 Potassium hydroxide is a stronger base that potassium carbonate, which is to say that the OH^- ion is a better hydrogen-ion acceptor than the CO_3^{2-} ion. Using KOH for the conversion of K_2HPO_4 to K_3PO_4 is therefore preferable.

23-10 From consideration of the conditions used in the two processes, the impurity would be $\boxed{\text{sodium chloride}}$.

23-12 (a) The oxidation of chloride ion by iodate ion is represented:

$$\boxed{3\,Cl_2(g) + I^-(aq) + 6\,OH^-(aq) \longrightarrow 6\,Cl^-(aq) + IO_3^-(aq) + 3\,H_2O(l)}$$

(b) The 0.216 L of $Cl_2(g)$ at STP is 0.00964 mol of Cl_2. According to the balanced equation, 1 mol of $I^-(aq)$ is consumed for every 3 mol of Cl_2. Therefore the solution originally must have contained 0.00321 mol of I^-. Dividing this chemical amount by the original volume of the solution in liters gives the original concentration of the solution. It was $\boxed{0.0643\ \text{mol L}^{-1}}$ in $I^-(aq)$.

23-14 (a) The oxidation state of the chlorine in perchloric acid is $\boxed{+7}$.

(b) The chemical formula of the anhydride of perchloric acid is $\boxed{Cl_2O_7}$.

(c)

$$\boxed{12\,HClO_4(l) + P_4O_{10}(s) \longrightarrow 4\,H_3PO_4(l) + 6\,Cl_2O_7(l)}$$

23-16 The hypochlorite ion is reduced in the first half-reaction and oxidized in the second. The standard potential for it to disproportionate is $\Delta\mathcal{E}^\circ = 0.90 - 0.48 = 0.42$ V. Hypochlorite ion is $\boxed{\text{unstable}}$ with respect to disproportionation in base.

23-18 The reaction to give phosphorus trifluoride from phosphorus and fluorine should be *more* exothermic (per mole of compound produced) than the reaction of phosphorus and chlorine to give phosphorus trichloride. The bond dissociation enthalpy for F_2 is less than for Cl_2, and the P—F bonds should be stronger (have a large dissociation enthalpy) than the P—Cl bonds.

23-20 (a) $SO_2(s) + 2\,F_2(g) \longrightarrow SOF_2(g) + 2\,O_2(g)$

(b) $MgO(s) + F_2(g) \longrightarrow MgF_2(s) + 1/2\,O_2(g)$

(c) $2\,Na(s) + F_2(g) \longrightarrow 2\,NaF(s)$

(d) $2\,GeO(s) + 4\,F_2(g) \longrightarrow 2\,GeF_4(g) + O_2(g)$

23-22 The graphite fluoride contains

$$\frac{2.53 \text{ g C}}{12.01 \text{ g mol}^{-1}} = 0.2106 \text{ mol C} \quad \text{for every} \quad \frac{1.000 \text{ g F}}{19.00 \text{ g mol}^{-1}} = 0.0526 \text{ mol F}$$

The ratio of these two chemical amounts is 4.001 to 1, so the empirical formula is $\boxed{C_4F}$.

23-24 The molar mass of the compound is 102.45 g mol^{-1}. Therefore, 225 g of it is 2.196 mol. One mole of an (ideal) gas occupies 22.4 L under STP conditions, so 2.196 mol occupies $\boxed{49.2 \text{ L}}$.

23-26 (a) The silicon has *SN* 6; the SiF_6^{2-} ion is octahedral about the central Si.
(b) The central chlorine has a steric number of 4; the ClF_2^+ ion is bent.
(c) The central iodine in IF_4^+ has *SN* 5. The structure is see-saw.
(d) The central As in AsF_6^- has a steric number of 6; the ion is octahedral.
(e) The central B has a steric number of 4; the BF_4^- ion is tetrahedral.
(f) The central Br has a steric number of 6; the BrF_6^+ ion is octahedral.

23-28 (a) The VSEPR theory predicts a $\boxed{\text{tetrahedral}}$ geometry about the central sulfur atom in sulfuryl difluoride. The *SN* of the sulfur is 4.

(b) The thionyl difluoride donates an electron pair to an oxygen atom. It acts as a $\boxed{\text{Lewis base}}$.

23-30 Reasoning by analogy to the case of water, one predicts that the entropy of vaporization of BrF_3 will be $\boxed{\text{larger}}$ than what is predicted by Trouton's rule.

23-32 (a) NaF (b) AsF_3 (c) C_5F_{12}

23-34 The reaction is $CF_4(g) + 2\,H_2O(l) \longrightarrow CO_2(g) + 4\,HF(aq)$. ·
for which

$$\Delta G^\circ = 4(-296.82) + (-394.36) - 2(-237.18) - 1(-879) = \boxed{-228\ \text{kJ}}$$

The reaction is spontaneous at standard conditions, but slow.

23-36 (a) The standard enthalpy change of the reaction in the problem is equal to the sum of the standard enthalpies of formation of the products minus those of the reactants. Let x equal ΔH_f° for $XeF_4(g)$. Then

$$\Delta H^\circ = -887\ \text{kJ} = 1\underbrace{(0)}_{Xe(g)} + 4\underbrace{(-271.1)}_{HF(g)} - 1\underbrace{(x)}_{XeF_4(g)} - 1\underbrace{(0)}_{H_2(g)}$$

Solving gives $x = \boxed{-197\ \text{kJ mol}^{-1}}$.

(b) From part (a), ΔH_f° of $XeF_4(g)$ is $-197\ \text{kJ mol}^{-1}$. Hence, for the reaction

$$Xe(g) + 2\,F_2(g) \longrightarrow XeF_4(g)$$

$\Delta H^\circ = -197$ kJ. This formation reaction can be imagined as proceeding with the production of 4 mol of $F(g)$ from $F_2(g)$ and then the creation of 4 mol of Xe—F bonds. It costs $4(79\ \text{kJ})$ to create $4\,F(g)$ from $2\,F_2(g)$. This means that the formation of the 4 mol of Xe—F bonds is exothermic to the extent of 513 kJ. The mean bond enthalpy is one-fourth of this, which is $\boxed{128\ \text{kJ}}$.

23-38 (a) Some new terminology is introduced, but it is not hard to figure it out. The ΔE_f° is simply the heat absorbed by a reaction at constant volume that goes to give a compound in its standard state from its constituent elements in their standard states. The ΔE_f° of 2.763×10^{-4} mol of $XeO_3(s)$ is therefore $+112$ J. The formation of one mole of the compound would absorb 405 kJ, that is, $\Delta E_f^\circ = \boxed{405\ \text{kJ mol}^{-1}}$.

(b) Start with the relationship $H = E + PV$ From this it follows that $\Delta H = \Delta E + \Delta(PV)$ and $\Delta H = \Delta E + (\Delta n_g)RT$ where Δn_g is the change in the chemical amount of gas and it is assumed that the volumes of the solids and liquids are negligible and that the temperature is constant. In the formation reaction, Δn_g is negative, so $\boxed{\Delta H_f^\circ < \Delta E_f^\circ}$.

23-40 (a) Perxenic acid is H_4XeO_6. Its large positive standard reduction potential means it is an excellent $\boxed{\text{oxidizing agent}}$.

(b) The XeO_3 is itself an excellent oxidizing agent, as shown by the large positive standard reduction potential of the second half-reaction. It is a very poor reducing agent and $\boxed{\text{stronger as an oxidizing agent}}$.

(c) Calculate $\Delta\mathcal{E}^\circ$ from the $\mathcal{E}^\circ$'s of the half-reactions in which the $XeO_3(aq)$ is reduced and oxidized. The result is $\Delta\mathcal{E}^\circ = 2.12 - 2.36 = -0.20$ V. Hence 1 M $XeO_3(aq)$ is $\boxed{\text{stable}}$ with respect to disproportionation in acid at pH 0.

23-42 You could generate chlorine to sterilize your water by reacting the potassium permanganate with sodium chloride in aqueous solution. Passing the $Cl_2(g)$ into the unclean water would kill the bacteria.

23-44 According to Table 23-1, the annual production of sodium carbonate in the U. S. is 9.5×10^9 kg yr^{-1}. Take this as the annual consumption in the U.S. as well. The Owens Lake formation of trona contains 3.5×10^{10} kg of sodium carbonate. Dividing the second number by the first gives $\boxed{3.7 \text{ years}}$.

23-46 (a) A current of 100,000 A corresponds to the passage of 100,000 C s^{-1}. There are 86,400 seconds in a day, so 8.64×10^9 C passes through each of the 250 cells daily. This equals 8.95×10^4 mol of electrons per cell (computed using the Faraday constant, 96,485 C mol^{-1}). For each mole of electrons that passes through a cell, 1/2 mol of Cl_2 is produced, according to the half-equation $Cl^- \longrightarrow Cl_2 + 2\,e^-$. Therefore, 4.48×10^4 mol of Cl_2 is produced by one cell in a day. This is 3.17×10^6 g of Cl_2 per cell, which means the production is 7.94×10^8 g of Cl_2 per day for the 250-cell plant. This is $\boxed{794 \text{ metric tons}}$.

(b) The power consumption in watts is the operating potential in volts multiplied by the current. Each cell uses 3.5 V $\times\ 100,000$ A $= 350,000$ W. The 250 cells use 250 times this: 8.75×10^7 W. Multiplying this rate of energy consumption by the 24 hours in a day gives 2.10×10^6 kW-hr as the energy consumption of the plant. Since a kilowatt-hour is 3600 J (see text page 515), this energy consumption is $\boxed{7.56 \times 10^9 \text{ J}}$.

(c) The electric bill for one day at this plant is $\boxed{\$105,000}$.

23-48 (a) By extension of the periodic trend, HAt should be $\boxed{\text{stronger}}$ than HI.

(b) Referring to the standard reduction potentials in Appendix E, the reduction potential for $At_2 + 2\,e^- \longrightarrow 2\,At^-$ must be algebraically greater than -0.7628 but less than 0.770 V.

(c) Bubbling Cl_2 through a solution of $At^-(aq)$ should generate At_2 by $\boxed{\text{oxidation}}$ since elemental chlorine should be a better oxidizing agent than elemental At_2.

(d) The At_2 should, like the other halogens, disproportionate:

$$\boxed{At_2(s) + 2\ OH^-(aq) \longrightarrow OAt^-(aq) + At^-(aq) + H_2O(l)}$$

23-50 Compute the quantity of electricity needed to oxidize the fluoride ion:

$$3.3 \times 10^3 \text{ g F}_2 \times \left(\frac{1 \text{ mol F}_2}{38.0 \text{ g F}_2}\right) \times \left(\frac{2 \text{ mol } e^-}{1 \text{ mol F}_2}\right) \times \left(\frac{96,485 \text{ C}}{1 \text{ mol } e^-}\right) = 1.676 \times 10^7 \text{ C}$$

This much electricity passes in 1 hour, so the quantity per second is 1/3600 as much or 4.655×10^3 C s^{-1}. This means the current is $\boxed{4700 \text{ A}}$.

23-52 (a) Convert the normal boiling point of HF to kelvins and proceed as follows:

$$d_{F_2} = \frac{P}{RT}\mathcal{M} = \left(\frac{1.00 \text{ atm}}{(0.08206 \text{ L atm mol}^{-1} \text{ K}^{-1})292.69 \text{ K}}\right) 38.0 \text{ g mol}^{-1}$$

$$= \boxed{1.58 \text{ g L}^{-1}}$$

(b) The measured density of $HF(g)$ at its boiling point is almost twice the ideal-gas value. The reflects strong non-bonded attractions (hydrogen bonds) among the molecules.

23-54 The most obvious effect of $F^-(aq)$ in the body would be to tie up calcium in an insoluble compound: $Ca^{2+}(aq) + 2F^-(aq) \longrightarrow CaF_2(s)$.

23-56 The reactions are

$$PuO_2 + 3O_2F_2 \longrightarrow PuF_6 + 4O_2$$

$$Pu + 3O_2F_2 \longrightarrow PuF_6 + 3O_2$$

23-58 (a) The molecule of S_2F_{10} consists of two octahedral sulfur atoms joined by a single bond with each S having 5 F's also single-bonded to it. Fluorine is not expect to form more than one bond.

(b) Both S's in this model of S_2F_{10} would have steric numbers of 6. The shape of the molecule would therefore be $\boxed{\text{two octahedra bonded at corners}}$.

23-60 The ionization energy of $Xe(g)$ is nearly equal to the ionization energy of $O_2(g)$. Further, the size of Xe^+ and O_2^+ are be about the same. These facts suggested to Bartlett that the compound $Xe^+PtF_6^-$ might form. In fact the product of the reaction is a complex mixture unless excess SF_6 is added as an inert diluent, in which case $XePtF_6$ can be isolated.

23-62 In this reaction, the AsF_5 is a fluoride-ion acceptor, which means it is an electron-pair acceptor and hence an acid. The XeF_2 is an electron-pair donor and hence a base.

Answers to Even-Numbered Problems, Chapter 24

24-2 Yes, it *is* possible for a gasoline to have an octane number that is less than zero. The gasoline would knock worse in an engine than *n*-heptane.

24-4 (a) The standard enthalpy of formation ΔH_f° of C_4H_8 is the enthalpy change in forming $C_4H_8(g)$ from $C(s)$ and $H_2(g)$ in their standard states: $4\,C(s) + 4\,H_2(g) \longrightarrow C_4H_8(g)$. Imagine this reaction to proceed by the atomization of 4 mol of $C(s)$ and 4 mol of $H_2(g)$ to form 4 mol of $C(g)$ and 8 mol of $H(g)$ followed by the formation of 4 mol of C—C bonds and 8 mol of C—H bonds. The sum of the ΔH°'s of these steps equals the desired estimate of ΔH°, by Hess's law.

$$\Delta H^\circ = 4\underbrace{(716.682)\ \text{kJ}}_{C(g)} + 8\underbrace{(217.96)\ \text{kJ}}_{H(g)} - 4\underbrace{(348)\ \text{kJ}}_{C-C} - 8\underbrace{(413)\ \text{kJ}}_{C-H} = \boxed{-86\ \text{kJ}}$$

(b) The standard enthalpy change for the combustion of cyclobutane is the standard enthalpy of formation of the products minus the standard enthalpy of formation of the reactants:

$$\Delta H^\circ = -2568\ \text{kJ} = 4(-393.51) + 4(-241.82) - 6(0) - x$$

where x is the exact ΔH_f° of cyclobutane. Solving gives $\Delta H_f^\circ = \boxed{-27\ \text{kJ mol}^{-1}}$.

(c) The actual ΔH_f° (part (b)) is less negative than the estimated ΔH_f° (part (a)). It follows that making the ring of four carbon atoms is endothermic by about $\boxed{59\text{kJmol}^{-1}}$. This is the "strain enthalpy."

24-6 (a) The catalytic cracking reaction is:

$$CH_3-\underset{\underset{\displaystyle CH_3}{|}}{\overset{\overset{\displaystyle CH_3}{|}}{C}}-\underset{\underset{\displaystyle H}{|}}{\overset{\overset{\displaystyle CH_3}{|}}{C}}-\underset{\underset{\displaystyle H}{|}}{\overset{\overset{\displaystyle CH_3}{|}}{C}}-\underset{\underset{\displaystyle CH_3}{|}}{\overset{\overset{\displaystyle CH_3}{|}}{C}}-CH_3 \longrightarrow H-\underset{\underset{\displaystyle CH_3}{|}}{\overset{\overset{\displaystyle H}{|}}{C}}-\underset{\underset{\displaystyle CH_3}{|}}{\overset{\overset{\displaystyle CH_3}{|}}{C}}-CH_3 \quad + \quad CH_3-\underset{\underset{\displaystyle CH_3}{|}}{\overset{\overset{\displaystyle CH_3}{|}}{C}}-\underset{\underset{\displaystyle H}{|}}{C}=CH_2$$

(b) The number of possible isomeric alkenes of the formula C_6H_{12} equals $\boxed{16}$. They include three butenes, five hexenes, and eight pentenes. The butenes are 3,3-dimethyl-1-butene (shown in the previous part) 2,3-dimethyl-1-butene, and 2,3-dimethyl-2-butene. The five hexenes are 1-hexene, 2-hexene, and 3-hexene of which the latter two exist in *cis* and *trans* forms. The eight pentenes include three 1-pentenes, which are 2-methyl-1-pentene, 3-methyl-1-pentene and 4-methyl-1-pentene, and five 2-pentenes, which are 2-methyl-2-pentene, 3-methyl-2-pentene and 4-methyl-2-pentene of which the last-two exist in *cis* and *trans* forms.

24-8 The structural formulas are:

(a) *[cyclobutene ring with H, H, CH$_3$, H on top carbons and H, CH$_3$ on bottom carbons]*

(b) $$CH_3-\underset{\underset{\displaystyle CH_3}{|}}{C}=\overset{\overset{\displaystyle H}{|}}{C}-CH_3$$

(c) $$\underset{\underset{\displaystyle H}{|}}{\overset{\overset{\displaystyle H}{|}}{C}}=\underset{}{\overset{\overset{\displaystyle H}{|}}{C}}-\underset{}{\overset{\overset{\displaystyle CH_3}{|}}{C}}=\underset{\underset{\displaystyle H}{|}}{\overset{\overset{\displaystyle H}{|}}{C}}$$

(d) $$CH_3-\underset{\underset{\displaystyle H}{|}}{\overset{\overset{\displaystyle CH_3}{|}}{C}}-\underset{\underset{\displaystyle CH_3}{|}}{\overset{\overset{\displaystyle C_2H_5}{|}}{C}}-\underset{\underset{\displaystyle H}{|}}{\overset{\overset{\displaystyle H}{|}}{C}}-\underset{\underset{\displaystyle H}{|}}{\overset{\overset{\displaystyle H}{|}}{C}}-CH_3$$

(e) $$CH_3-\underset{\underset{\displaystyle H}{|}}{\overset{\overset{\displaystyle H}{|}}{C}}-\underset{\underset{\displaystyle H}{|}}{\overset{\overset{\displaystyle H}{|}}{C}}-\underset{\underset{\displaystyle C_2H_5}{|}}{\overset{\overset{\displaystyle H}{|}}{C}}-\underset{\underset{\displaystyle H}{|}}{\overset{\overset{\displaystyle C_2H_5}{|}}{C}}-\underset{\underset{\displaystyle H}{|}}{\overset{\overset{\displaystyle H}{|}}{C}}-\underset{\underset{\displaystyle H}{|}}{\overset{\overset{\displaystyle H}{|}}{C}}-CH_3$$

(f) *[cyclooctene ring]*

(g) $H_2C=C=CH_2$

(h) $CH_3-CH_2-C\equiv C-CH_3$

24-10 The isomers of 4-octene are:

$$\begin{matrix} H_7C_3 \\ \diagdown \\ C=C \\ \diagup \quad\diagdown \\ H \qquad H \end{matrix} \qquad\qquad \begin{matrix} H_7C_3 \\ \diagdown \\ C=C \\ \diagup \quad\diagdown \\ H \qquad C_3H_7 \end{matrix}$$

24-12 (a) 2,3-dimethyl-1,3-butadiene   (b) 2,4-hexadiene   (c) 2,2-dimethylbutane
(d) methylpropene

24-14 (a) All of the carbon atoms except the two methyl carbon atoms are sp^2-hybridized. The two methyl carbons are sp^3-hybridized.

(b) The two carbon atoms on the extreme ends of the chain are sp^3-hybridized. The others are all sp^2-hybridized.

(c) All six carbons are sp^3-hybridized.

(d) The two methyl carbons are sp^3-hybridized; the other two carbons are sp^2-hybridized.

24-16 (a)

$$\text{(cyclopropanol)} \;+\; 4\,O_2 \longrightarrow 3\,CO_2 + 3\,H_2O$$

(b)

$$H_3C-\overset{O}{\overset{\|}{C}}-O-\underset{H}{\overset{CH_3}{\underset{|}{\overset{|}{C}}}}-CH_3 \;+\; HOH \longrightarrow H_3C-\overset{O}{\overset{\|}{C}}-OH \;+\; HO-\underset{H}{\overset{CH_3}{\underset{|}{\overset{|}{C}}}}-CH_3$$

(c)

$$H-\underset{H}{\overset{H}{\underset{|}{\overset{|}{C}}}}-\underset{H}{\overset{H}{\underset{|}{\overset{|}{C}}}}-OH \;\longrightarrow\; HOH \;+\; \begin{matrix} H \\ \diagdown \\ C=C \\ \diagup \quad\diagdown \\ H \qquad H \end{matrix}$$

(d) $CH_3CH_2CH_2CH_2I + H_2O \longrightarrow CH_3CH_2CH_2CH_2OH + HI$

24-18 (a) React acetic acid, ethylene, and oxygen in the presence of a catalyst: $CH_3COOH + CH_2CH_2 + 1/2\,O_2 \longrightarrow CH_3COOCHCH_2 + H_2O$.

(b) React formic acid and ammonia: $HCOOH + NH_3 \longrightarrow CHO(NH_2) + H_2O$.

(c) React ethylene with fluorine: $H_2CCH_2 + F_2 \longrightarrow FCH_2CH_2F$.

24-20 A tertiary amine has no hydrogen atoms on the nitrogen atom. Hence, it cannot contribute hydrogen to a water molecule as do primary and secondary amines when they condense with a carboxylic acid.

24-22 Compute the amount of ethylene dichloride required to make the 3.73×10^9 kg of vinyl chloride. It is clear without writing an equation that the molar ratio of the vinyl chloride to the ethylene dichloride is 1 to 1. Also, 1 mol of HCl is produced for every 1 mol of vinyl chloride. Then:

$$3.73 \times 10^9 \text{ kg } C_2H_3Cl \times \left(\frac{1 \text{ mol } C_2H_3Cl}{0.062453 \text{ kg } C_2H_3Cl} \right) \times \left(\frac{1 \text{ mol } C_2H_4Cl_2}{1 \text{ mol } C_2H_3Cl} \right)$$

$$\times \left(\frac{0.0989 \text{ kg } C_2H_4Cl_2}{1 \text{ mol } C_2H_4Cl_2} \right) = 5.91 \times 10^9 \text{ kg } C_2H_4Cl_2$$

This is $\boxed{0.944}$ of the total production of ethylene dichloride. The mass of the by-product HCl is $\boxed{2.18 \times 10^9 \text{ kg}}$, by a similar computation.

24-24 (a) The two compounds are 2,4-dichlorophenol and glycolic acid (hydroxyacetic acid).

(b) If its two chlorine atoms are replaced by hydrogen atoms, the compound with the structure on the left becomes phenol. The compound on the right becomes ethanol if its —COOH group is changed to a —CH$_3$ group.

24-26 (a) The structure of acetaminophen (N-(4-hydroxyphenyl)acetamide), the active ingredient in Tylenol, is:

$$\underset{\text{OH}}{\diagup}\!\!\!-\!\!\!\bigcirc\!\!\!-\!\!\!\overset{\overset{\displaystyle H}{|}}{N}\!\!-\!\!\overset{\overset{\displaystyle O}{\|}}{C}\!\!-\!CH_3$$

This substance has the molecular formula $C_8H_9NO_2$.

(b)

$$0.500 \text{ g } C_8H_9NO_2 \times \left(\frac{1 \text{ mol } C_8H_9NO_2}{151.17 \text{ g}}\right) = \boxed{3.31 \times 10^{-3} \text{ mol } C_8H_9NO_2}$$

24-28 To make testosterone ($C_{19}H_{28}O_2$), the C=O at C-11 in cortisone ($C_{21}H_{28}O_5$) becomes —CH_2—, and the —$COCH_2OH$ side-group at C-17 is replaced by —H.

24-30 (a) The critical temperature of ethylene is 9.50°C, which is $\boxed{49.10°F}$. The critical pressure is $\boxed{736.21 \text{ psi}}$.

(b) At 54 atm and 15°C, the ethylene in the pipeline is a $\boxed{\text{supercritical fluid}}$. Both the critical temperature and critical pressure are exceeded.

(c) The volume of the pipeline is its interior cross-sectional area times its length. Expressing all dimensions in meters gives:

$$V = \pi r^2 l = 3.14159(125 \times 10^{-3} \text{ m})^2(1.00 \times 10^5 \text{ m}) = 4.91 \times 10^3 \text{ m}^3$$

This is 4.91×10^6 L. An ideal gas at 54 atm and 15°C contains 2.284 mol L^{-1}. If ethylene (molar mass 28.054 g mol^{-1}) behaves ideally, its density is 64.07 g L^{-1}. The product of this density and the volume of the pipeline in liters is 3.15×10^8 g, or $\boxed{315{,}000 \text{ kg}}$.

(d) The actual density of the ethylene in the pipeline is 0.20 g cm^{-3}, which is 200 g L^{-1}. The supercritical ethylene is 3.12 times denser than the ideal-gas ethylene; hence the pipeline holds 3.12 times more. Its contents weigh $\boxed{982{,}000 \text{ kg}}$.

(e) The operator should $\boxed{\text{increase the pressure}}$ to keep the ethylene all liquid.

24-32 Thermal cracking proceeds at a higher temperature than catalytic cracking and uses no catalyst. The products are short-chain alkenes, which are good feedstocks for chemical industry but too volatile to be good components of gasoline. Catalytic cracking gives longer-chain alkenes that are desirable in gasoline.

24-34 (a) There are five candidate structures of benzene. Structures (i), (ii), and (iv) would form only one C_6H_5Cl. The other two proposed structures accommodate Cl in place of H in at least two distinct locations.

(b) Only structure (ii) gives exactly three isomers of formula $C_6H_4Cl_2$. Structures (i), (iii), and (v) give more than three isomers, and structure (iv) gives only two isomers.

24-36 Make an alcohol that contains a radioactive isotope of oxygen, for example, $R(^{18}O)H$. Prepare the ester and determine how much, if any, of the radioactivity is incorporated in it. It is found that almost all of the labeled oxygen is incorporated in the ester.

24-38 (a) The two equations are both dehydrations:

$$CH_3CH_2OH \longrightarrow H_2C{=}CH_2 + H_2O \quad \text{over alumina, 400 °C}$$

$$2\,CH_3CH_2OH \longrightarrow CH_3CH_2{-}O{-}CH_2CH_3 + H_2O \quad \text{over alumina, 230°C}$$

(b) The reaction is a dehydrogenation: $CH_3CH_2OH \longrightarrow CH_3CH{=}O + H_2$.

24-40 (a) The structure of stanolone is

(b) The molecular formulas of stanalone is $C_{19}H_{30}O_2$.

Answers to Even-Numbered Problems

25-2 The addition polymerization of tetrafluoroethylene is written:

$$n\,C_2F_4 \longrightarrow (CF_2CF_2)_n$$

25-4 Polymethyl methacrylate forms by addition polymerization of methyl methacrylate, which has the structure:

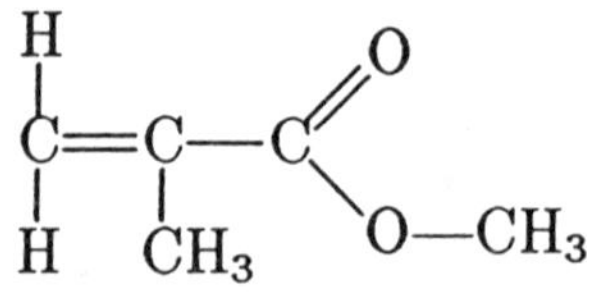

$$\begin{array}{ccc}
H & & O \\
| & & \diagup\!\diagup \\
C\!=\!C\!-\!C & & \\
| & | & \diagdown \\
H & CH_3 & O\!-\!CH_3
\end{array}$$

25-6 The starting monomer, which is alanine, has the structure:

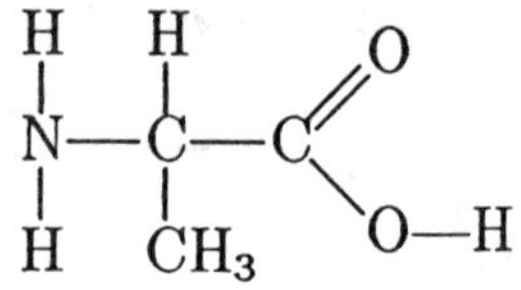

$$\begin{array}{ccc}
H & H & O \\
| & | & \diagup\!\diagup \\
N\!-\!C\!-\!C & & \\
| & | & \diagdown \\
H & CH_3 & O\!-\!H
\end{array}$$

25-8 The initiation is

$$I^- - I^+ + CH_2 = CHCH_3 \longrightarrow I - CH_2CH(CH_3)^+ \, I^-$$

25-10 The two ends of a peptide chain are distinguishable. Any of the 20 amino acids can be at the amino end and any of the 20 can be at the carboxylic acid end. The answer is $20 \times 20 = 400$.

25-12 The pentapeptide is:

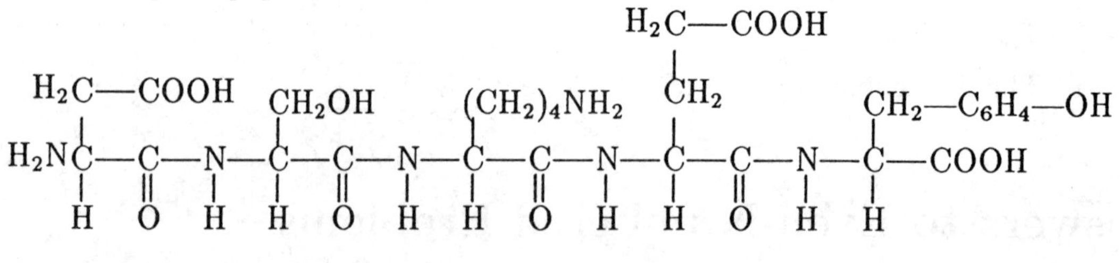

$$\begin{array}{c}
H_2C\!-\!COOH \\[2pt]
| \\
CH_2 \\[2pt]
| \\
\end{array}$$

$$H_2NC\!-\!C\!-\!N\!-\!C\!-\!C\!-\!N\!-\!C\!-\!C\!-\!N\!-\!C\!-\!C\!-\!N\!-\!C\!-\!COOH$$

with side groups $H_2C\!-\!COOH$, CH_2OH, $(CH_2)_4NH_2$, CH_2, and $CH_2\!-\!C_6H_4\!-\!OH$.

The pentapeptide has mostly polar side-groups. It should be more soluble in water than in *n*-octane.

25-14 The ring form of D-ribose has four asymmetric carbon atoms:

25-16 A nucleotide consists of a molecule of deoxyribose ($C_5H_{10}O_4$), a molecule of phosphoric acid (H_3PO_4) and a molecule of one of the four bases (adenine ($C_5H_5N_5$), cytosine ($C_4H_5N_3O$), thymine ($C_5H_6N_2O_2$), and guanine ($C_5H_5N_5O$)) linked as a unit by the loss of two molecules of water. The average molar mass of a nucleotide is 327 g mol^{-1}. The links between nucleotides form with the loss of still another molecule of water, so the average molar mass of a repeating unit in DNA is 309 g mol^{-1}. If the molar mass of the DNA is 4×10^9 g mol^{-1}, there are 1×10^7 nucleotides along the DNA chain. Each link in DNA contains a base, so this DNA is said to have 10^4 kilobases.

25-18 (a) Cellulose is a condensation polymer of glucose ($C_6H_{12}O_6$). A molecule of water is lost at every condensation stage so that $(C_6H_{10}O_5)_n$ is the molecular formula of cellulose and $C_6H_{10}O_5$ is its empirical formula. In guncotton, three —OH groups are replaced by three —ONO$_2$ groups in every monomer unit. The empirical formula is therefore $C_6H_7N_3O_{11}$.

(b) In cellulose acetate, three —OH groups in cellulose are replaced by three —OOC—CH$_3$ (acetate) groups in every monomer unit. The formula of the monomer unit is therefore $C_{12}H_{16}O_8$, and the empirical formula is $C_3H_4O_2$,

25-20 The repeating unit in the polyester has the formula $C_{10}H_8O_4$. This formula is the sum of the molecular formulas of terephthalic acid and ethylene glycol minus twice the formula of water, a relationship that derives from the fact that the diacid and diol combine by condensation with loss of water. The molar mass of the repeating unit is 192.17 g mol^{-1}. Then the number of moles of repeating unit needed is:

$$(10.0 \times 10^3) \text{ g polymer} \times \left(\frac{1 \text{ mol units}}{192 \text{ g polymer}} \right) = 52.04 \text{ mol}$$

This means that the synthesis needs 52.04 mol of terephthalic acid (molar mass 166.1 g mol^{-1}) and 52.04 mol of ethylene glycol (molar mass 62.06 g mol^{-1}). These chemical amounts convert to 8.64 kg of terephthalic acid and 3.23 kg of ethylene glycol.

25-22 An *n*-alkane is a straight chain of —CH_2— groups. In linear polyethylene, the chain may have hundreds of thousands of such links. Hence the term "ultimate."

25-24 The mass of polystyrene is 2.84×10^{12} g, and the mass of a monomer unit of styrene (C_7H_8) is 104.2 g mol^{-1}. In addition polymerization, no mass is lost by the splitting out of small molecules. Hence, 2.72×10^{10} mol of styrene were incorporated, which is 1.64×10^{34} molecules (by multiplication by Avogadro's number).

25-26 The polymerization reaction must be driven by an increase in entropy in the surroundings, because it itself involves a local decrease in entropy. Heat must therefore be evolved into the surroundings; the reaction is exothermic.

25-28 In graft polymerization, one fully polymerized structure is first formed, after which a second polymeric structure may be bonded to the first through unsaturated bonds that are present in the first. Neither polyvinyl chloride nor polystyrene has such bonds, so the answer is no. However, vinyl chloride and styrene might graft together if hydrogen-atom transfer to a growing polystyrene chain from a pre-existent polyvinyl chloride chain occurred. This would give some product, but not as a result of graft polymerization in the usual sense.

25-30 Hair and fingernails are proteins that are cross-linked by disulfide bonds, which give them strength. Oxidation of these sulfur-rich compounds gives the same odorous sulfur-containing products.

25-32 (a) The equation for the degradation of glucose to alcohol and carbon dioxide is $C_6H_{12}O_6(aq) \longrightarrow 2\,CO_2(aq) + 2\,C_2H_5OH(aq)$.

Compute $\Delta G°$ of this reaction from the $\Delta G_f°$'s of the reactants and products:

$$\Delta G° = 2(-385.98) + 2(-181.64) - 1(-917.2) = -218.0 \text{ kJ}$$

25-34 Since thymine (T) is complementary to adenine (A), and cytosine (C) is complementary to guanine (G), the sequence in the complementary strand is TGAACTGGC.

25-36 If the only difference between natural rubber and gutta-percha is the geometrical relationship (*cis* in natural rubber, but *trans* in gutta-percha) at the recurring double bonds, then conversion of the double bonds to single bonds by reaction with H_2 should give identical products.

Answers to Even-Numbered Problems, App. A

A-2 The trailing zeros in (b), (c), and (e) must not be omitted when the number is put into scientific notation. The two negative signs in answer (b) sometimes trouble students.
(a) 4.579×10^3 (b) -5.020×10^{-2} (c) 2.134560×10^3
(d) 3.825 (e) 4.50×10^{-5} (f) 9.814

A-4 (a) 0.003333 (b) $-12,000,000$ (c) 0.0000279 (d) 30 (e) 0.06700.

A-6 Remind students of how to locate numbers on the number line:

$$-2.17 \times 10^3 < -14.8 \times 10^{-4} < 11.7 \times 10^{-7} < 3.19 \times 10^{-2}$$

A-8 The amount of vanadium is 4.6×10^{-6} g.

A-10 (a) Statistical methods for deciding whether to omit a outlier are not developed in the Appendix. Instead the appeal is to use good judgment. Although there is plenty of scatter, none of the values is grossly out of line (as in problem A-9), and all should be retained.

(b) The average volume is 555 cm^3.

(c) The standard deviation is $\sigma = 27$ cm^3. The unit is part of the answer.

A-12 The average of the determinations of mass in problem A-9 differs by 6.4 percent from the "true" value (the one obtained with a better balance). The average volume in problem A-10 differs by only 0.36 percent from the "true" volume. The second measurement is more accurate despite being less precise.

A-14 (a) two   (b) ambiguous (one, two, three or four significant figures)   (c) seven
(d) four   (e) three.

A-16 (a) -0.0025 (no change needed)   (b) 7.0×10^3 g   (c) 1.4×10^2 s
(d) 2.7×10^7 Pa   (e) 2.0×10^{-19} J

A-18 Eight, 2,997,215.5; seven, 2,997,216; six, 2.99722×10^6; five, 2.9972×10^6; four, 2.997×10^6; three, 3.00×10^6; two, 3.0×10^6; one, 3×10^6.

A-20 (a) 250.89 (b) -77.7 (c) 6.552×10^{19} (d) -1.467×10^{-13}

A-22 (a) 4189 (b) 0.257 (c) 2.948×10^{-26} (d) 6.8×10^{-4}

A-24 The length of the table is 198.88 in. Five significant figures appear in the answer (not three) because the value "2.54" is a definition and, despite its appearance, has an infinite number of significant figures, not three.

Answers to Even-Numbered Problems, App. B

B-2 (a) 6.6×10^{-5} K   (b) 1.59×10^7 kg m^{2}s^{-2} (c) 1.3×10^{-7} kg
(d) 6.2×10^{10} kg m^{-1}s^{-2}

B-4

$$\mathrm{JT}^{-1} = \frac{\mathrm{J}}{\mathrm{T}} = \frac{\mathrm{kg\ m^2\ s^{-2}}}{\mathrm{kg\ s^{-2}\ A^{-1}}} = \frac{\mathrm{m^2}}{\mathrm{A^{-1}}} = \mathrm{A\ m^2}$$

B-6 Convert all to the same unit for comparison
$10^6\mu\mathrm{m} < 10^{-2}\mathrm{km} < 1000 \mathrm{\ in} < 10^{-4} \mathrm{\ Mm} < 10^6 \mathrm{\ mm}$.

B-8 (a) The answer is 9032°F assuming four significant figures in the given temperature. It is hard to measure such a high temperature to within one degree Celsius. If the given temperature has two significant figures, the answer is 9000°F. The addition or subtraction of 32°F can often be omitted in conversions involving high temperatures.

(b) 104°F   (c) 414°F   (d) −40°F.

B-10   (a) 4727 K   (b) 313.2 K   (c) 485 K   (d) 233 K.

B-12 Translate the "C⟶F" routine to the following mathematical form:

$$(t_{\circ\mathrm{C}} + 40) \times \frac{9}{5} - 40 = t_{\circ\mathrm{F}}$$

Completing the indicated multiplication and addition gives the standard formula for $t_{\circ\mathrm{F}}$ in terms of $t_{\circ\mathrm{C}}$:

$$\frac{9}{5}(t_{\circ\mathrm{C}}) + 72 - 40 = t_{\circ\mathrm{F}}$$
$$\frac{9}{5}(t_{\circ\mathrm{C}}) + 32 = t_{\circ\mathrm{F}}$$

The approach is similar for the "F⟶C" routine:

$$(t_{\circ\mathrm{F}} + 40)\frac{5}{9} - 40 = t_{\circ\mathrm{C}}$$
$$(t_{\circ\mathrm{F}} + 40)\left(\frac{5}{9}\right) - 72\left(\frac{5}{9}\right) = t_{\circ\mathrm{C}}$$
$$\left(\frac{5}{9}\right)(t_{\circ\mathrm{F}} + 40 - 72) = t_{\circ\mathrm{C}}$$
$$\left(\frac{5}{9}\right)(t_{\circ\mathrm{F}} - 32) = t_{\circ\mathrm{C}}$$

B-14 (a) 6.82 MPa (b) 1.0×10^6 V m^{-1} (c) 2.3×10^{-17} m s^{-1} (d) 2.24×10^{-2} m^3 mol^{-1}
(e) The answer is 1.03×10^4 kg m^{-2}. Part (e) may be harder for those with experience than for beginners. If "14.7 lb in^{-2}" is recognized as ordinary atmospheric pressure, then the answer 1.01×10^5 N m$^{-2} = 1.01 \times 10^5$ Pa may be written immediately. But the use of the pound as a unit of force has been avoided in the Appendix.

B-16 The mile per hour and inch per day are obviously suitable to measure speed. Replacement of the N by kg m s^{-2} and cancelation shows that the N s kg^{-1} is the same as the m s^{-1}, and is therefore a possible unit of speed. The kg per J reduces to s^2 per m^2, and is *not* a possible unit for speed.

B-18 The unit-factor method gives:

$$1 \frac{\text{mile}}{\text{gal}} \times \left(\frac{1 \text{ gal}}{3.785 \text{ dm}^3} \right) \times \left(\frac{1609.344 \text{ m}}{1 \text{ mile}} \right) = 425.2 \frac{\text{m}}{\text{dm}^3}$$

Therefore, 30.0 miles per gallon is 1.28×10^4 m dm^{-3}. Fuel consumption could also be given in the sleek but opaque unit "m^{-2}".

B-20 (a) The conversion factor between miles and meters is in problem B-13. The answer is 1.86×10^5 miles per second.

(b)

$$3.00 \times 10^8 \frac{\text{m}}{\text{s}} \times \frac{86400 \text{ s}}{1 \text{ day}} \times \frac{14 \text{ days}}{1 \text{ fortnight}} \times \frac{1 \text{ in}}{0.0254 \text{ m}} \times \frac{1 \text{ ft}}{12 \text{ in}} \times \frac{1 \text{ furlong}}{660 \text{ ft}}$$

$$= 1.80 \times 10^{12} \text{ furlongs per fortnight}$$

Answers to Even-Numbered Problems, App. C

C-2 The slope is 0.0146 atm (°C)$^{-1}$.

C-4 (a) The equation is in the required form with m the slope equal to -2 and b the intercept equal to -8.

(b) The equation is $y = 3/4x + 7/4$. The slope is 3/4 and the intercept is 7/4.

(c) The equation is $y = 16/7x - 53/7$. The slope is 16/7 and the intercept is $-53/7$.

C-6 The graph of y versus x for the equation

$$y = \frac{8 - 10x - 3x^2}{2 - 3x}$$

is linear. Note that the numerator can be factored into $(2 - 3x)(4 - x)$ so that the equation is equivalent to $y = 4 - x$.

C-8 (a) $x = 3/4$ (b) $x = -5/2$ (c) $x = 2$.

C-10 The answers are given to 5 significant figures. (a) $x = -0.14132, -2.3587$ (b) $x = -0.47178, 1.2718$ (c) $x = 1.0000, 2.3333$.

C-12 (a) Assuming that x is small compared to 2.00 gives $x = 5.27 \times 10^{-17}$. The other roots are $x = -2.00, x = -2.50, x = 3.00$, as can be confirmed by approximating the right side of the equation as zero.

(b) The best method of solution is graphical. There are three roots because this is a third-degree equation: $x = 0.0827, 0.967, -3.05$.

(c) The only real root is $x = 0.287$. It can be arrived at graphically. The other two roots are imaginary: $x = 0.023155 \pm 1.7028i$. They are of little interest in chemical applications.

C-14 (a) 2.96×10^{-17}; three significant figures in the mantissa (the 528 in the exponent) mean three significant figures in the answer. (b) 31.392 (c) 1.5×10^6 (d) -10.3025

C-16 The answer is 1.3419×10^{-7}. The number of significant digits given can be shown to be correct by recasting the natural logarithm as a base-10 logarithm (by dividing by 2.30258509) and using the rule for significant figures given on text page A-24.

C-18 Few calculators accommodate a number with an exponent exceeding 99 in absolute value. To answer this problem, write $10^{-107.8} = 10^{-108} \times 10^{+.2}$. Then, compute the parts of the product separately. The answer is 2×10^{-108}.

C-20 The two logarithms are 192.428 and -288.572.

C-22 The problem is to find x in the equation $x = 1/\ln x$. One way to proceed is to guess an x, put it into the right side of the equation and see on a calculator if the indicated operations gives back the guess. The answer is 1.763.

C-24 The volume of xenon is 9.74×10^{-3} L.